How and why do males become male and females become female? And what are the consequences if the decision is not incisive? Drawing upon interests in animal genetics and molecular biology, the author endeavours to answer these difficult yet topical questions. Genetic determination of mammalian sex is explained with examples from the latest molecular findings and with particular reference to the rôle of the gene *SRY/Sry*. How sex determination leads to differentiation of male or female gonads and a corresponding genital tract is then discussed. As inferred, these clear-cut decisions can be compromised in several ways leading to intersexuality, a significant problem in domestic farm animals and also well documented in laboratory species and man. How can genetic females without Y-chromosome-related DNA sequences generate organs such as a testis or an ovotestis, frequently on a unilateral basis? The author explains the various components of genetic instruction that may prompt such developmental anomalies. Using examples taken from intersexuality, chimaeras and asymmetries, this book describes the underlying molecular basis of sex determination and sexual differentiation, and focuses on the critical rôle of the rate of embryonic development in these vital processes. Male precocity is a recurrent theme, as is the involvement of Sertoli cells and their secretion of anti-Müllerian hormone.

This is a highly pertinent book for reproductive physiologists, geneticists and developmental biologists whose interests may extend from animal science through veterinary medicine to human clinical medicine.

SEX DETERMINATION, DIFFERENTIATION AND INTERSEXUALITY IN PLACENTAL MAMMALS

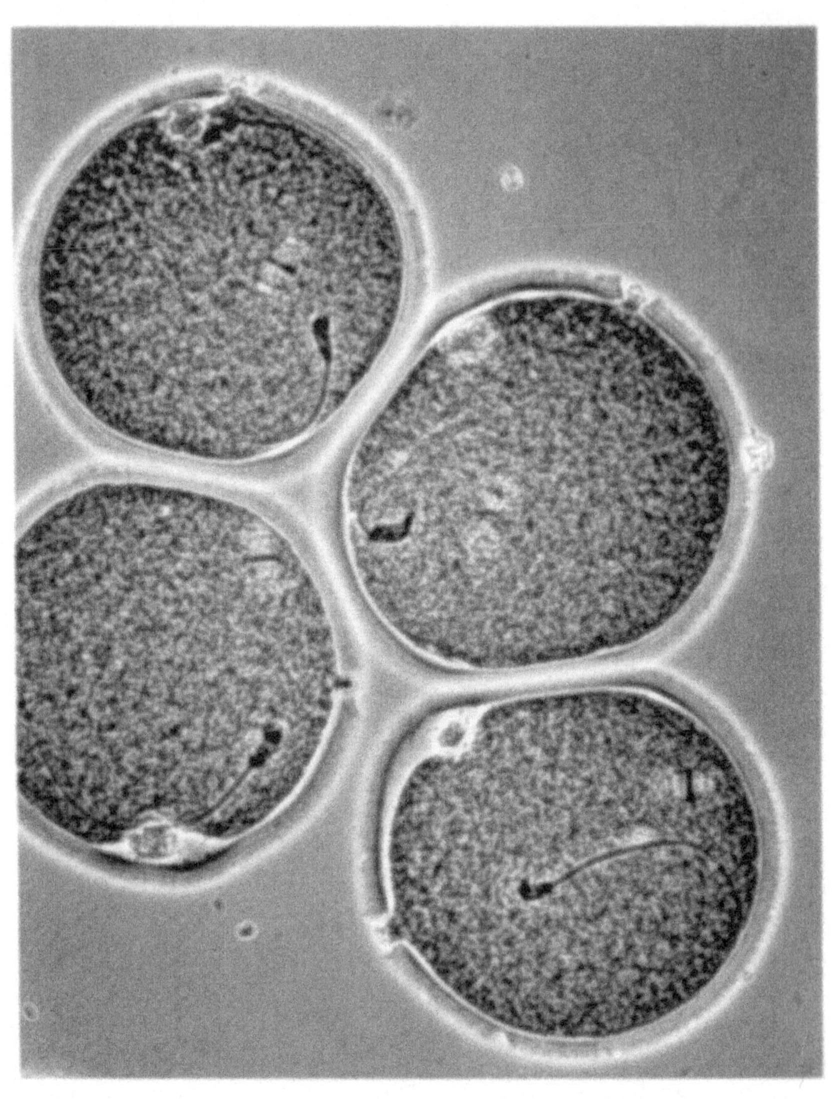

The moment of sex determination in four eggs of the golden hamster that have recently been penetrated and activated by either an X-bearing or a Y-bearing spermatozoon. Note that the chromosomes of the second meiotic spindle are arranged at telophase, that each sperm head is already swollen and that a portion of the flagellum is being incorporated into the vitellus. (Unpublished photograph by the author.)

This image is available for download in colour from www.cambridge.org/9780521182294

Contents

Plate 5.1 will be found between pp. 144 and 145; Plates 5.2 and 5.3 are between pp. 158 and 159

Preface

This book has been written from the perspective of a university teacher whose research has been in the field of mammalian reproductive physiology but who is also much interested by developments in animal genetics and molecular biology. Preparing the individual chapters offered an exciting opportunity for bringing these disciplines together in various ways. The result is seen primarily as a text for advanced (Honours) undergraduates in Schools of Biological Sciences, Medicine, Veterinary Medicine and Animal Science. It should also appeal to those on taught MSc courses and to PhD students interested in both developmental biology and reproductive physiology in the higher mammals.

The book was planned during my time in the Faculty of Veterinary Medicine, University of Montréal, but the administrative commitment there – together with lecturing and activities in the operating theatre – meant that a serious spell of writing had to await my return to Edinburgh. In fact, the chapters were prepared in draft between May 1991 and June 1993, and then brought up to date as far as December 1993 on the basis of the extensive journal coverage available in the University of Edinburgh libraries. A small number of 1994 references has also been included.

As to the origins of this work, they almost certainly date back to the author's post-doctoral days in Paris (1968–1970) listening to lectures on sexual differentiation by the late Professor A. Jost and observing the studies of his assistants, Drs J. Prépin and B. Vigier, on freemartin calves at the Station de Physiologie Animale, Jouy-en-Josas. In similar vein, Professor R. V. Short, FRS, gave a memorable lecture on sexual differentiation in September 1970 at an Anglo-French colloquium held in Nouzilly, just outside Tours. However, the present work was also prompted by observations at the laboratory bench and during abdominal surgery. In extensive studies on normal and abnormal fertilisation in domestic farm animals, our species of choice was the pig because of the number of eggs shed at ovulation. The inbred females used in these surgical studies often revealed unusual gonads in the form of an ovotestis or ovary on one side and a testis on the other. Initially, such intersex animals were simply seen as a major inconvenience, interrupting a carefully planned programme of research, but in due course they became a subject of research in their own right. How could apparently genetic females generate an ovary and an ovotestis within the same animal, and what form of genetic instruction would prompt the unilateral appearance of testicular tissue? Attempts to answer these questions will be found in the chapters that follow.

Although the text examines many of the latest findings on sex determination and then sexual differentiation of the gonads and genital tract, it does not dwell on dimorphisms elsewhere in the organ systems. Except in a context of intersexuality, brain sex is not considered in any specific way nor are the resulting patterns of behaviour, nor indeed sexual dimorphisms in organs such as the liver. This is not an oversight. Despite the fascination of such material and the many new discoveries, the scope of the present volume had to be kept within reasonable limits, and a systematic treatment of these other fields will have to await a future endeavour. One other limitation concerns use of the term 'placental mammals'. In reality, the text focuses on a small number of eutherians – man, mouse and several domestic species – although there is occasional reference to marsupials (which, of course, have a placenta) where they serve to illustrate major new findings or divergences from the eutherian model.

The treatment of abnormal reproductive tissues or conditions is also limited. The objective has certainly not been to document as many bizarre conditions as possible but rather to seek the underlying genetic lesions. A molecular explanation has been offered wherever possible, although research at this level is moving so fast that the text cannot be completely up to date. Nonetheless, what comes through from the writing is of major significance – the fact that a point mutation, a single base change, can wreak such havoc. Evolution may very well thrive on mutations but the cost to individuals of our own species can be physically taxing and emotionally devastating, especially in a reproductive context.

Each chapter has been intended as an essay in its own right. This has inevitably resulted in some repetition. Nowhere is this extensive except in the final chapter, for this is presented in the form of an overview. As to a policy on references, I have tried to be reasonably comprehensive and to cite both new work and old: new for there is much excitement in arriving at molecular descriptions of diverse clinical conditions, old since it is sobering to note the extent to which many of the more fashionable ideas in today's literature have a remarkably ancient pedigree. Finally, as to conventions, the possessive adjectival form of referring to syndromes has been maintained (e.g. Turner's) rather than adopting the American approach that omits the apostrophe 's'. The original format of describing such aneuploidy has also been adhered to (i.e. 45,XO) rather than simply 45,X. And such classical and attractive spellings as chimaera, disulphide and foetus have been preferred and used consistently throughout the text; such is an author's privilege!

Edinburgh
February, 1994

R.H.F. Hunter

Acknowledgements

The burden of preparing any modern academic text is always eased if one has the support of people one admires and respects. This has been very much the case in the present instance. Professor T. R. R. Mann, CBE, FRS, Emeritus Professor of the Physiology of Reproduction in the University of Cambridge and my professor since first attending his lectures in 1961, maintained an active interest in the writing up to the time of his death in November 1993. Professor E. J. C. Polge, CBE, FRS, Honorary Professor of Animal Reproduction in the University of Cambridge, has also been enormously supportive and helpful, both directly and in more subtle ways, and has been a source of advice and assistance since I joined him as a PhD student in 1962. Similarly, Professor C. Thibault, Emeritus Professor of Reproduction in the University of Paris VI and still exceedingly active at the Station de Physiologie Animale, Jouy-en-Josas, has been a very considerable source of encouragement. Indeed, during a period of six months working with him and his close colleague Marie-Claire Levasseur on a rather difficult project in 1992, there was an opportunity for much discussion and also free access to his wonderful library and collection of reprints. To all of these friends and mentors, I wish to record my warmest words of appreciation.

Many other friends and colleagues have helped me in diverse ways. At the level of molecular biology, Dr Corinne Cotinot of the Centre de Recherches, Jouy-en-Josas, welcomed me to her laboratory during 1992 and instructed me in the latest techniques whilst we were pursuing the question of whether intersex pigs possess the so-called testis-determining gene, *Sry*. This visit was facilitated by Dr F. Grosclaude, Director of Animal Production at the Institut National de la Recherche Agronomique. A little earlier, the Society for the Study of Fertility had asked Dr Suzanne Ullmann and myself to organise a Symposium on 'Sex Determination and

Differentiation', and this was duly held during the Annual Conference at the University of Glasgow in July, 1992. Particularly beneficial was the associated exhibition of relevant ancient literature organised from the University Library's collections and highlighted in the excellent booklet produced by Dr Brian Cook, entitled: *Contributions of the Hunter Brothers to our understanding of reproduction.*

Concerning preparation of the text, Frances Anderson tackled my handwriting with cheerfulness and energy, and maintained her good spirits and willingness throughout, despite the continued addition of references and of new experimental material. My very best thanks go to her for all this help and for producing an immaculate final copy. Of course, this stage was reached only after many close friends and colleagues had generously given of their time to read and comment on draft chapters. In this connection, I particularly wish to thank Professor T. G. Baker and Drs K. P. Bland, P. S. Burgoyne, A. Chandley, M. Clinton, B. Cook, K. W. Jones, R. F. O. Kemp, A. A. Macdonald, C. Price, T. Taketo, S. Ullmann, R. Webb and J. D. West. Most of these specialists read one or two chapters but Dr Brian Cook was thoughtful enough to cast an eye over the whole text, even though some considerable distance from his own field of expertise. Responsibility for any errors that remain must be that of the author alone.

The following people were kind enough to send me reprints, preprints, specific information or specialist advice: N. Affara, B. Afzelius, T. G. Baker, J. B. L. Bard, R. S. P. Beddington, J. Bell, F. G. Biddle, J. D. Biggers, K. P. Bland, S. Blecher, N. A. Brown, P. S. Burgoyne, A. G. Byskov, B. M. Cattanach, A. Chandley, H. M. Charlton, B. Cook, C. Cotinot, S. Dolci, P. J. Dziuk, D. S. Falconer, M. de Felici, M. Fellous, J. E. Fléchon, R. L. Gardner, F. W. George, A. K. Goff, I. Gordon, R. G. Gosden, J. E. Griffin, J. P. Hardelin, R. P. Harrison, C. F. Heyns, W. G. Hill, S. H. Hillier, C. E. Hinks, M. Hooper, C. A. Hunter, A. Illius, P. Jacobs, R. L. Jäger, R. Jimenez, N. Josso, C. J. H. Kellnar, R. F. O. Kemp, W. A. King, M. Kirzenbaum, M. C. Levasseur, M. F. Lyon, A. McLaren, S. Magre, P. Malet, U. Mittwoch, H. D. M. Moore, C. Nagamine, C. Nixey, J. W. Overstreet, S. J. Palmer, T. Paull, C. Petit, C. Pilgrim, P. Popescu, G. Rappold, J. Rossant, P. Rossi, W. Schempp, R. V. Short, D. W. Silversides, J. M. W. Slack, D. G. Szollosi, T. Taketo, C. Thibault, S. Ullmann, P. van der Schoot, B, Vigier, H. B. Ward, H. Wartenberg, R. Webb, C. J. G. Wensing, J. D. West, J. D. Wilson and U. Wolf. Illustrative material or permission to copy figures and diagrams, was generously provided by N. Affara, T. G. Baker, C. E. Bishop, R. L. Brinster, N. A. Brown, P. S. Burgoyne, A. Chandley, Y. Clermont, B. Cook, C. Cotinot, M. Dym, N.

Ellis, D. W. Fawcett, F. W. George, C. F. Heyns, I. J. Jackson, N. Josso, M. C. Levasseur, A. McLaren, S. Magre, P. Malet, U. Mittwoch, C. Nagamine, D. C. Page, T. Paull, P. Popescu, G. Rappold, J. Rossant, J. W. R. Schwabe, R. V. Short, E. M. Simpson, T. Taketo, C. Thibault, A. A. Travers, P. van der Schoot, H. Wartenberg, C. J. G. Wensing, I, Wilmut, J. D. Wilson and U. Wolf.

Photographic work used in illustrating the text was enthusiastically prepared by Mr Ian Goddard of Edinburgh University Library in George Square and by Mr Jack Cable of the Department of Anatomy, Edinburgh University Medical School. The latter arrangement received the blessing and support of Professor M. H. Kaufman. Mr Nicholas Smith of Cambridge University Library, Mrs Jo Currie of Edinburgh University Library and Dr Tim Hobbs of Glasgow University Library were all helpful in providing copies of rare material. To each of these specialists, I am most grateful.

In connection with citations from published work, my best thanks are extended to the Editors of the following journals and reviews: *Advances in Developmental Biology*, *American Journal of Anatomy*, *Anatomy and Embryology*, *Animal Breeding Abstracts*, *Annual Review of Physiology*, *Annals of Human Genetics*, *BioEssays*, *Biology of Reproduction*, *British Medical Journal*, *Cell*, *Current Biology*, *Development*, *Developmental Biology*, *Differentiation*, *Genomics*, *Hormone Research*, *Human Genetics*, *Journal of Anatomy*, *Journal of Experimental Biology*, *Journal of Experimental Zoology*, *Journal of Genetics*, *Journal of Reproduction and Fertility*, *Journal of Theoretical Biology*, *Molecular Reproduction and Development*, *Nature*, *Oxford Reviews of Reproductive Biology*, *Proceedings of the National Academy of Sciences*, *Proceedings of the Royal Society of London*, *Recent Progress in Hormone Research*, *Reviews in Physiology*, *Biochemistry and Pharmacology*, *Trends in Endocrinology and Metabolism*, and *Trends in Genetics*; and to the following publishing houses: Academic Press, Cambridge University Press, CRC Press, Ellipses, Elsevier-North Holland Inc., Heinemann Medical, Macmillan, Oxford University Press, Pitman and Sons, Raven Press, W. B. Saunders Company, Springer-Verlag, John Wiley & Sons and Williams and Wilkins.

Finally, I very much wish to record in print the support of my family. My father, who long years ago had his own Scottish academic connections, generously sponsored production of the text, and to him I shall always remain warmly grateful. The debt to my wife is so considerable that she would be seriously embarrassed to see it committed to paper. However, a domestic situation that permits one to indulge an urge to write and study in complete freedom whilst, at the same time, taking over much of the family home with collections of books, reprints, and manuscript material is one that cannot pass unmentioned. Nor can the endless help in checking, rechecking and editing the whole of the text, and for supporting this undertaking with good humour and affection throughout. I am immensely grateful to her.

Abbreviations

ACTH	adreno-corticotrophic hormone
AMH	anti-Müllerian hormone
AMP	adenosine 3′,5′-monophosphate
cAMP	cyclic AMP
CRF	corticotrophin releasing factor
cDNA	complementary DNA
DNA	deoxyribonucleic acid
ELISA	enzyme-linked immunoabsorbent assay
FSH	follicle stimulating hormone
GnRH	gonadotrophin releasing hormone
HMG	high mobility group (proteins)
H-Y	H-Y antigen
i.u.	international units
IVF	*in vitro* fertilisation
LH	luteinising hormone
LIF	leukaemia inhibitory factor
MIS	meiosis inducing substance
MPS	meiosis preventing substance
Mr	relative molecular mass
NCAM	neural cell adhesion molecule
ORF	open reading frame
PCR	polymerase chain reaction
PG	prostaglandin
PGC	primordial germ cell
PGE_2	prostaglandin E_2
PMDS	persistent Müllerian duct syndrome
PMSG	pregnant mare serum gonadotrophin
RNA	ribonucleic acid
mRNA	messenger ribonucleic acid
RPS4	ribosomal protein S4

TGF	transforming growth factor
TGF-β	transforming growth factor-β
Yp	short arm of the Y chromosome
Yq	long arm of the Y chromosome

Gene abbreviations

AMH	anti-Müllerian hormone gene in man
Amh	mouse homologue of *AMH*
AZF	azoospermia factor
Bkm	Banded Krait minor (banded Krait is a snake)
Hox	homeobox genes
hpg	hypogonadal mouse
Hya	male-specific histocompatability antigen
iv	situs inversus viscerum mutation in mouse
KAL	Kallmann's syndrome gene
KALIG-I	Kallmann's syndrome (interval) gene
P	polled (hornlessness) in goats
RPS4Y	ribosomal protein S4 gene on the Y chromosome
SOX	*SRY* box genes
Sox	mouse homologue of *SOX*
Spy	spermatogenesis gene
SRY	sex-determining gene of the Y chromosome in man
Sry	mouse homologue of *SRY*
Sxr	sex reversal mutation (factor) in mouse
Sxr'	an H-Y antigen-negative variant of *Sxr*
Tas	T-associated sex reversal in mouse
Tda-1	testis-determining autosomal-1 in mouse
TDF	testis-determining factor (gene) in man
Tds	testis-determining sequences in mouse
Tdy	testis-determining gene on Y chromosome in mouse
TFM	testicular feminisation in man
Tfm	mouse homologue of *TFM*
WT	Wilms' tumour gene in man
WTI	Wilms' tumour suppressor gene

Wt1	mouse homologue of *WT1*
XIST	X-inactive specific transcript
Xist	mouse homologue of *XIST*
ZFX	zinc finger gene on the X chromosome
Zfx	mouse homologue of *ZFX*
ZFY	zinc finger gene on the Y chromosome
Zfy	mouse homologue of *ZFY*

1
Historical landmarks in studies of reproduction and sex determination

Introduction

These opening pages represent no more than the briefest survey of highlights upon a canvas extending from the speculations of classical antiquity to the findings of contemporary molecular genetics. The viewpoint offered is, of course, a personal one and there is little doubt that other authors would – in part at least – have focused on different events and different material. In this regard, inspection of some of the volumes cited as references should give access to alternative interpretations and indeed to modified historical perspectives. Readers may perhaps detect a certain partiality towards contributions from the University of Edinburgh in the last portion of this short chapter. Taking into consideration the author's earlier connections with this institution (see Preface), and its extensive influence in the field of mammalian reproductive physiology, then it is hoped that the significance of the studies mentioned will not be judged as completely inappropriate.

Ancient Greek philosophers

As discussed by Short (1969), the dual problems of sex determination and sexual differentiation have fascinated mankind since the dawn of history, and sexual abnormalities of one sort or another have become the centrepiece of many a legend and fable. Primitive man undoubtedly had a keen appreciation of the vital parts of the anatomy, this being well illustrated in cave paintings, carvings, and frequently in the nature of mutilations inflicted upon rivals. Anomalous sexual anatomy certainly attracted attention and sometimes even social esteem, a situation which has persisted in some cultures up to the present day. Discourses on the origin of the two sexes occupy a prominent niche in the early medical literature. The possibility of controlling the sex of offspring seems to have been in the mind of many authors, and prescriptions as to the manner in which this might be achieved have entered into mythology. In addition to their application in domestic animals, such prescriptions very much concerned our own species, and one is never far from the inference that being able to produce males at will would in diverse ways be advantageous.

In this context, the teachings of Democritus of Abdera (460–370 BC) deserve attention. Although his writings have been lost, his message concerning determination of sex retained a prominent place in ancient philosophy. Democritus believed that the two testes played quite separate rôles in procreation, and that females in some manner originated from the left testis, males from the right. Such a distinction between left and right can only have been intuitive and yet, as pointed out by Mittwoch (1977), it led to the general concept of 'rightness' being associated with maleness, with the corresponding attributes of being adroit, dexterous, righteous. Females fared less generously in this school of thought, although the uterus did find a place in left–right considerations for the determination of sex. On the basis of observations in pigs in which it was claimed that male foetuses were found predominantly in the right horn of the uterus, it was suggested that this situation was due to semen flowing into the right uterine horn during the prolonged copulation. This somewhat fanciful notion was then extended to recommendations for the position to be adopted during human coitus, orientation of the two bodies depending on the desired sex of the child. Perhaps rather too generously, Mittwoch (1977) indicates that the Greeks had produced an explanation for sex determination which contains an element of truth. In the embryo, the relationship of male to female bears a dominance relationship which is essentially similar to that manifested by the right and left sides (Mittwoch, 1977). However, in that there are but two sexes and two gonads, it would not have been a profound speculation to associate one of the sexes predominantly with one of the gonads.

Other views abroad at the time of the Ancients concerned a more specific rôle of fluids in the determination of sex or of foetal composition. Pythagoras (580–500 BC) thought that male semen gave rise to the noble parts of the foetus whereas female semen gave rise to the gross parts. Hippocrates (460–377 BC) considered that each parent emits a liquor, and that such liquors when intermingled played a joint part in formation of the new being. Each contained a vital principle referred to as the *sperma* or 'seed'. If the *sperma* from each parent were strong, then a male would be generated whereas, if weak, a female would arise. However, if one *sperma* were strong and the other weak, then the *sperma* present in larger quantity would determine the sex of the offspring. As a further consideration the *sperma* was presumed to be derived from the entire body, each tissue contributing in some unique way, thereby enabling the gonads to contain ingredients from every part of the body. Such was the theory of *pangenesis*.

Also concerned with fluids, Aristotle (384–322 BC) refined the earlier notion of Pythagoras by suggesting that the male transmits the spirit, the soul, the principle of life and movement. His view was that an egg is generated by means of an intermingling of the male semen (the 'seed') with the *catamenia* or menstrual coagulum (the 'soil'), the latter being regarded as female semen. The 'seed' of the male was again referred to as the *sperma*, and an egg was thus the product of *sperma* interacting with *catamenia* to mould or shape it. The sex of the resultant offspring was thought to be determined environmentally, that is by conditions within the uterus – a view not without relevance when we come to Chapter 2. A male embryo reflected the principles of heat and dryness whereas a female was associated with cold and damp conditions. Embryonic sex might therefore be an expression of the influence of warm or cold winds, the side of the womb in which the infant was presumed to have been generated, or even the side of the male from which the *sperma* was derived. Interaction between these diverse factors was seemingly not viewed as a complication!

Afzelius & Baccetti (1991) drew attention to one other consideration concerning fluids. They remind us that the talmudic thinker and instructor Rabbi Simlai also believed that there were maternal as well as paternal contributions to the formation of a foetus. The Rabbi was much struck by the contrast between the red colour of menstrual blood and the whitish colour of seminal fluid. This led him to propose that the embryo's skin, flesh, blood, hair and dark parts of the eye were derived from the mother whereas bones, tendons, nails and the white part of the eye come from the father. No explanation was offered as to precisely how the mixing of male and female fluids led to this rather strict segregation of tissues.

Hermaphrodites also occupy a significant place in views on the determination of sex. According to Greek mythology Hermaphroditus, the son of Hermes and Aphrodite, merged with the nymph Salmacis to form one body with male and female characteristics. This then gave rise to the further view that human males and females arose by uniform bisection or cleavage of hermaphrodites. Why this notion of cleavage should have arisen is uncertain, but it may have been associated with the fear of aggression – the splitting by an axe – or perhaps the influence of lightning. Merging or fusion, on the other hand, was probably indicative of affection or protection (see Mittwoch, 1986). Be that as it may, Plato (427–347 BC) in his

Symposium presents descriptions of three sexes – males, females and androgynes – each split in half by an angry Zeus and each forever seeking its partner. Apollo was bidden to heal their wounds and, once this was accomplished, the two halves came together in mutual embrace – a neat and appealing manner in which to explain homosexual and heterosexual attraction. Mittwoch (1986) also makes reference to an Oriental legend existing in Mesopotamia, Persia and India, in which a primitive hermaphrodite was bisected by the Deity. She then goes on to link these myths with the nature of the embryonic gonad in mammals, a not completely convincing parallel since, as will be discussed in due course, the bipotentiality of the mammalian gonad is effectively lost as the sex of the embryo expresses itself by means of gene action.

Pre-nineteenth century landmarks

The span from ancient Greece until the end of the 1800s is chronologically enormous, but progress in terms of a realistic understanding of reproductive processes was slight during most of the intervening centuries. Although the Greek physician and biologist Galen (AD 130–200) had been much influenced by the teachings and traditions of the medical school at Alexandria, he nonetheless went on to propose novel views concerning reproduction. Basing his observations on the bicornuate uterus of domestic farm animals, he correctly noted ducts that terminated in the tips of the uterine horns and interpreted these paired ducts as the female counterpart of the male seminal ducts; their rôle, he mistakenly supposed, would be to carry 'female semen' filtered by the ovaries from the bloodstream down into the uterus to interact with male semen, the latter representing the concept of Aristotle. The doctrines of Galen – with their echoes of Aristotle – remained prominent for almost fourteen centuries after his death, undoubtedly a mark of his stature in medical history but also a rather clear reflection of the dearth of intellectual vitality in a turbulent Europe. Despite this latter remark, various universities were firmly established by the twelfth and thirteenth centuries and, in a medical context, the Italian schools in particular assumed prominence in the succeeding centuries.

The wonderful drawings of Leonardo da Vinci (1452–1519) that focused on the human reproductive system might have acted as a catalyst to research and scholarship, but regrettably these were not published systematically for the benefit of the slowly proliferating scientific world until the late nineteenth century. By contrast, the numerous and meticulous observations of Andreas Vesalius (1514–1564) were published in 1543 in his

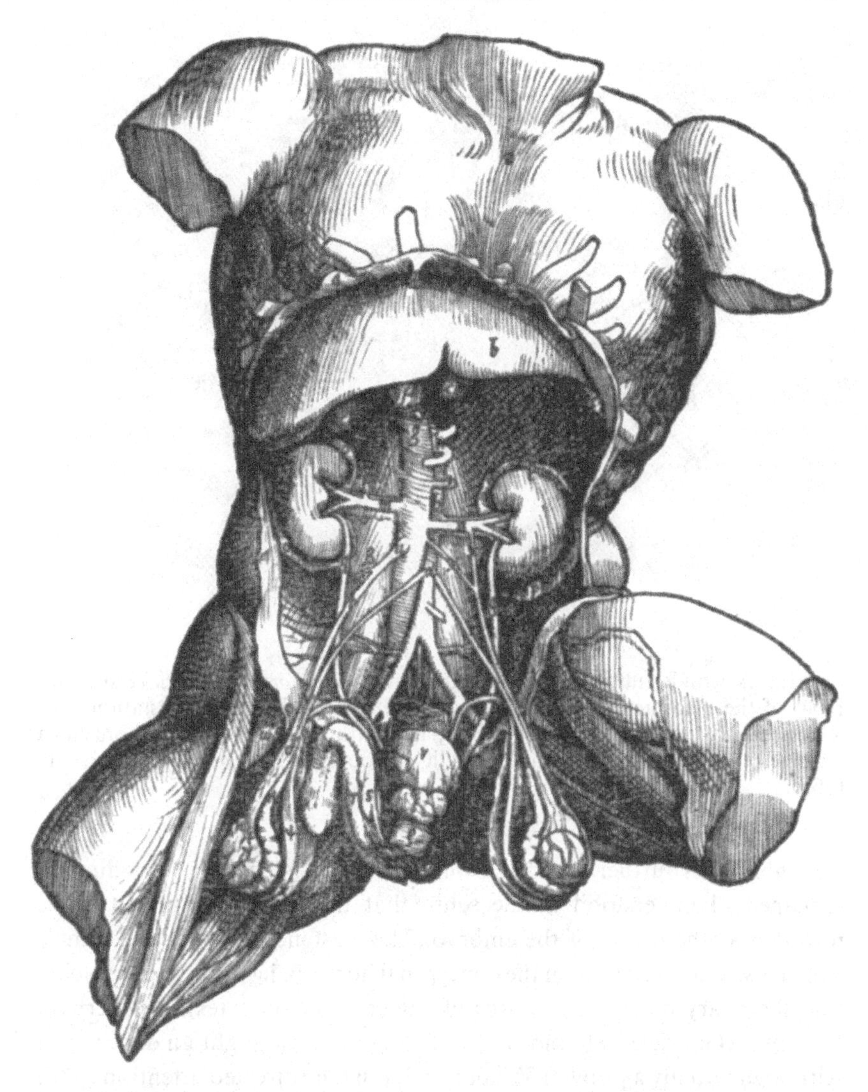

Fig. 1.1. In any modern essay bearing on reproduction in mammals, this superb illustration from Vesalius (1543) is an excellent starting point. It depicts the organs of generation, and also highlights some of the asymmetries in the disposition of the major blood vessels and kidneys. (Courtesy of Edinburgh University Library.)

masterpiece entitled *De Humani Corporis Fabrica Libri Septem*, the seven volumes usually being known together as the *Fabrica*. They are generally considered to form the basis of modern systematic research in anatomy and contain extensive illustrations of the organs of generation (Fig. 1.1), and doubtless influenced numerous successors in his adopted Padua, as well as

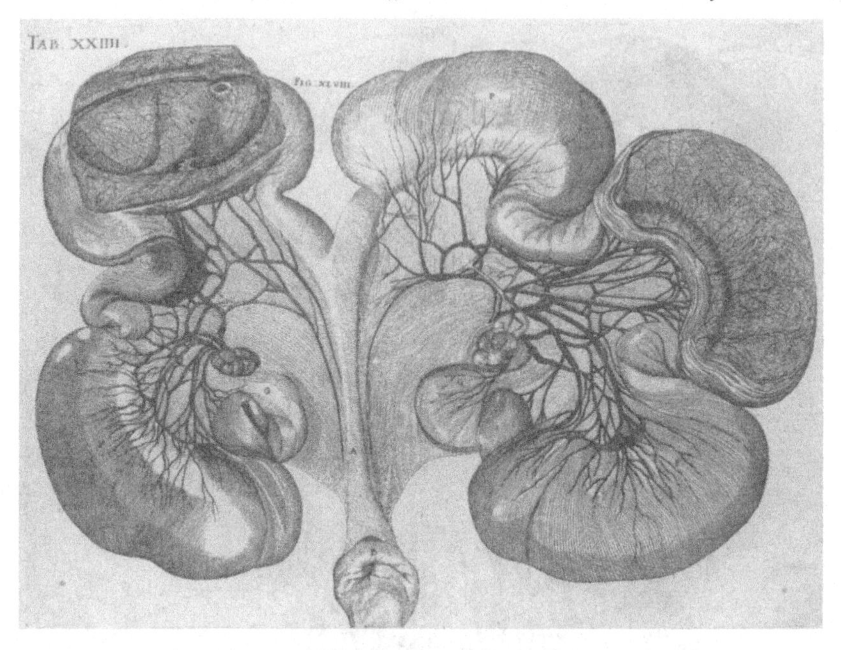

Fig. 1.2. A wonderfully precise illustration from *De Formato Foetu* (Fabricius, 1604) of the bicornuate uterus of a pig at an advanced stage of gestation. The relatively small number of foetuses compared with the number of corpora lutea indicates considerable embryonic mortality. (Courtesy of Cambridge University Library.)

students much further afield. However, the first stages of reproduction remained ill understood in the sense that the uterus continued to be regarded as the origin of the embryo. The existence of male and female gametes was not yet even on the conceptual horizon, let alone the suspicion that the ovary might be a source of egg cells (i.e. oocytes). But perhaps Fallopius (Gabriele Fallopio, 1523–1562) gave some slight guidance here. Although initially a pupil of Vesalius in Padua, he directed attention to the duct that now bears his name and in which, it was later appreciated, the process of fertilisation takes place. The definitive description of the Fallopian tube appears in his *Observationes anatomicae* (1561) and is rather widely cited in its English translation.

Also in Padua, a pupil of Fallopius called Fabricius (1537–1619) concluded that the Fallopian tube or duct was an organ of secretion, having produced a reasonably accurate account of its function in the formation of chicken eggs. As a gifted anatomist, Fabricius also provided accurate descriptions of the uterus and round ligaments (Fig. 1.2), and likewise of the ovarian and uterine arteries and their anastomoses. As is widely known, the Englishman William Harvey (1578–1657) was a student of Fabricius and much interested in reproduction as well as in the circulation of blood. He examined the contents of the uterus in various mammals after mating and, based largely on his studies in deer with delayed implantation and hence delayed embryonic development, concluded that the existence of 'female semen' was a myth and indeed that male semen was not involved in formation of the foetus. His positive contribution to reproduction was the inspired proposition that 'Ex ovo omnia' or 'All living things come from eggs' (Harvey, 1651), a view that incidentally was not supported by actual observation for a further 200 years. Even so, Harvey's pronouncement represented the ovist school of thought, the one in which the offspring is considered to be already preformed in the egg.

A Dutchman called de Graaf (1641–1673) kept the focus on eggs, even though he was quite unable to see them as distinct cells. Basing his studies initially on avian gonads, he identified the tertiary follicles that were later to take his name in mammalian ovaries (Table 1.1). In his publication of 1672, de Graaf maintained that the eggs originated in the ovaries, but was confronted by Harvey's (and Galen's) notion of conception, that of an ovum formed *in utero*, which still prevailed. There was also the fact that the transport function of the Fallopian tubes, that is between the ovaries and the uterus, was not yet generally appreciated, although de Graaf himself had supposedly traced the passage of eggs through the tubes to the uterus. He believed that the follicles were eggs and had been misled by noting their disappearance after copulation in the rabbit, an induced ovulator, followed by the appearance of blastocysts in the uterus a few days later. However, as he himself was willing to admit, the relative size of ovarian follicles compared with that of the Fallopian tubes could not be reconciled with the passage of eggs, unless in reality it was the contents of the follicles that entered the tubes.

As improved models of microscope became available, investigators gradually – very gradually – drew closer to a factual understanding of the first steps of reproduction. Another Dutchman, Anton van Leeuwenhoek (1678), was able to report to the Royal Society of London how semen examined under a microscope that he himself had devised contained millions of little animalcules. The observations were made in 1677 on the semen of fish, frogs and mammals. This is usually thought to have been the first factual description of spermatozoa, and Leeuwenhoek (1683) went on to postulate that de Graaf's 'egg' required to be impregnated by one of the animalcules for pregnancy to occur, although of course no such

Table 1.1. *Vital landmarks since the seventeenth century in understanding reproductive events in mammals*

de Graaf	1672	Asserted homology of mammalian ovary with that of bird, and focused on tertiary (Graafian) follicles
van Leeuwenhoek	1678	Spermatozoa recorded as animalcules in semen of fish, frogs and mammals
Cruickshank	1797	Identified rabbit eggs in the Fallopian tubes after coitus; (induced ovulation in rabbits)
von Baer	1827	Described origin of the mammalian egg in the Graafian follicle
Bischoff	1854	Early details of sperm penetration into egg
Van Beneden	1875	Arguably the first systematic description of fertilisation in mammals (rabbit)
Waldeyer	1888	Key description and naming of chromosomes following the observations of Hertwig (1876)
Heape	1890	Successful embryo transplantation between donor and recipient rabbits
Marshall	1910	Summarised much of the available knowledge concerning the physiology of reproduction
Pincus	1936	Stimulated the development of techniques of *in vitro* fertilisation
Jost	1947	Demonstrated local humoral influence of embryonic gonads on development (and regression) of neighbouring genital ducts
Ford *et al.*; Jacobs & Strong	1959	Key rôle of the Y chromosome in sex determination of mammalian embryos
Tarkowski	1961	Generation of mammalian chimaeras to study differentiation
Wachtel *et al.*	1975	H-Y antigen proposed as the elusive testis-determining factor
Steptoe & Edwards	1978	First full-term human offspring born as a consequence of IVF and embryo transplantation
Koopman *et al.*	1991	Production of sex-reversed transgenic mice by *Sry* injection into pronucleate eggs

observation was made. In any event, the alternative version of the preformationist theory had come to life with the animalculist or spermist school of believers (Fig. 1.3). During the time needed for Leeuwenhoek's letter to be translated from Dutch to Latin and then printed in the Royal Society (1678) volume that actually appeared in 1679, others were able to submit comparable reports on spermatozoa to the French Academy and gain apparent priority (i.e. Hartsoeker, 1678; Huguens, 1678), whereas they had

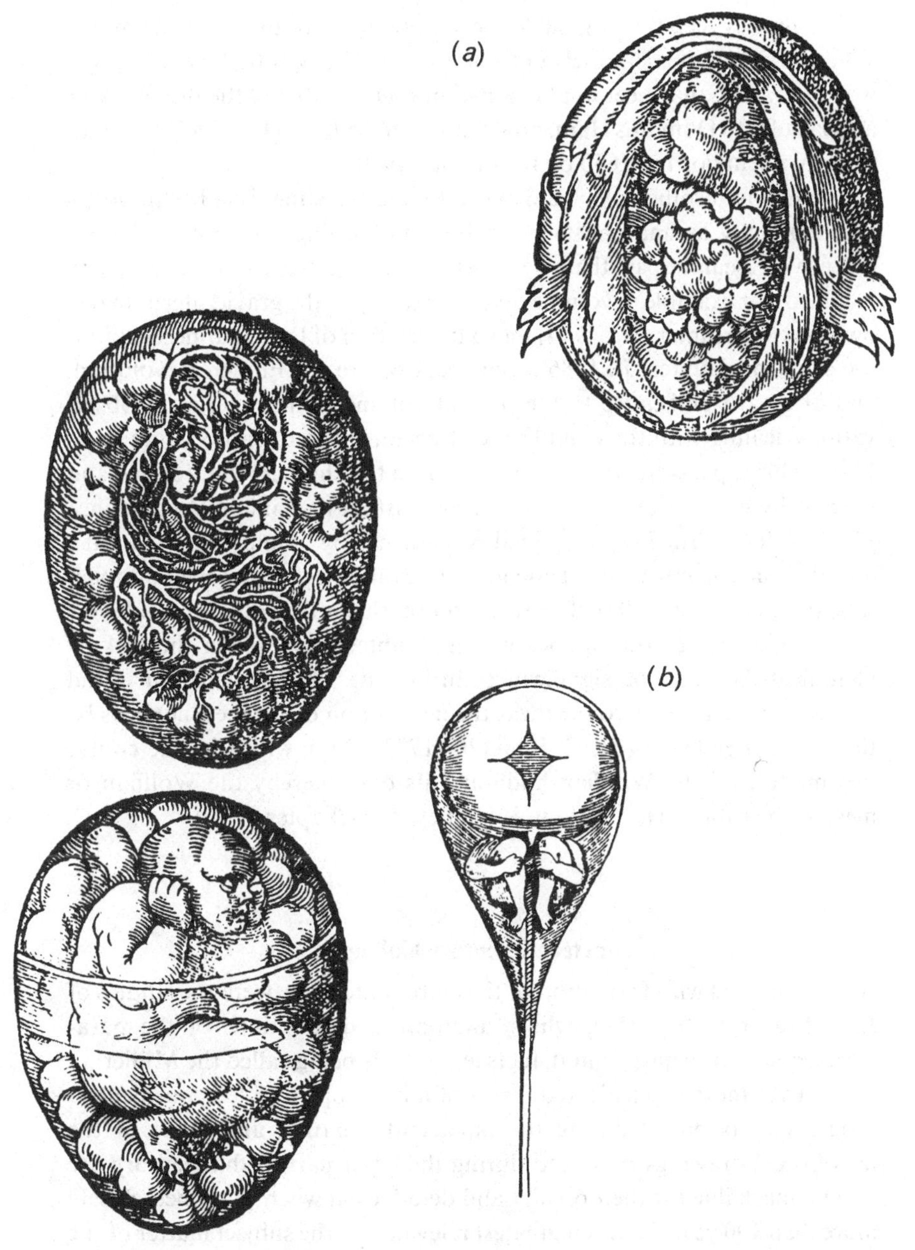

Fig. 1.3. Drawings to portray some imaginative ideas on the formation of a new individual taken from (*a*) Rueff (1554) and (*b*) Hartsoeker (1694). The drawings of Rueff represent an early version of the ovist view, which stands in marked contrast to the subsequent spermist view of generation. (Courtesy of Glasgow University Library.)

in fact already seen the original letter. Before drawing any moral from this situation, which has parallels not unknown in the twentieth century, it is worth recalling that Leeuwenhoek had himself attributed the discovery of animalcules to Dominus Ham, now thought to be Johan Ham, a Dutchman from Arnhem born in 1650 or 1651 (Cole, 1930).

The Hunter brothers, William and John, both distinguished as surgeons and especially as anatomists, were busy publishing the results of their extensive researches in the later part of the eighteenth century. These included, for example, a volume on the anatomy of the gravid uterus in our own species (Hunter, W., 1774) and a description of testicular descent into the scrotum (Hunter, J., 1786). Ten years before the French Revolution, and of great significance to the present volume, came a series of observations dealing with the condition of freemartinism in cattle (Hunter, J., 1779). This topic is treated at some length in the chapters that follow, and specifically in Chapter 5. As anatomical assistant to William Hunter since 1771, William Cruickshank gained prominence by amplifying de Graaf's observations, having isolated and identified rabbit eggs from the Fallopian tube (Cruickshank, 1797). The secretion of mucin that develops in layers around the eggs of this species was undoubtedly helpful in the task of identification. Also of significance during the eighteenth century and relevant to the present volume was the description of the mesonephros by the German embryologist C. F. Wolff (1733–1794), which in due course became termed the Wolffian body and its duct thereby the Wolffian or mesonephric duct. This is discussed in detail in Chapter 4.

Nineteenth century highlights

Also concerned with formation of the embryonic ducts were the studies of J. P. Müller (1801–1858), whose anatomical descriptions of the paramesonephric or female genital ducts led to their being called the Müllerian ducts. Ever more sophisticated forms of microscope enabled conspicuous progress to be made during the nineteenth century, and some of the histological drawings produced during the latter part of that century are quite remarkable for their quality and detail, even when examined closely more than 100 years later. Of greatest relevance to the subject matter of the present monograph are the studies on eggs, on the process of fertilisation, and on embryo transplantation. As is generally appreciated by biologists, it was von Baer (1827) who was given credit for discovering the mammalian egg or, more correctly, its source in the Graafian follicle, thereby overcoming the confusion existing between eggs and follicles. This was achieved by

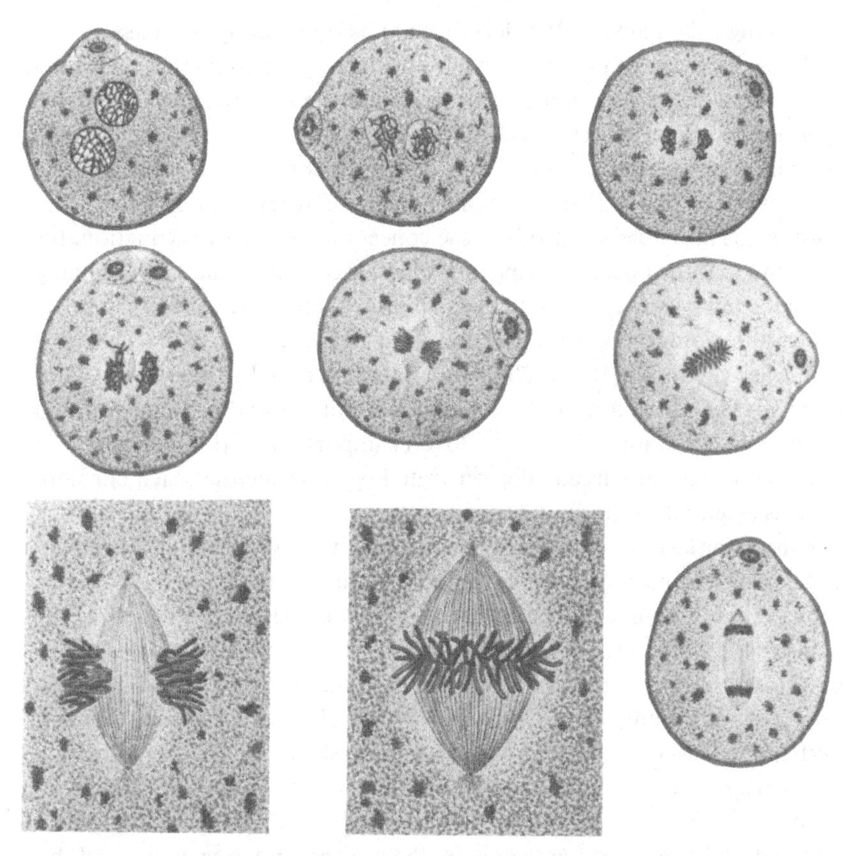

Fig. 1.4. A selection of illustrations from the extensive series published by Sobotta (1895) showing condensation of the haploid groups of chromosomes within the male (lower) and female pronuclei, and arrangement of chromosomes on the first mitotic (cleavage) spindle. Even the microtubules of the latter structure are represented in outline.

dissecting open Graafian follicles and examining the liberated contents within their cumulus masses. von Baer recorded the triumphant moment:

Led by curiosity . . . I opened one of the follicles and took up the minute object on the point of my knife, finding that I could see it very distinctly and that it was surrounded by mucus. When I placed it under the microscope I was utterly astonished, for I saw an ovule just as I had already seen them in the tubes, and so clearly that a blind man could hardly deny it. (Translation from Corner, 1933.)

This was followed by the publications of Barry (1843) and Bischoff (1854) dealing with the entry of sperm into mammalian eggs, and then came the detailed and systematic report of Van Beneden (1875) on fertilisation in the rabbit egg and of Sobotta (1895) on early embryonic development in the mouse (Fig. 1.4).

Although the name of Waldeyer (1888) has frequently been associated with the discovery of mammalian chromosomes, Henking (1871) working with invertebrates had noted a chromosomal element present in half the spermatids of the insect *Pyrrhocoris apterus*, and Van Beneden (1883) studying the nematode *Ascaris* had described chromosomes in both the sperm nucleus and egg nucleus which combine at fertilisation. The latter author was also able to introduce the concept of chromatin reduction, for he found that the nucleus of ripe reproductive cells contained half as many chromosomes as the germinal cells to which they gave birth. This key observation on chromatin reduction, that is the reduction division, was confirmed by Boveri (1887). These foundations led to Wilson's (1905) work on meiosis, and the notion that combination of haploid gametes restored the diploid condition. Moreover, Boveri's important work on dispermy in sea urchin eggs had already demonstrated by this time that each chromosome is essential for normal development.

During the last twenty years of the nineteenth century, Walter Heape was developing his interest in mammalian reproductive biology, and covering topics ranging from fertility and infertility in sheep (Heape, 1899*a*,*b*) to the menstrual cycle of monkeys (Heape, 1894). He was certainly interested in questions of sex determination and was even rumoured to have achieved a means of controlling sex. If there was any truth in such a rumour, then it is perhaps surprising that details were never published – a significant omission for such a prolific author. However, his experiments on embryo transplantation in rabbits were successful and were published (Heape, 1890, 1897), and not only are they amazing in themselves and one source of his distinguished reputation, but they also enabled an early appreciation of the distinction between maternal influences on the embryo or foetus and expression of its own genetic constitution. The technique of embryo transplantation, or egg transfer as it is frequently termed, has been of outstanding value in a wide variety of experiments and is an essential step for testing the viability of an embryo after diverse *in vitro* procedures or manipulations. Indeed, this brief historical survey concludes below with reference to transgenic mice and sheep that required the specific step of embryo transplantation for their production after genetic manipulation at the pronuclear stage.

Before turning to the twentieth century, it is worth summarising the views of the preformationists and contrasting these with the epigeneticists. In essence the preformationists, be they ovists or spermists, believed that the new individual already existed in minute form in the egg or in the spermatozoon. An early version of the ovist view is illustrated in Fig. 1.3

taken from the work of Rueff (1554) and, as will be realised from previous sections of this chapter, it existed long before the spermist view. In large measure, it stemmed from the beliefs of Aristotle, although it came to be modified during the seventeenth century with Harvey's pronouncement and as attention focused on the Graafian follicle. By contrast, the spermist view only materialised with the identification of Leeuwenhoek's animalcules, or spermatozoa, and the preformationist impression here comes across clearly in Hartsoeker's depiction of an homunculus, a complete preformed foetus that merely had to grow (Fig. 1.3). However, it will be clear that long before the turn of the nineteenth century, the preformationist school had lost all validity. The contrasting school, that of epigenesis, was progressively taking to the stage, but few of its proponents had even begun to suspect the quite enormous complexity and subtlety of the process of differentiation, let alone its regulation by chemical units that came to be called genes.

Twentieth century developments

Findings of significance during the last 100 years or so that bear directly on the substance of this book have been many, various and profound, and represent a quantum leap in accumulated knowledge. They include identification of the sex chromosomes during the first decade of the century, the fact that males and females differ in one of their chromosome pairs (see McClung, 1902; Morgan, 1910; Wilson, 1911), and definition of the relationship between chromosomes and genes (Morgan, 1926); extraction, characterisation and synthesis of key steroid hormones by the late 1930s (see Parkes, 1962); the pivotal rôle of the testis and the so-called permissive rôle of the ovary in mammalian sexual differentiation (Jost, 1947); the genetic basis of heredity residing in nuclear DNA arranged as a double helix (Watson & Crick, 1953) and the subsequent emergence of molecular biology as a distinct discipline (see Pollock, 1965); and appreciation of the involvement of the Y chromosome in sex determination in mammals (Ford *et al.*, 1959; Jacobs & Strong, 1959; Welshons & Russell, 1959). The last discovery in turn led to clinical conditions such as Klinefelter's or Turner's syndromes in man being ascribed to anomalies of the sex chromosome complements (Chapter 7). Mammalian chimaeras were generated in the laboratory by Tarkowski (1961) and by Mintz (1962), and soon thereafter experimental chimaeras were being used by a series of biologists to analyse diverse problems of development (see Chapter 8).

Notable landmarks in molecular genetics and its application include

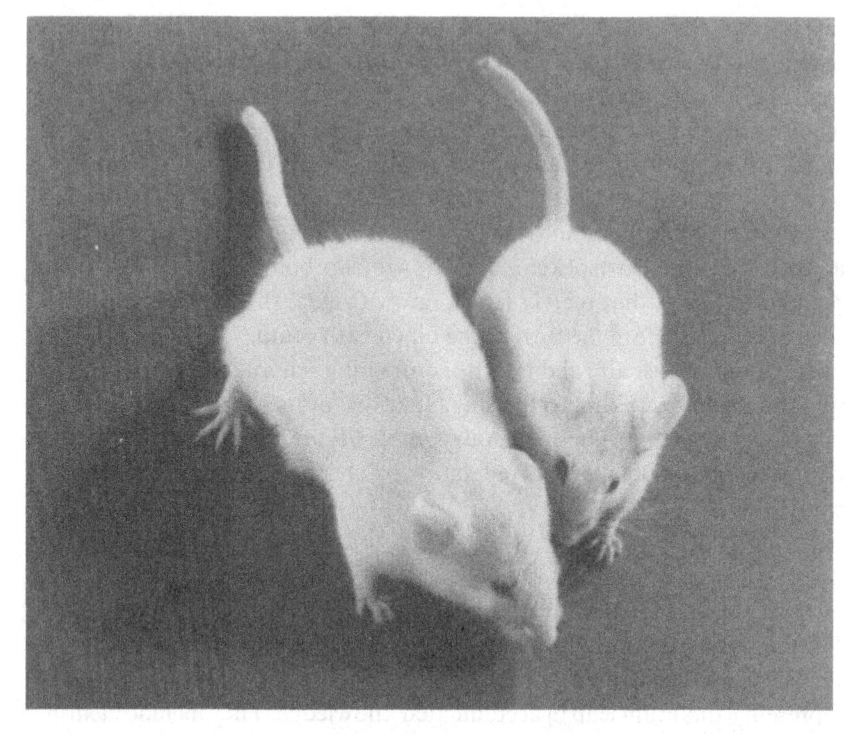

Fig. 1.5. A dramatic illustration of the impact of molecular biology techniques on animal growth and development. The mouse on the right is of normal mature size whereas that on the left is a transgenic 'giant' mouse weighing almost twice as much as its sibling. It was produced by introducing a growth hormone gene construct into a pronucleus by the approach of microinjection, followed by transplantation of the manipulated embryo into a recipient animal. (Courtesy of Dr R. L. Brinster.)

identification of a number of candidates for the putative male or testis-determining factor such as H-Y antigen, snake satellite (*Bkm*) DNA sequences, zinc finger proteins and, most recently, the tiny DNA sequence termed *Sry* (Chapter 2). Transgenic giant mice (Fig. 1.5) bearing a gene construct for growth hormone were engineered by Palmiter *et al.* (1982), and gene transfer also enabled modified milk proteins to be secreted in mice and sheep (Simons *et al.*, 1987, 1988). Equally impressively, transgenic mice have been produced to demonstrate sex reversal after a minimum sex-specific DNA sequence (*Sry*) was microinjected into female embryos or, more strictly, eggs at the pronuclear stage (Koopman *et al.*, 1991). Molecular biological techniques have also been used to achieve separation of X- and Y-bearing spermatozoa prior to artificial insemination (Johnson, Flook & Hawk, 1989; Métézeau *et al.*, 1991) and to identify the sex of bovine blastocysts prior to transplantation (Leonard *et al.*, 1987; Cotinot *et al.*, 1991). Comments on these latter technologies are to be found in Hunter (1988).

An ever-increasing number of texts and major publications bearing on mammalian reproduction underline the establishment of reproductive physiology as a mature and fascinating discipline in its own right. British universities such as Cambridge and Edinburgh became key centres for reproductive studies (Table 1.2), and the formation of specialist research groups such as the ARC Unit of Reproductive Physiology and Biochemistry in Cambridge (1955) and the MRC Unit of Reproductive

Table 1.2. *Some prominent contributors to reproductive studies in or associated with the University of Edinburgh (studies in clinical and experimental endocrinology are not cited)*

Barry, M.	1802–1855	Practised medicine in Edinburgh. Published key paper on fertilisation of mammalian eggs (1843)
Darwin, C.	1809–1882	Sometime medical student (1825–1827). Not impressed by surgical demonstrations. Later to offer views on evolution
Simpson, J. Y.	1811–1870	Professor of Midwifery 1840–1870. Introduced chloroform anaesthetic in childbirth. University Maternity Pavilion named after him
Ewart, J. C.	1851–1933	Professor of Natural History 1882–1927. Experiments on telegony. Author of *The Penycuik Experiments* (1899)
Marshall, F. H. A.	1878–1949	Worked in Edinburgh 1900–1908, initially as Assistant to Cossar Ewart, latterly as first University Lecturer in reproductive physiology (1905–1908). Wrote first edition of *The Physiology of Reproduction* (published 1910)
Crew, F. A. E.	1886–1973	Head of Institute of Animal Breeding (1920). First Buchanan Professor of Genetics (1928) and Head of Institute of Animal Genetics (1930). Much interested in sex determination and intersexuality
Walton, A.	1897–1959	Graduated in agriculture (1923) and commenced research on artificial insemination in Edinburgh. Moved to Cambridge to become Sir John Hammond's senior colleague. Much interested in sperm physiology

Table 1.2. (*cont.*)

Donald, H. P. Hancock, J. L. Land, R. B.	Relevant studies from 1958	ARC Animal Breeding Research Organisation (1950). Reproductive experiments such as selection for fertility, heterospermic insemination and embryo transplantation
Jacobs, P. Chandley, A. Evans, J.	1959	MRC Unit of Clinical and Population Cytogenetics Research. Key work on human chromosomes, especially the sex chromosomes, and anomalous chromosome constitutions in clinical cases
Beatty, R. A. McLaren, A. Edwards, R. G.	Late 1950s and especially 1960s	Working in Institute of Animal Genetics. R. G. Edwards was graduate student in Department of Genetics (PhD 1955). The group's interests included male and female gametes, embryonic development in mammals, and chimaeras
Short, R. V. Baird, D. T.	1972	First Director and Deputy Director of newly-formed MRC Unit of Reproductive Biology. Founder members already established in Edinburgh were D. T. Baird, R. A. Beatty and A. McLaren
Jones, K. W. Bishop, J. O.	1980s	Particular interest in molecular genetics, working in the Department of Genetics. K. Jones involved in early work with DNA probes for candidate genes involved in mammalian sex determination
Land, R. B. Clark, A. J. Simons, P.	1982	Established molecular biology programme for the Animal Breeding Research Organisation. Produced transgenic sheep yielding novel milk proteins
Hooper, M. L. Patek, C. E. West, J. D.	1990	Use of DNA probes to analyse cell origin in chimaeras and the sex chromosome constitution of different cell lines. Interest in stem cells

Biology in Edinburgh (1972) also testify to this fact. Concerning publications, it is seldom easy to offer a completely balanced selection yet many people in the field would probably agree that the following texts give a useful historical flavour of reproductive developments during the present century.

1. *The Physiology of Reproduction* (Marshall, 1910).
2. *The Physiology of Reproduction in the Cow* (Hammond, 1927).

3. *The Internal Secretions of the Ovary* (Parkes, 1929).
4. *The Eggs of Mammals* (Pincus, 1936).
5. *Patterns of Mammalian Reproduction* (Asdell, 1946).
6. *Neural Control of the Pituitary Gland* (Harris, 1955).
7. *Reproductive Physiology* (Nalbandov, 1958).
8. *Sex and Internal Secretions* (Young, 1961).
9. *The Biochemistry of Semen and of the Male Reproductive Tract* (Mann, 1964).
10. *Reproduction in Mammals* (Austin & Short, 1972).
11. *Handbook of Physiology* (Greep, 1973).
12. *Reproduction in Farm Animals* (Hafez, 1974).
13. *The Ovary* (Zuckerman & Weir, 1977).
14. *Conception in the Human Female* (Edwards, 1980).
15. *The Physiology of Reproduction* (Knobil & Neill, 1988).

The reference work that many still regard as the reproductive bible (*Marshall's Physiology of Reproduction*) is slowly emerging from the press in its fourth edition (Lamming, 1984), the comprehensive French volume *La Reproduction chez les Mammifères et l'Homme* (Thibault & Levasseur, 1991) is now available in an up-dated English edition entitled *Reproduction in Mammals and Man* (1993), and more specialised monographs abound, as do volumes emanating from conferences, symposia and workshops. As to the processes of sex determination and differentiation in mammals, selected milestones would include the writings of Crew (1927, 1965), Ford (1963, 1970), Mittwoch (1967, 1973), Ohno (1967, 1979) and Wachtel (1983, 1989). The two concise works of McLaren (1976, 1981) are also highly pertinent and a pleasure to read on the grounds of clarity of style and elegance of exposition – important aspects in works of scholarship not always conspicuous in the latter part of the twentieth century. And not to be overlooked as the reader becomes immersed in the detail of the following chapters is the classical treatment of Maynard Smith (1978) entitled *The Evolution of Sex* and a subsequent volume on *The Origin and Evolution of Sex* edited by Halvorson & Monroy (1985).

References

Afzelius, B. A. & Baccetti, B. (1991). History of spermatology. In *Comparative Spermatology 20 Years After*, vol. 75, ed. B. Baccetti, Serono Symposia Publications, pp. 1–10. New York: Raven Press.

Asdell, S. A. (1946). *Patterns of Mammalian Reproduction*. Ithaca, New York: Comstock.

Austin, C. R. & Short, R. V., ed. (1972). *Reproduction in Mammals*, 1st edn. Cambridge: Cambridge University Press.

Baer, K. E. von (1827). *De ovi mammalium et hominis genesi*. Leipzig.
Barry, M. (1843). Spermatozoa observed within the mammiferous ovum. *Philosophical Transactions of the Royal Society, Series B*, **133**, 33.
Bischoff, T. L. W. (1854). *Entwicklungsgeschichte des Rehes*. Giessen.
Boveri, T. (1887). Zellenstudien I: Die Bildung der Richtungskörper bei Ascaris megalocephala und Ascaris lumbricoides. *Zeitschrift für Naturwissenschaften*, **21**, 423. Jena.
Cole, F. J. (1930). *Early Theories of Sexual Generation*. Oxford: Clarendon Press.
Corner, G. W. (1933). The discovery of the mammalian ovum. In *Lectures on the History of Medicine, 1926–1932*. Philadelphia: Mayo Foundation Lectures.
Cotinot, C., Kirszenbaum, M., Leonard, M., Gianquinto, L. & Vaiman, M. (1991). Isolation of bovine Y-derived sequence: potential use in embryo sexing. *Genomics*, **10**, 646–53.
Crew, F. A. E. (1927). *The Genetics of Sexuality in Animals*. Cambridge: Cambridge University Press.
Crew, F. A. E. (1965). *Sex-Determination*, 4th edn. London: Methuen.
Cruickshank, W. (1797). Experiments in which, on the third day after impregnation, the ova of rabbits were found in the Fallopian tubes; and on the fourth day after impregnation in the uterus itself; with the first appearance of the foetus. *Philosophical Transactions of the Royal Society*, **87**, 197–214.
De Graaf, R. (1672). *De mulierum organis generationi inservientibus tractatus novus*. Leyden.
Edwards, R. G. (1980). *Conception in the Human Female*. London & New York: Academic Press.
Ewart, J. C. (1899). *The Penycuik Experiments*. London: Adam & Charles Black.
Fabricius, H. (1604). *De Formato Foetu*. Padua.
Fallopius, G. (1561). *Observationes anatomicae*. Venice.
Ford, C. E. (1963). The cytogenetics of human intersexuality. In *Intersexuality*, ed. C. Overzier, pp. 86–117. London: Academic Press.
Ford, C. E. (1970). Cytogenetics and sex determination in man and mammals. *Journal of Biosocial Science (Supplement)*, **2**, 7–30.
Ford, C. E., Jones, K. W., Polani, P. E., De Almeida, J. C. & Briggs, J. H. (1959). A sex-chromosome anomaly in a case of gonadal dysgenesis (Turner's syndrome). *Lancet*, **i**, 711–13.
Greep, R. O., ed. (1973). *Handbook of Physiology*. Section 7. Endocrinology II. Female Reproductive System, Part II. Washington DC: American Physiological Society.
Hafez, E. S. E., ed. (1974). *Reproduction in Farm Animals*, 3rd edn. Philadelphia: Lea & Febiger.
Halvorson, H. O. & Monroy, A., ed. (1985). *The Origin and Evolution of Sex*. New York: Alan Liss.
Hammond, J. (1927). *The Physiology of Reproduction in the Cow*. Cambridge: Cambridge University Press.
Harris, G. W. (1955). *Neural Control of the Pituitary Gland*. London: Edward Arnold.
Hartsoeker, N. (1678). Extrait d'une lettre de M. Nicolas Hartsoeker écrite à l'Auteur du Journal touchant la manière de faire les nouveaux microscopes, dont il a été parlé dans le Journal il y a quelques jours. *Journal des Scavans, Paris*, **30**, 355–6, le 27 août, 1678.
Hartsoeker, N. (1694). *Essay de Dioptrique*. Paris.
Harvey, W. (1651). *Exercitationes de generatione animalium*. London & Amsterdam.

Heape, W. (1890). Preliminary note on the transplantation and growth of mammalian ova within a uterine foster-mother. *Proceedings of the Royal Society*, **48**, 457–8.

Heape, W. (1894). XI. The menstruation of *Semnopithecus entellus*. *Philosophical Transactions of the Royal Society of London, Series B*, **185**, 411–71.

Heape, W. (1897). Further note on the transplantation and growth of mammalian ova within a uterine foster-mother. *Proceedings of the Royal Society of London, Series B*, **62**, 178–83.

Heape, W. (1899*a*). Note on the fertility of different breeds of sheep, with remarks on the prevalence of abortion and barrenness therein. *Proceedings of the Royal Society of London, Series B*, **65**, 99–111.

Heape, W. (1899*b*). Abortion, barrenness and fertility in sheep. *Journal of the Royal Agricultural Society, Series III*, **10**, 1–32.

Henking, H. (1871). Untersuchungen über die ersten Entwicklungsvorgänge in den Eiern der Insekten. II Über Spermatogenese und deren Beziehung zur Entwicklung bei *Pyrrohocoris apterus*. *Zeitschrift für Wissenschaftliche Zoologie, Abt A*, **51**, 685–736.

Hertwig, O. (1876). Beiträge zur Kenntniss der Bildung, Befruchtung und Theilung des thierischen Eies. *Morphologisches Jahrbuch*, **1**, 347–434.

Huguens, C. (1678). Extrait d'une lettre de M. Huguens de l'Academie R. des Sciences à l'Auteur du Journal, touchant une nouvelle manière de microscope qu'il a apporté de Hollande. *Journal des Scavans, Paris*, **28**, 331–2, le 15 août, 1678.

Hunter, J. (1779). An account of the free martin. *Philosophical Transactions of the Royal Society*, **69**, 279–93.

Hunter, J. (1786). A description of the situation of the testis in the foetus, with its descent into the scrotum. In *Observations on Certain Parts of the Animal Œconomy*, pp. 1–26. London.

Hunter, R. H. F. (1988). *The Fallopian Tubes: Their Rôle in Fertility and Infertility*. Berlin: Springer-Verlag.

Hunter, W. (1774). *The Anatomy of the Human Gravid Uterus*. Birmingham: Baskerville.

Jacobs, P. A. & Strong, J. A. (1959). A case of human intersexuality having a possible XXY sex-determining mechanism. *Nature (London)*, **183**, 302–3.

Johnson, L. A., Flook, J. P. & Hawk, H. W. (1989). Sex preselection in rabbits: live births from X and Y sperm separated by DNA and cell sorting. *Biology of Reproduction*, **41**, 199–203.

Jost, A. (1947). Recherches sur la différenciation sexuelle de l'embryon de lapin. III. Rôle des gonades foetales dans la différenciation sexuelle somatique. *Archives d'Anatomie Microscopique et de Morphologie Expérimentale*, **36**, 271–315.

Knobil, E. & Neill, J. *et al.*, ed. (1988). *The Physiology of Reproduction*. New York: Raven Press.

Koopman, P., Gubbay, J. , Vivian, N., Goodfellow, P. & Lovell-Badge, R. (1991). Male development of chromosomally female mice transgenic for *Sry*. *Nature (London)*, **351**, 117–21.

Lamming, G. E., ed. (1984). *Marshall's Physiology of Reproduction*, 4th edn, vol. 1. Edinburgh & London: Churchill Livingstone.

Leeuwenhoek, A van (1678). Observationes de Anthonii Leeuwenhoek, de natis è semine genitali animalculis. *Philosophical Transactions of the Royal Society*, **12**, 451–2.

Leeuwenhoek, A. van (1683). An abstract of a letter from Mr Anthony Leeuwenhoek of Delft about Generation by an Animalcule of the Male seed.

Animals in the seed of a frog. Some other observables in the parts of a Frog. Digestion, and the motion of the blood in a Feavor. *Philosophical Transactions of the Royal Society*, **13**, 347–55.

Leonard, M., Kirszenbaum, M., Cotinot, C., Chesné, P., Heyman, Y., Stinnakre, M., Bishop, C., Vaiman, M. & Fellous, M. (1987). Sexing bovine embryos using Y-chromosome specific DNA probe. *Theriogenology*, **27**, 248.

McLaren, A. (1976). *Mammalian Chimaeras.* Cambridge & London: Cambridge University Press.

McLaren, A. (1981). *Germ Cells and Soma: A New Look at an Old Problem.* New Haven & London: Yale University Press.

McClung, C. E. (1902). The accessory chromosome-sex determinant? *Biological Bulletin, Marine Biological Laboratory (Woods Hole, Mass.)*, **3**, 43–84.

Mann, T. (1964). *The Biochemistry of Semen and of the Male Reproductive Tract.* London: Methuen.

Marshall, F. H. A. (1910). *The Physiology of Reproduction.* London: Longmans, Green.

Maynard Smith, J. (1978). *The Evolution of Sex.* Cambridge: Cambridge University Press.

Métézeau, P., Cotinot, C., Colas, G., Azoulay, M., Kiefer, H., Goldberg, M. E. & Kirszenbaum, M. (1991). Improvement of flow cytometry analysis and sorting of bull spermatozoa by optical monitoring of cell orientation as evaluated by DNA specific probing. *Molecular Reproduction and Development*, **30**, 250–7.

Mintz, B. (1962). Formation of genotypically mosaic mouse embryos. *American Zoologist*, **2**, 432.

Mittwoch, U. (1967). *Sex Chromosomes.* New York & London: Academic Press.

Mittwoch, U. (1973). *Genetics of Sex Differentiation.* New York & London: Academic Press.

Mittwoch, U. (1977). To be right is to be born male. *New Scientist*, **73**, 74–6.

Mittwoch, U. (1986). Males, females and hermaphrodites. *Annals of Human Genetics*, **50**, 103–21.

Morgan, T. H. (1910). Chromosomes and heredity. *American Naturalist*, **44**, 449–96.

Morgan, T. H. (1926). *The Theory of the Gene.* New Haven, Connecticut: Yale University Press.

Nalbandov, A. V. (1958). *Reproductive Physiology.* San Francisco & London: W. H. Freeman.

Ohno, S. (1967). *Sex Chromosomes and Sex-Linked Genes.* Berlin: Springer-Verlag.

Ohno, S. (1979). *Major Sex-Determining Genes.* Berlin: Springer-Verlag.

Palmiter, R. D., Brinster, R. L., Hammer, R. E., Trumbauer, M. E., Rosenfeld, M. G., Birnberg, N. C. & Evans, R. M. (1982). Dramatic growth of mice that develop from eggs microinjected with metallothionein-growth hormone fusion genes. *Nature (London)*, **300**, 611–15.

Parkes, A. S. (1929). *The Internal Secretions of the Ovary.* London: Longmans.

Parkes, A. S. (1962). Prospect and retrospect in the physiology of reproduction. *British Medical Journal*, **ii**, 71–5.

Pincus, G. (1936). *The Eggs of Mammals.* New York: Macmillan.

Pollock, M. R. (1965). 1965 Nobel Prize for Medicine. *Nature (London)*, **208**, 1250–2.

Rueff, J. (1554). *De conceptu et generatione hominis.* Zürich.

Short, R. V. (1969). An introduction to some of the problems of intersexuality. *Journal of Reproduction and Fertility, Supplement*, **7**, 1–8.

Short, R. V. (1977). The discovery of the ovaries. In *The Ovary*, 2nd edn, ed. S. Zuckerman & B. J. Weir, ch. I, pp. 1–39. New York & London: Academic Press.
Simons, J. P., McClenaghan, M. & Clark, A. J. (1987). Alteration of the quality of milk by expression of sheep β-lactoglobulin in transgenic mice. *Nature (London)*, **328**, 530–2.
Simons, J. P., Wilmut, I., Clark, A. J. , Archibald, A. L., Bishop, J. O. & Lathe, R. (1988). Gene transfer into sheep. *Bio Technology*, **6**, 179–83.
Sobotta, J. (1895). Die Befruchtung und Furchung der Eies der Maus. *Archiv fur Mikroskopische Anatomie und Entwicklungsmechanik*, **45**, 15–93.
Steptoe, P. C. & Edwards, R. G. (1978). Birth after the reimplantation of a human embryo. *Lancet*, **ii**, 366.
Tarkowski, A. (1961). Mouse chimaeras from fused eggs. *Nature (London)*, **190**, 857–60.
Thibault, C. & Levasseur, M. C. (1991). *La Reproduction chez les Mammifères et l'Homme*. Paris: Ellipses.
Thibault, C., Levasseur, M. C. & Hunter, R. H. F. (1993). *Reproduction in Mammals and Man*. Paris: Ellipses.
Van Beneden, E. (1875). La maturation de l'oeuf, la fécondation et les premières phases du développement embryonnaire des mammifères d'après recherches faites chez le lapin. *Bulletin de l'Académie Belgique, Classe Science*, **40**, 686–9.
Van Beneden, E. (1883). Recherches sur la maturation de l'oeuf, la fécondation et la division cellulaire. *Archives de Biologie*, **4**, 610–620.
Vesalius, A. (1543). *De humani corporis fabrica libri septem*. Basel.
Wachtel, S. S. (1983). *H-Y Antigen and the Biology of Sex Determination*. New York: Grune & Stratton.
Wachtel, S. S., ed. (1989). *Evolutionary Mechanisms in Sex Determination*. Boca Raton, Florida: CRC Press.
Wachtel, S. S., Ohno, S., Koo, G. C. & Boyse, E. A. (1975). Possible role for H-Y antigen in the primary determination of sex. *Nature (London)*, **257**, 235–6.
Waldeyer, W. (1888). Über Karyokinese und ihre Beziehungen zu den Befruchtungsvorgängen. *Archiv für Mikroskopische Anatomie und Entwicklungsmechanik*, **32**, 1–122.
Watson, J. D. & Crick, F. H. C. (1953). Molecular structure of nucleic acids. A structure for DNA. *Nature (London)*, **171**, 737–8.
Welshons, W. J. & Russell, L. B. (1959). The Y-chromosome as the bearer of male determining factors in the mouse. *Proceedings of the National Academy of Sciences, USA*, **45**, 560–6.
Wilson, E. B. (1905). The chromosomes in relation to the determination of sex in insects. *Science*, **22**, 500–2.
Wilson, E. B. (1911). The sex chromosomes. *Archiv fürMikroskopische Anatomie und Entwicklungsmechanik, Abt II*, **77**, 249–71.
Young, W. C. (1961). *Sex and Internal Secretions*, 3rd edn. Baltimore: Williams & Wilkins.
Zuckerman, S. & Weir, B. J. (1977). *The Ovary*, 2nd edn. New York, San Francisco & London: Academic Press.

2

Mechanisms of sex determination

Introduction

Diverse mechanisms are involved in the determination of sex in lower animals, such as the ratio of the number of X chromosomes to autosomes in *Drosophila* (Hodgkin, 1990) or the ambient temperature during incubation of the embryo in turtles, alligators, crocodiles and some lizards (Charnier, 1966; Bull, 1980, 1983, 1987; Head, May & Pendleton, 1987; Deeming & Ferguson, 1988; Ewert & Nelson, 1991). However, the dogma for eutherians remains that it is the genetic nature of the spermatozoon penetrating and activating the oocyte that is the primary determinant of the sex in the resultant zygote. Such a genetic mechanism is taken to indicate evolutionary progress from the lower vertebrates in which environmental influences exercise such a large and often decisive rôle in directing the path of sexual differentiation. The genetic mechanism for sex determination in mammals is viewed as relatively stable, inferring that mammals have thereby gained an increased control over their environment (Mintz, 1968).

Because females are the homogametic sex in mammals, bearing two X chromosomes in diploid cells, it falls to the heterogametic population of spermatozoa to impose the decision and developmental programme according to the type of spermatozoon fusing with the oocyte and delivering its haploid complement of chromosomes into the vitellus. Activation of the egg by a Y-bearing spermatozoon is classically held to give rise to a male embryo, if the events of fertilisation and early development proceed successfully. Fertilisation by an X-bearing spermatozoon, by contrast, normally heralds formation of a female embryo. A major presumption in these statements is that the primary sex-determining mechanism is located uniquely or overridingly on the sex chromosome of a highly specialised cell, the male gamete. It would certainly seem worth considering at the very outset of this essay whether this latter notion is unequivocally correct on all occasions in placental mammals. If it is, then this would indicate a remarkable degree of chromosomal integrity during the mitotic and meiotic events of spermatogenesis within the highly vulnerable location of the scrotal testis, a precision not always demonstrable during cell division in non-germinal tissues. On the other hand, putative errors during the process of cell division might conceivably involve the translocation of sex-determining information from its native (Y) chromosome to another sex-determining (X) chromosome or even the production by spontaneous mutation of sex-determining information on an autosome.

Rôle of chromosomes in sex determination

Despite the mildly speculative reservations in the preceding paragraph concerning the chromosomal specificity of primary sex determining mechanisms, the sex chromosome constitution of the mammalian zygote is accepted as having been imposed at fertilisation by the penetrating spermatozoon. It therefore follows from the above that a female would be expected to have an XX sex chromosome constitution and a male an XY constitution. In essence, what is being stated is that all tissues of individuals recognisable phenotypically as male or female should match these predictions although, occasionally, chromosomal sex and phenotypic sex do not correspond or the sexual phenotype may be uncertain.

Whilst it was appreciated by the turn of the century that hereditary information resided in the chromosomes, the chromosomal basis for sex determination was first established between 1910 and 1916, largely by T. H. Morgan and his American colleagues. Its basis was deduced especially from a consideration of sex ratios, then sex linkage of diverse traits, hybridisation occurring in inversed sexual phenotypes and, more recently, by direct visualisation of the sex chromosomes using the technique of karyotyping developed in the late 1950s and early 1960s. These topics have been reviewed extensively during the last thirty years (e.g. Beatty, 1960, 1970; Crew, 1965; Gallien, 1965; Mittwoch, 1967; Ohno, 1967; Maynard Smith, 1978; Wachtel, 1983).

Although the rather small Y chromosome was originally believed not to transmit sex-determining information (two X chromosomes leading to female sex determination), it has been accepted since 1959 that (*a*) the Y chromosome is male determining (Table 2.1) and (*b*) there is a vital segment on the Y chromosome that acts as the primary signal for male development in both mouse and man (Ford *et al.*, 1959; Jacobs & Strong, 1959; Welshons

Table 2.1. *Selected key events in the search for the male determining factor(s) in mammals*

Year	Nature of observation	References
1959	Y chromosome shown to be necessary for testis determination in mouse and man	Ford *et al.* (1959); Jacobs & Strong (1959); Welshons & Russell (1959)
1966	Testis determining factor(s) located on short arm of the human Y chromosome	Jacobs & Ross (1966)
1975	H-Y antigen proposed as the testis-inducing substance (putative *Tdy* gene product)	Wachtel *et al.* (1975)
1981	*Bkm* sequences suggested as candidate gene for male determination in mice	Jones & Singh (1981)
1987	*Zfy* proposed as candidate gene	Page *et al.* (1987)
1990	*SRY* proposed as candidate gene in man	Sinclair *et al.* (1990)
	Sry proposed as candidate gene in mouse	Gubbay *et al.* (1990)
1991	*Sry* confirmed as a male-determining gene in transgenic mice	Koopman *et al.* (1991)

& Russell, 1959). Irrespective of the number of X chromosomes detected in the embryo, testicular development will be prescribed as long as a Y chromosome is present, as in XY, XXY, XXYY, XXXY, etc. (Davis, 1981), resulting in phenotypic males. In the absence of a Y chromosome and its primary signal (a presumptive gene or sequence of genes), development will proceed in a female direction. However, despite apparent development of a testis in animals containing a Y chromosome, the process of spermatogenesis is impaired in the presence of two X chromosomes (e.g. in an XXY constitution), probably because inactivation of the X chromosome during normal spermatogenesis is essential for fertility (see below). Some X-coded gene product must presumably act to prevent development of male germ cells (McLaren & Monk, 1981).

Mammalian sex chromosomes are markedly different in size and genetic content but there is nonetheless a pairing of X and Y chromosomes during meiosis which ensures that they segregate appropriately. Pairing of the X and Y chromosomes does not occur at the centromere but rather takes place in a small region of homology between the short arm of the X and the short arm of the Y (Burgoyne, 1986). Following the initial suggestion and illustration of Koller & Darlington (1934), derived from cytogenetic observations in the rat (Fig. 2.1), Burgoyne (1982) postulated that a single obligatory crossing-over event would take place between the homologous pairing regions of the X and Y chromosomes in every male meiosis. He also

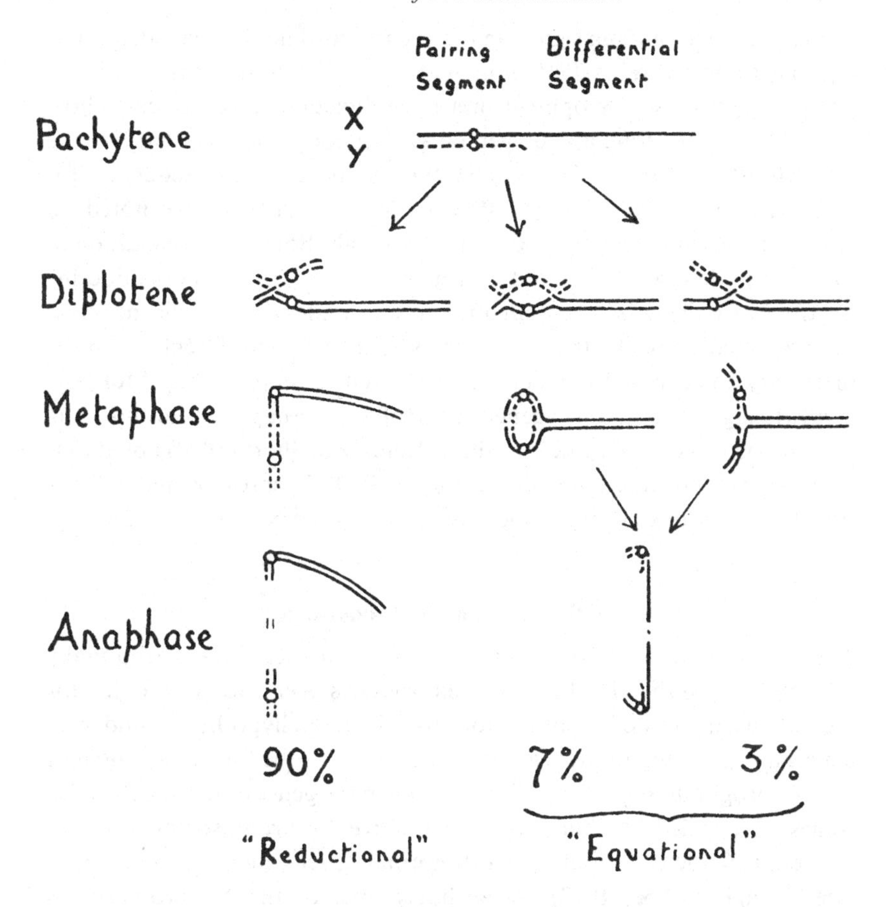

Fig. 2.1. The original diagram of Koller & Darlington (1934) showing the behaviour of the X and Y chromosomes during the first stages of meiosis in the rat.

pointed out that any gene located distal to the cross-over point would be transmitted to 50% of the XX and 50% of the XY progeny, that is it would be transmitted like an autosomal gene and would not show sex linkage. Burgoyne therefore termed this the pseudoautosomal region, the area of normal X–Y interchange. Various lines of evidence, especially cytogenetic analysis of structural abnormalities of the Y chromosome in humans, suggested that the primary testis-determining information is situated on the short arm of the Y (Jacobs & Ross, 1966). Loss of the fluorescent segment of the long arm of the Y does not interfere with the formation of a normal testis (Davis, 1981). Even so, additional genetic information on the short or long arms may be essential for a fully functional testis demonstrating normal spermatogenesis. The above discussion infers, correctly, that the primary sex-determining signal on the Y chromosome is a testis-determining signal referred to as *TDF* in man and *Tdy* in mouse. Formation of a testis then ensures development in a male direction since testicular hormones act to elaborate a male genital system and male secondary sexual characteristics (Jost, 1953). Whilst this is the accepted sequence for eutherian mammals, O, Short, Renfree & Shaw (1988) have noted an apparent and important exception in marsupials. Somatic sexual differentiation was deduced to be under primary genetic rather than endocrine control, since the gubernaculum and scrotum undergo a significant phase of differentiation whilst the gonads are still sexually indifferent, as do the mammary anlagen and pouch in females (Renfree *et al.*, 1992). McLaren (1991*a*) suggested that direct genetic control may perhaps be determined by X chromosome dosage, whereas Shaw, Renfree & Short (1990) noted that scrotal development occurs where only a single X chromosome is functional whilst two X chromosomes are necessary for pouch formation.

Inactivation of an X chromosome

The concept that one of the X chromosomes is rendered transcriptionally inactive early in the life of a female embryo has been widely accepted for more than a quarter of a century, for this – the Lyon hypothesis – underlies a mechanism of quantitative genetic adjustment. If the full complement of cellular programming information residing in the genes of the sex chromosomes were called into play, then two active X chromosomes in extra-gonadal somatic cells could perturb normal development of the embryo (see Takagi & Abe, 1990). Accordingly, one of the X chromosomes becomes inactivated in both the soma and germ line of females and can be visualised in suitable cell preparations as a highly-condensed, pyknotic lump of chromatin, the Barr body (Lyon, 1961, 1962). Females (XX) thus become the equivalent of males (XY) in terms of X chromosome dosage. In instances of supernumerary X chromosomes, all X chromosomes except one become inactivated.

The specific proposal of Lyon (1961) was that a dosage compensation mechanism is achieved by X chromosome inactivation in order to avoid the aneuploidy effect that would stem from the presence of more than one X chromosome. Such inactivation was thought initially to be at random in the sense that, in individual cells, either the maternal-or paternal-derived X chromosome would remain active. This supposedly random inactivation was considered to involve all somatic cells of the female. In female germ cells, however, one X chromosome is thought to be inactivated during the phase of migration (Ohno, Klinger & Atkin, 1962), whereas reactivation has been noted by the time the germ cells reach the ovarian anlagen (Teplitz & Ohno, 1963; Ohno, 1967; Migeon & Jelalian, 1977) or as they enter meiosis (Kratzer & Chapman, 1981; McLaren, 1981; Monk & McLaren, 1981). In human and mouse oogonia, only one X chromosome is active but the silent X is reactivated at about the time of entry into meiosis. Both X chromosomes are thought to be active during oogenesis itself. In the mouse, reactivation of the inactive X chromosome occurs in the germ line at about 12.5 days *post coitum*, when the germinal vesicle of diploid oocytes contains four times the haploid amount of DNA. In the male, reactivation is close to the time of mitotic arrest in the testis (Monk & McLaren, 1981). These observations, which are discussed in a stimulating chapter by Jones (1989), infer the influence of vital regulatory genes. In somatic cells, X inactivation occurs during embryonic life (Lyon, 1974; Cattanach, 1975; Gartler, Rivest & Cole, 1980), although there are distinct differences in the pattern of inactivation between different cell lineages in, for example, young rodent embryos (West, 1982; Gardner *et al.*, 1985).

Monk & Harper (1979) highlighted temporal differences in the pattern of X chromosome inactivation in developing embryos. Such inactivation is thought to occur in most cells of the female mouse embryo at the blastocyst stage when there are two principal cell types: the outer sphere of trophectoderm cells and the inner cell mass. X inactivation occurs at 3.5 days *post coitum* in the trophectoderm and at 4.5 days in the primary endoderm. However, because there is genetic evidence that both X chromosomes are potentially active in inner cell mass cells, this led to the supposition that X chromosome inactivation may have occurred initially in only the trophectoderm cells. Furthermore, Monk & Harper (1979) proposed that X chromosome inactivation is linked to cellular differentiation, occurring at different times in different cell populations as they depart or differentiate terminally from a pluripotent foetal stem line. Analysis of large numbers of inner cell mass cells from female blastocysts yielded data consistent with the presence of two active X chromosomes. In the post-implantation embryo, two active X chromosomes would be expected in the pluripotent epiblast region before gastrulation but only one in the corresponding extra-embryonic ectoderm (trophectoderm derivative) and primary endoderm (inner cell mass derived); as noted, this is apparently the situation. Inactivation is complete in the epiblast between 6.0 and 6.5 days of gestation at the onset of gastrulation (Monk & Harper, 1979). As far as cell lineages are concerned, the process appears to be essentially random with respect to parental origin of the X chromosomes in primitive ectoderm-derived tissues, but preferential inactivation of the paternally inherited X chromosome has been demonstrated unequivocally in derivatives of the trophectoderm and primitive endoderm (Takagi & Sasaki, 1975; West *et al.*, 1977; Gardner *et al.*, 1985).

Since the above views were summarised, there has been the challenging suggestion of Chandra (1985, 1986) that the primary function of X chromosome inactivation in mammals might be one of sex determination rather than of dosage compensation. Ferguson-Smith & Affara (1988) found this suggestion highly plausible. Lyon (1986) accepted that this could have been so in a hypothetical ancestor having homomorphic sex chromosomes, but holds to her view that the requirement for dosage compensation in modern mammals is vital. Nonetheless, the original prediction of Lyon (1962) that the human X chromosome might have a pairing segment which was non-inactivated is not discarded. On the contrary, evidence in favour of this prediction is marshalled from human and mouse studies, being essentially that a pairing segment of the sex chromosomes carrying homologous genes does indeed escape inactivation.

The precise nature of the inactivation process, its control mechanisms and sequence of imposition still require full elaboration, but inactivation appears to involve travel of some form of signal along the chromosome from an inactivation centre (see Cattanach, 1975; Lyon, 1986). One suggested explanation for X inactivation involves chemical modification of the DNA or, more precisely, methylation of the X chromosome and its condensation as heterochromatin, thereby repressing its gene activity (Yen *et al.*, 1984; Jagiello *et al.*, 1987). Demethylation would be associated with the onset of expression of specific genes. Details for an involvement of methylation in the differential programming of cell lineages are given in thoughtful papers by Monk, Boubelik & Lehnert (1987) and Brandeis, Ariel & Cedar (1993). Even so, as inferred by Bird (1986) in his discussion of CpG-rich islands and pointed out in the excellent review on DNA methylation and chromatin structure by Selker (1990), it is still not fully clear whether increased methylation is a cause or a consequence of gene inactivity. What is clear is that methylation patterns of specific genes undergo dynamic changes in the germ line and early embryo (Kafri *et al.*, 1992).

Recent findings in man have focused on the so-called X inactivation centre. A starting point here is the long-standing observation that in patients with Turner's syndrome having structural X chromosome aberrations, the abnormal X is always the inactivated X. Because these structural X abnormalities invariably included a proximal part of the long arm of the X chromosome termed Xq13, this suggested the presence of an active locus in Xq13 which was involved in the process of X inactivation (Ferguson-Smith, 1991). This putative locus was designated the X inactivation centre. A gene has recently been isolated which could be a candidate for the inactivation centre: it is referred to as *XIST* (X-inactive specific transcript) and is expressed from the inactive but not the active X chromosome (Brown *et al.*, 1991). Using structural aberrations in somatic cell hybrids, the gene has been mapped to Xq13. Its exons seem to contain a number of stop codons, which suggests that the gene product is a structural RNA which may act directly on chromosomal DNA (Brockdorff *et al.*, 1992; Brown *et al.*, 1992). Even so, the specific function of *XIST* in the inactivation process is unclear (Ferguson-Smith, 1991), although its suggested mode of action has been endorsed in the mouse. Quite exceptionally, it appears that *Xist* messenger RNA does not make a protein or travel from the nucleus to the cytoplasm. Rather it remains trapped in the nucleus to associate with the Barr body, and may be specifically involved in the initiation of X chromosome inactivation (Kay *et al.*, 1993), being switched on in the very young mouse embryo just before the occurrence of X inactivation. In human females whose karyotype includes tiny ring X chromosomes, there is a level of *XIST* transcription comparable to that of an active X – hence the strong suggestion that the inability of these rings to inactivate is responsible for the multiple congenital malformations and severe mental retardation (Migeon *et al.*, 1993).

Sex-determining genes as gonadal determinants

In the lucid and wide-ranging essay of Short (1982), the reader is informed that sex determination in mammals is a simple event, in which perhaps only a single gene is involved in deciding whether the indifferent gonad is to develop into a testis or an ovary. This viewpoint, which features in the introduction to many modern papers and derives from the proposal of Fisher (1930), is in essence arguing that a master gene acts directly to control a hierarchy or cascade of secondary regulatory genes. Whilst the present treatment will dwell in due course on the notion that there may indeed be a single sex-determining gene in placental mammals, perhaps the reader should not yet be completely persuaded by such an interpretation. A number of reservations could be offered, but one that paves the way for considerations of intersexuality later in the text will be presented here. It concerns plasticity or flexibility in differentiation and development of the mammalian gonad. Irrespective of the chromosomal constitution, the very fact that a gonad may develop as an ovotestis could be interpreted to suggest the primary involvement of more than one sex-determining gene. A closely associated argument might be that since the very young vertebrate embryo is sexually indifferent, having a double set of rudiments permitting production of a male or female genital system, this bipotentiality might in itself suggest the involvement of more than one sex-determining gene. However, in view of the overriding rôle of the male, the concept of an ovary-determining gene pathway is seldom given a reasonable or sufficient airing, yet specific ovarian programming information is undoubtedly required in genetic females.

Nonetheless, there is a proposal stemming from observations by Eicher & Washburn (1983, 1986) on sex-reversed mice that, in addition to the testis-determining gene on the Y chromosome, autosomal genes are necessary for normal gonadal differentiation and that there are indeed also ovary-determining genes (Eicher & Washburn, 1986). In their hypothesis, testis-determining genes are activated prior to – and themselves function to inactivate – ovary-determining gene(s) in XY animals. Male development is therefore precocious. XX animals lack the initial gene in the testis-determining cascade, and hence the ovary-determining gene(s) dominates. Mutations of the testis-determining loci that interfere with the timely and coordinated expression of these genes may result in failure of suppression of ovarian determinants, with subsequent development of ovarian tissue in XY individuals.

Conventional teaching remains that the presence of a Y chromosome in mammals is normally associated with development of testes in the embryo, leading to the formation of a male. As noted above, individuals with sex chromosome constitutions such as XY, XXY or even XXXXY develop as phenotypic males whereas those with XO, XX or XXX constitutions develop as females (e.g. Byskov & Hoyer, 1988). These observations have been interpreted to indicate that the Y chromosome carries what are referred to as testis-determining sequences (*Tds*). The classical view holds that this gene or genes may function by encoding or controlling the production of a specific determining factor, such as a protein, that acts to induce testicular differentiation. In the absence of the Y-related gene, ovaries are formed and female characteristics develop. Only since 1990 has a specific testis-determining gene been identified with confidence. As discussed below, the search for such a gene or genes produced a number of potential candidates, and there were similarly periodic reports of a putative testis-inducing substance. Disappointingly for those involved, many of the initial claims failed to stand up to rigorous analysis. Before discussing the

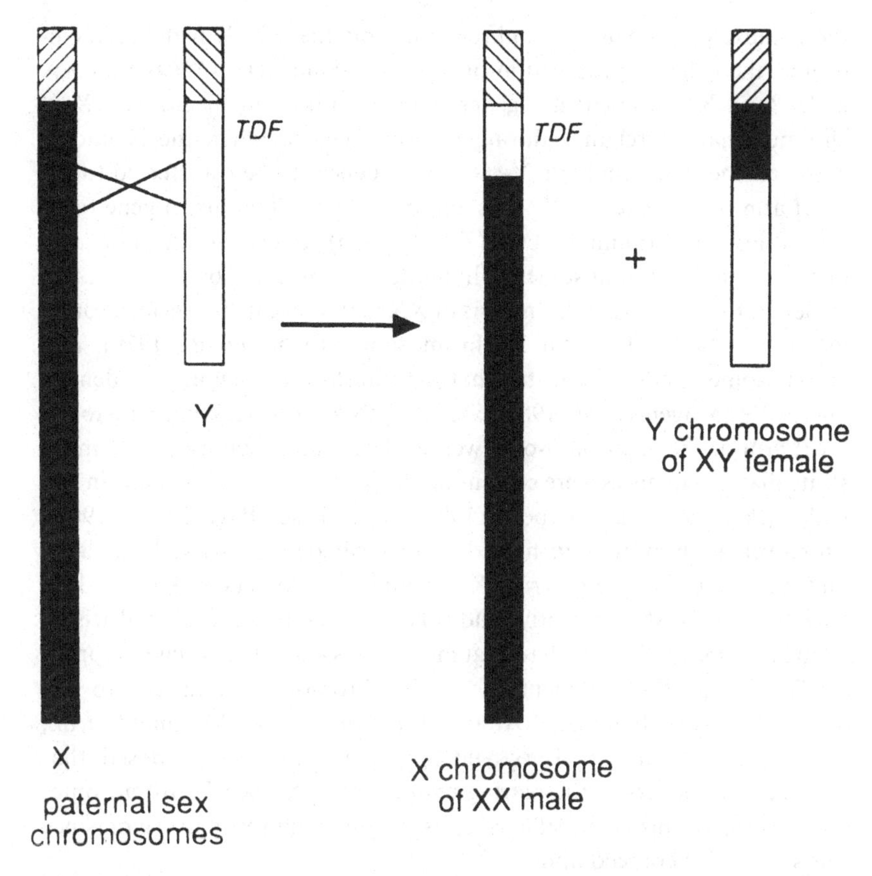

Fig. 2.2. To illustrate the manner of accidental interchange of the testis-determining region or factor (*TDF*) from the Y to X chromosome during errors in paternal meiosis that can thereby generate XX males and XY females. (Adapted from Affara (1991) and Wolf *et al.* (1992).)

more recent evidence for candidate genes, some relevant information from observations on 'sex reversal' will be summarised.

Involvement of the Y chromosome: evidence from sex reversal

One of the more elegant lines of evidence for testis determination being effected through the Y chromosome comes from mammals that are referred to as being sex reversed (*Sxr*), that is when genetically XX *Sxr* individuals develop as phenotypically normal males. However, the X chromosome of these presumptive males carries a segment that has been translocated from the Y chromosome during errors in paternal meiosis (Fig. 2.2). Indeed, in the context of human hermaphroditism and the XX Klinefelter's syndrome, the original proposition of Ferguson-Smith (1966) was that XX males (and XY women) are generated by a rare and illegitimate X–Y chromosomal interchange during paternal meiosis when the X and Y chromosomes pair, enabling Y-specific sequences to be transferred to the short arm of the paternal X. This led to the hypothesis that a gene to be termed testis determining factor (*TDF* in man) resided on the short arm (Yp) of the Y chromosome, a hypothesis confirmed by two lines of evidence. First, cytogenetic analysis of XX males revealed translocation of material onto the paternal X chromosome (Evans *et al.*, 1979), and chromosome banding indicated that the translocated region was identifiable as Yp (Magenis *et al.*, 1982). Second, DNA sequences mapping to the short arm of the Y chromosome were isolated, and then used to demonstrate that Y sequences were present on the paternal X chromosome in XX males (Page & de la Chapelle, 1984; Page, 1986; Page *et al.*, 1987). Moreover, analysis by *in situ* hybridisation located the Y-derived sequences on the short arm of the paternal X chromosome (Anderson, Page & de la Chapelle, 1986). More recently, study of cases of human sex reversal arising from abnormal X–Y interchange at meiosis has led to the precise mapping of *TDF* to a short segment of the Y chromosome adjacent to the pseudoautosomal boundary (Affara *et al.*, 1986, 1987; Vergnaud *et al.*, 1986). As noted above, Burgoyne (1982) had already proposed that crossing over occurs in the pairing region of the X and Y chromosomes during meiosis, this being referred to as the pseudoautosomal region since genes located there recombine.

In mice, the corresponding anomaly of sex reversal is attributed to a gene, *Sxr*, that converts chromosomal females (XX or XO) into phenotypic males by inducing development of testes able to produce androgens and thereby masculinisation (Cattanach, Pollard & Hawkes, 1971; Evans, Burtenshaw & Cattanach, 1982). A Y-specific DNA fragment was detected on the distal end of an X chromosome in XX_{Sxr} metaphase chromosomes (Singh & Jones, 1982). Adult XX mice carrying *Sxr*, termed *Sxr* pseudomales by LeBarr, Blecher & Moger (1991), lack the initial segment of the epididymis although they are not deficient in androgens. The testes have absent (in XX animals) or defective (in XO animals) spermatogenesis. *Sxr* is believed to have arisen as a duplication of the testis-determining segment of the mouse Y chromosome (Fig. 2.3), repositioned as the result of an inversion to the distal end of the Y beyond the pairing region (McLaren, 1988 *a*, *b*). In meiosis, *Sxr* is translocated from this chromosome, designated Y_{Sxr}, to a paternal X chromosome, now designated X_{Sxr}, that on

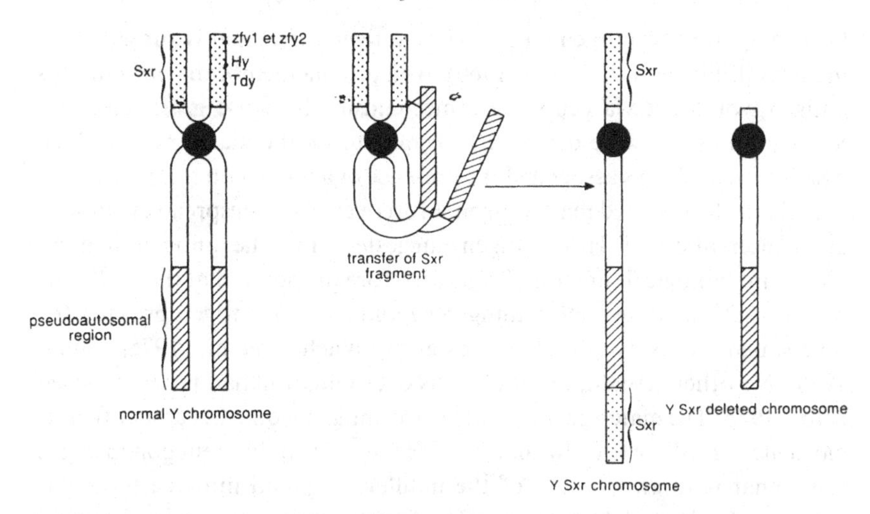

Fig. 2.3. Diagram to illustrate the manner of generating the *Sxr* translocation in mice by duplicating the testis-determining region and repositioning this by means of an inversion to the distal end of the Y chromosome beyond the pairing region. (From Cotinot, McElreavey & Fellous, 1993.)

fertilisation produces an XX *Sxr* zygote (Evans *et al.*, 1982). In other words, males carrying the mutation are able to sire male mice with an apparent XX karyotype.

H-Y antigen: a putative testis-inducing substance

If it is accepted that the Y chromosome carries a major sex-determining gene(s), then how might this gene act at the molecular level to prescribe and/or regulate initial formation and organisation of a testis from a sexually-indifferent gonad? In other words, what might be the primary determinant of testis formation? One early proposal was the involvement of a protein termed H-Y antigen, a male-specific cell-surface transplantation antigen originally thought to be controlled by the Y chromosome and demonstrated most clearly by rejection of skin grafts. In certain inbred strains of mice, male tissue grafts were rejected when transplanted onto female mice of the same strain, but such males do not reject skin grafts from females nor from males (Eichwald & Silmser, 1955). Male-specific antibodies were detected in the serum of female mice with male skin grafts, and the graft rejection in females was therefore attributed to a transplantation antigen present only in males. Since the sole genetic difference anticipated between the sexes in an inbred line would be the presence of a Y chromosome in males, a gene on this chromosome was suggested to code for a minor histocompatibility antigen termed H-Y antigen, presumably present only in males (Billingham & Silvers, 1960). As a plasma membrane protein, this antigen would act via cell-surface mechanisms to cause graft rejection. Serological assays were developed for measuring the male antigen. Such assays almost always associated the presence of a testis with the presence of a serologically detected male antigen. Moreover, the widespread evolutionary conservation of H-Y antigen suggested that the antigen had an important biological function. It was therefore proposed that the *H-Y* gene was, in fact, the testis-determining gene and its H-Y antigen product the long-sought testis-determining substance (Wachtel *et al.*, 1975; Ohno, 1976). No other actual inducer of testis determination had been proposed before 1975. The male-specific cell-surface antigen would act as a diffusible molecule to mediate transformation of the sexually indifferent gonad into a testis, that is to switch cells of the indifferent gonad into the testicular pathway. Lack of H-Y antigen in the female, so it was reasoned, would result in differentiation of ovaries.

How might transformation into a testis be achieved? Ohno (1977) proposed that a first step would be induction of seminiferous tubule formation, initiated by the binding of H-Y antigen to its receptor site. H-Y antigen is a component of the cell surface in many species. This binding would result in increased synthesis of H-Y antigen, which would cause displacement of other organogenesis-directing proteins from the 'anchorage sites' of the surrounding gonadal cells. Ohno (1977) further suggested that once the receptor sites were saturated with H-Y antigen, even XX cells in a foetal gonad would organise into a typically masculine structure and, even with a minority of XY cells in the indifferent gonad, neighbouring XX cells would be induced to engage in testicular differentiation. In line with her long-standing interest in differential rates of gonadal growth within the same foetus, Mittwoch (1977) then linked H-Y antigen expression with enhanced growth of the dominant gonad.

Tentative evidence for the involvement of H-Y antigen in testis formation came from several experiments. Suspensions of disaggregated rat testicular cells subjected to an anti-H-Y antiserum in culture reorganised into 'follicular-like' structures typical of an ovary. In the absence of antiserum, tubular structures were formed. Addition of H–Y antigen of epididymal fluid origin caused rat ovarian cells in suspension to re-associate into what appeared as testicular structures (Zenzes *et al.*, 1978). And bovine foetal ovarian cells revealed testicular-like organisation *in vitro* with the addition of purified H-Y antigen (Ohno *et al.*, 1979). Experiments of this nature served to indicate the presence of receptors for H-Y antigen on both ovarian and testicular cells, the antigen in some way programming the organisation of testicular structures irrespective of the sex chromosome constitution of the cells (i.e. XY or XX). H-Y antigen receptors were not demonstrable by binding reactions in non-gonadal tissues such as liver, kidney, brain or epidermis of either sex whereas ovarian and testicular cells would bind exogenous H-Y antigen. Whether a cell was intrinsically H-Y positive or not (i.e. expression) was seemingly confined to male cells; and all male cells tested up to that time, including gonadal cells, typed H-Y positive (Müller *et al.*, 1978). Exceptions were immature germ cells, which do not express the antigen. Secretion of H-Y antigen is confined to the Sertoli cells (Zenzes *et al.*, 1978).

Despite the body of circumstantial evidence presented above, there were always reservations as to a primary involvement of H-Y antigen in the process of testis determination. If H-Y antigen is required for the organisation of testicular tissue in the undifferentiated gonad, then it follows that individual mammals possessing testes should invariably type as H-Y antigen positive. Although sex-reversed 'male' mice (XX *Sxr*) usually type positive, McLaren *et al.* (1984) noted that some T16H/*Sxr* males (XX *Sxr'*) were H-Y negative, thereby indicating a dichotomy between testis determination and the expression of H-Y antigen and thus different genes for *Tdy* and H-Y (Simpson *et al.*, 1986, 1987). The *Sxr* symbol has now been subdivided, for a variant of *Sxr* was discovered in the early 1980s (McLaren *et al.*, 1984) and part of the translocated region was found to have been deleted (Fig. 2.4: Bishop *et al.*, 1988). The terminology now adopted in respect of these two versions is Sxr^{a} (referred to as *Sxr'* above) with at least seven genes and the abbreviated version known as Sxr^{b} retaining only three of the seven genes (see McLaren, 1991*a*). The relevance here is that XX Sxr^{b} male mice show normal testis formation, comparable to the situation in XX Sxr^{a} males, yet they lack the gene controlling expression of H-Y antigen. Such mice are demonstrably negative for H-Y whether by an *in vivo* skin grafting test or by means of T-cell mediated *in vitro* assays (McLaren, 1991*a*). Moreover, some XX human males also type H-Y antigen negative, and XY females are found who are H-Y positive, endorsing the view that H-Y antigen alone cannot be responsible for testis determination (Simpson *et al.*, 1987). In the face of such experimental evidence, the H-Y hypothesis could no longer be sustained.

Human females with Turner's syndrome (XO karyotype) have been found to be H-Y positive using a transplantation assay (Wiberg, 1985), as

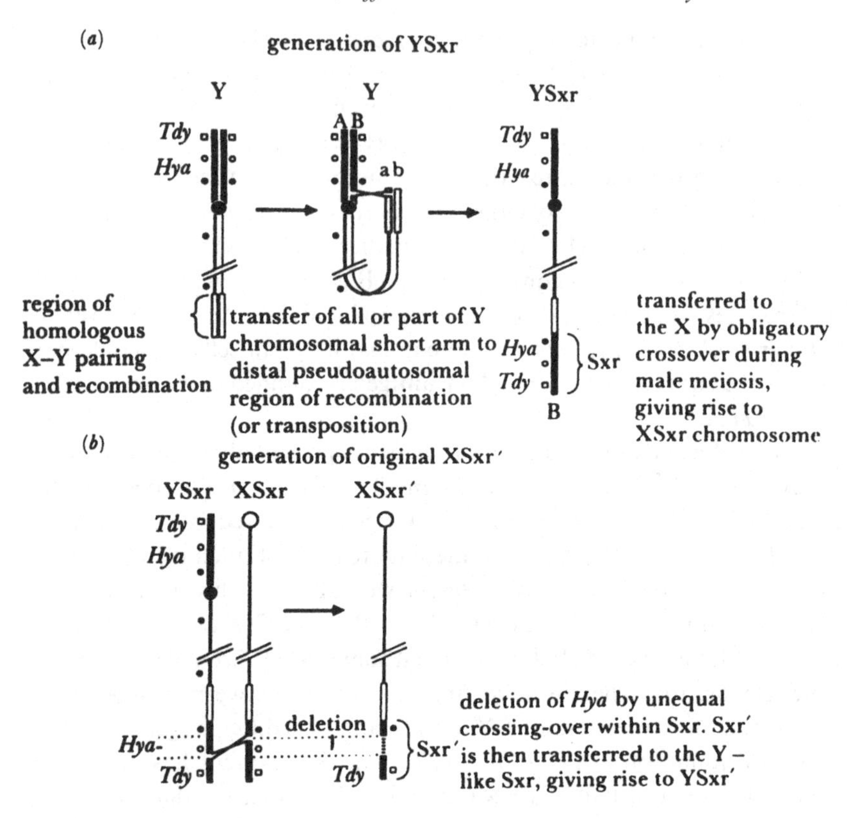

Fig. 2.4. A further illustration of the proposed origin of *Sxr* and its *Sxr′* variant, indicating loss of the H-Y antigen gene (*Hya*) from the chromosome carrying *Sxr*. This model was presented by Bishop *et al.* (1988).

have female XO wood lemmings (Wiberg & Gunther, 1985), indicating an autosomal or X chromosome location for the gene in these instances. This point is endorsed by the observation that XX true hermaphrodites who lack detectable Y chromosome sequences also type H-Y positive (Waibel *et al.*, 1987). An earlier speculation was that if the H-Y gene is autosomal yet under the control of X- and Y-linked genes, then a testis might develop when a threshold titre of H-Y antigen is reached (Wolf *et al.*, 1980*a*). Although this concept of thresholds may be relevant, the gene controlling expression of H-Y antigen is situated in the proximal part of the long arm of the human Y chromosome (Fig. 2.5) whereas the testis-determining gene (*TDF*) maps to the short arm of the Y. Nor would a concept of thresholds easily explain the unilateral formation of a testis or ovotestis in XX pigs bearing one ovary (Hunter, Baker & Cook, 1982; Hunter, Cook & Baker,

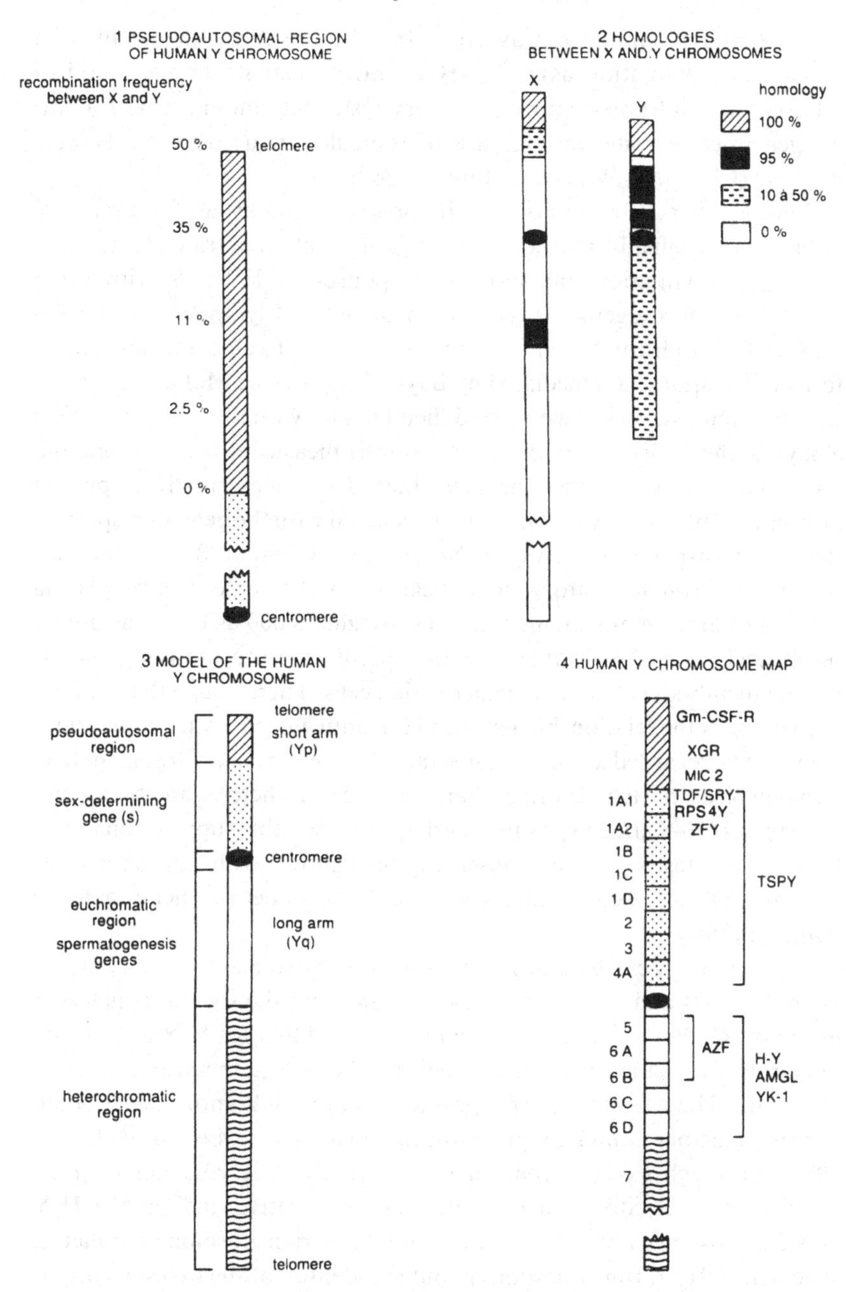

Fig. 2.5. A model of the human Y chromosome to illustrate the location of *SRY* adjoining the pseudoautosomal region on the short arm and that of the gene for H-Y antigen on the proximal portion of the long arm of the chromosome. (From Cotinot *et al.*, 1993.)

1985; Hunter, Chalmers & Cavazos, 1988). In any event, the results of a specific transplantation assay for H-Y antigen activity in the blood of intersex pigs did not support a primary testis-determining rôle for this antigen since, despite the presence of testicular tissue, the results were negative (Chalmers, Wiberg & Hunter, 1988).

Some further reservations may be presented briefly as follows. H-Y antigen was originally defined on the basis of rejection of transplanted skin (i.e. a transplantation antigen) and responses to H-Y, as with other transplantation antigens, proved to be mediated by T lymphocytes. Serological tests search for an antigen on the surface of cells, and antibodies formed in response are mediated by B lymphocytes (see McLaren, 1991*a*). In most circumstances, it would be difficult to know whether the serological assay and the cytotoxic lymphocyte assay were measuring precisely one and the same property. Thus, the gene that determines the H-Y plasma membrane antigen may not, in fact, be identical with the gene that specifies the H-Y transplantation antigen. Nonetheless, Ohno (1989) has stressed that in his own laboratory, identification of the male-specific plasma membrane antigen of subunit molecular weight 18 000 as H-Y was not by serological means, but rather via the specific receptor residing on the plasma membrane of bovine foetal ovarian cells. There is also the question as to whether the relationship between H-Y antigen and testicular development can be regarded as one of cause and effect. A troubling argument here is that in birds, the female rather than the male is the heterogametic sex and the female is H-Y antigen positive. Perhaps a reasonable supposition would be that any testis inducer conserved throughout evolution should be present in all testis-bearing males, whether homogametic or heterogametic (Ohno, 1978).

Other contrary evidence has been summarised in the excellent reviews of Wiberg (1987) and George & Wilson (1988), and the overall conclusion must now be that the testis-determining gene and that for H-Y antigen are not one and the same, even though they may be located in reasonably close proximity. There remains the proposition that the H-Y antigen gene could code for a factor essential to spermatogenesis (Burgoyne, Levy & McLaren, 1986), although exactly what such a spermatogenic rôle might entail remains to be clarified. There is also the provocative finding that H-Y antisera cross-react specifically with anti-Müllerian hormone (Müller & Wachtel, 1991), raising questions about the identity of the two substances. Finally, as alluded to on p. 56, there is a suggestion that H-Y antigen could be contributing to an accelerated rate of cell division in male embryos.

Specific candidates for the testis-determining gene

The so-called Y-linked, testis-determining factor responsible for initiating development as a male mammal assumed a more specific focus with the application of molecular techniques and in particular, the use of Y-specific DNA probes (reviewed by Wolf, Schempp & Scherer, 1992). Surprisingly, perhaps, repetitive DNA sequences isolated from the W chromosome of a snake gave the lead. One of the earliest candidates for the rôle of regulator of primary sex determination was a sex-specific Banded Krait minor (*Bkm*) satellite sequence. This satellite DNA cross-hybridised with DNA from organisms as diverse as *Drosophila* and humans, indicating either a conservation of such sequences or a completely independent evolution (Singh, Purdom & Jones, 1976, 1980; Singh, Phillips & Jones, 1984). The conserved nucleotide sequences of *Bkm* were concentrated at each end of the Y chromosome in XY male mice carrying *Sxr* (see above; Singh & Jones, 1982), were found to be transcribed, and such transcription was noted to be male specific and developmentally regulated (Singh *et al.*, 1984). Even so, the sex-specificity of these transcripts was not confirmed by other investigators (Epplen *et al.*, 1982; Schäfer, Ali & Epplen, 1986), nor was there any evidence using single-stranded probes that the *Bkm* nucleotide repeat sequences are directly or critically involved in mammalian sex determination (Durbin, Stalvey & Erickson, 1989). Indeed, the sequence was barely represented on the human Y chromosome. The rôle of *Bkm* repeat sequences remains unknown, although the suggestion has been made that they are non-functional accumulations on the Y chromosome and that, rather than being evolutionarily conserved, they may well have arisen independently in the wide spectrum of organisms examined (Levinson, Marsh & Epplen, 1985; Durbin *et al.*, 1989).

Considerable attention was attracted in 1987 when Page and colleagues reported evidence for a 'zinc-finger' gene on the human Y chromosome, with the suggestion that the zinc finger sequence, *ZFY*, could be implicated in sex determination (Page *et al.*, 1987). The strategy for identifying the putative *TDF* gene had been progressively to narrow down its location by analysing XX males carrying the testis-determining region of the Y at the tip of one of their X chromosomes and XY females having a deletion in this region. The region of overlap between the translocated segment and the deletion would be expected to carry the gene provoking male development. Such deletion mapping in sex-reversed humans showed that a 140 kilobase region on the short arm of the Y chromosome was necessary and sufficient

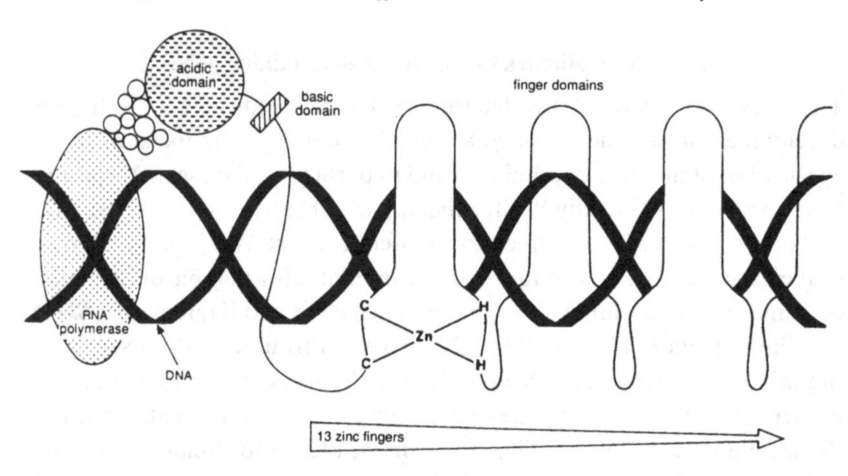

Fig. 2.6. To illustrate certain characteristics of a zinc-finger protein, the gene *ZFY* being considered a strong candidate for the testis-determining sequence in the late 1980s. (From Cotinot *et al.*, 1993.)

to induce testis formation. Within this region, Page *et al.* (1987) detected a sequence with an open reading frame corresponding to 404 amino acids. The deduced protein sequence was typical of a zinc-finger protein (Fig. 2.6), which would be expected to bind to DNA (or RNA) in a sequence specific manner and might regulate transcription. The identification and involvement of this *ZFY* gene appeared the more compelling since the gene is strongly conserved on the Y chromosome of various placental mammals, with a corresponding and closely related gene, *ZFX*, on the X chromosome of several eutherians examined. Although such observations were accepted by many at the time as evidence for the elusive testis-determining factor, the sequences most closely related to *ZFY* in marsupials are located on autosomes and not on the Y chromosome (Sinclair *et al.*, 1988), despite the Y chromosome being regarded as male-determining in marsupials. And *ZFX*, the homologue of *ZFY*, was shown in humans to escape inactivation (Schneider-Gadicke *et al.*, 1989; Palmer *et al.*, 1990), seemingly excluding *ZFY* as a candidate.

Moreover, since the cloning of this potential sex-determining gene by Page & colleagues in 1987, the results of further research had failed to substantiate the rôle of *ZFY* as the testis-determining factor. Erickson & Verga (1989) presented arguments against *ZFY* being the sex-determining gene and other authors had reported results that underlined this negative viewpoint. Indeed, Palmer *et al.* (1989) reported three XX men, and one XX intersex with ovarian and testicular tissue, who lacked *ZFY* and yet who

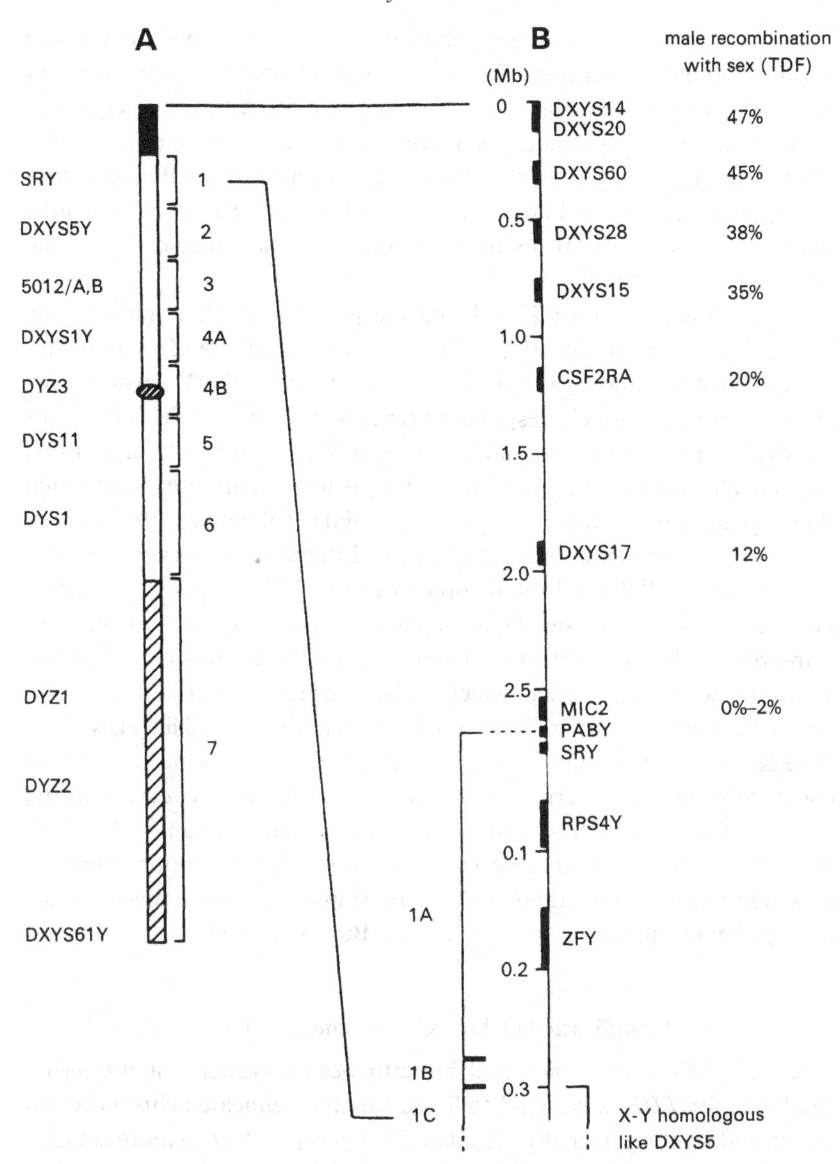

Fig. 2.7. A detailed map of a portion of the human Y chromosome to illustrate the relative positions of *SRY* and *ZFY* and the anticipated frequency of recombination for various genes on the short arm. (From Wolf *et al.*, 1992.)

had Y-chromosome-specific DNA sequences mapping to the region very close to the pseudoautosomal border. Hence, *ZFY* could not be the primary testis-determining gene, which itself would seem to be situated very close to the pseudoautosomal boundary (Fig. 2.7). In a *Nature* article entitled

'Thumbs down for zinc finger', Burgoyne (1989) reasoned that if the original anomaly giving rise to familial cases of XX sex reversal involved an X–Y interchange, then they should be positive for Y-specific sequences adjacent to the pseudoautosomal boundary. At the conclusion of his article, Burgoyne suggested that the human Y-chromosomal testis-determining factor (*TDF*) could be equated with *TDY* + *ZFY*. In other words, full programming of testis tissue formation in humans would require the influence of both *TDY* and *ZFY*.

Other evidence against *ZFY* being the primary sex-determining gene comes from mice, in which there are two copies of the *ZFY* sequence designated *Zfy-1* and *Zfy-2* within the *Sxr* region of the Y chromosome (Mardon *et al.*, 1989). Corresponding genes were found on the X chromosome (*Zfx*) and as an autosomal homologue (*Zfa*). Mardon & Page (1989) were unable to detect transcription of *Zfy-1* in newborn mouse testes, which thus argued against a rôle for *Zfy -1* in gonadal development. On the other hand, post-meiotic expression of *Zfy-1* in adult mouse testes was noted by Kallikin *et al.* (1989), whilst Koopman *et al.* (1989) using the sensitive polymerase chain reaction (PCR) reported *Zfy-1* as in fact having three transcripts in differentiating foetal mouse testes. But expression of *Zfy* was associated with the germ cells, which are believed to play no essential part in testis determination, rather than with the crucial somatic cell lineages of the developing gonad. Furthermore, in W^e/W^e mutant mice, which are almost completely lacking in germ cells, testicular development occurred in the absence of detectable *Zfy-1* and *Zfy-2* expression (Koopman *et al.*, 1989). An overwhelming primary rôle for *Zfy-1* in sex determination therefore appeared improbable, although *Zfy* may be involved in spermatogenesis, perhaps being equivalent to the *Spy* gene (Burgoyne, 1992).

Identification of *Sry*/*SRY* in mouse and man

Since the *ZFY* gene, which had hitherto been considered a promising candidate for *TDF*, was found to lie outside the delineated chromosomal region and was thus formally excluded (Palmer *et al.*, 1989), a more detailed search for breakpoints in XX males was conducted (Sinclair *et al.*, 1990). The search was thereby focusing within a 60 kilobase region of the human Y chromosome proximal to the pseudoautosomal boundary (Fig. 2.8). Using further DNA probes, the minimal portion of Y chromosome DNA able to induce maleness was noted to extend just 35 kilobases adjacent to the pseudoautosomal boundary. A specific *TDF* gene would thus be presumed

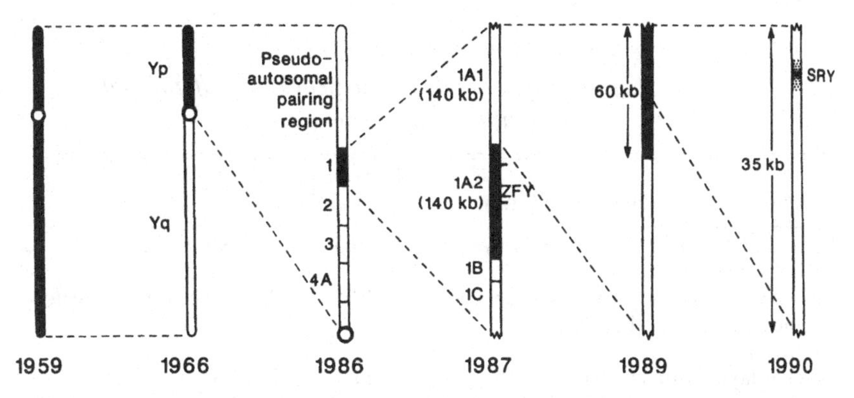

Fig. 2.8. A valuable historical depiction of the manner in which the search for the testis-determining region on the Y chromosome was progressively narrowed down between 1959 and 1990, being reduced from the 30–40 million bases originally under consideration to less than 250 bases encoding the conserved 80-amino acid motif of *SRY*. kb, kilobases. (From McLaren, 1990.)

to lie somewhere within this small region. Sinclair *et al.* (1990) tested this hypothesis by searching for such a gene, which they termed the human Y-located gene *SRY* (sex-determining region of the Y chromosome). A paper published simultaneously by Gubbay *et al.* (1990) pursued essentially the same approach in mice, searching for a putative *Sry* gene in the homologous region that would encode a testis-specific transcript. Both the Sinclair and Gubbay (overlapping) research teams in London were successful in their quest, discovering a gene with the anticipated hallmarks of the key to maleness.

In summary, Sinclair *et al.* (1990) and Gubbay *et al.* (1990) found that such a novel transcribed gene was present in the sex-determining region of man and mouse. Genomic clones covering 35 kilobases of the region adjacent to the pseudoautosomal boundary on the human Y chromosome were used to look for the presence of evolutionarily conserved sequences. One of these clones detected conserved sequences on the Y chromosome of a wide range of eutherian mammals tested – a strongly conserved motif of 80 amino acids – sequences therefore homologous to *SRY*. *SRY* encodes a protein with a potential DNA-binding domain, which is shared with the Mc protein of the mating-type locus of fission yeast (*Saccharomyces pombe*) and several non-histone (nuclear) proteins related to the high mobility groups, HMG1 and HMG2. In July 1990, *SRY* became the best candidate for *TDF*, the male sex-determining gene in humans. As noted by Sinclair *et al.* (1990) in their conclusion, proof of identity between *SRY* and *TDF*

Table 2.2. *The timing of* Sry *expression in the somatic cells of the urogenital ridge in relation to morphological stages of gonadal differentiation in the mouse*

Stage of gonadal differentiation	Days *post coitum*[a]	*Sry*
Mesonephros distinct but no trace yet of genital ridge	9.5	Not expressed
Thin layer of mesenchymal cells between the mesonephros and coelomic epithelium	10.5	Transcripts first detected
Thicker layer with active cell proliferation but testis differentiation only just beginning	11.5	Expressed
Sertoli cells have differentiated and aggregated to form testis cords	12.5	Expression declining
Appearance of functional Leydig cells	13.5	Transcription no longer detectable

[a] Figures depend in part on the genetic strain of mouse and the precise convention for timing the onset of pregnancy.
Note the relatively short period of *Sry* expression in the urogenital ridge. Note also that *Sry* transcripts have been reported from the 2-cell stage to that of the blastocyst (Zwingman *et al.* 1993).

would require mutational analysis of XY females or the production of sex-reversed transgenic animals. Such persuasive evidence was not long in appearing.

In the work of Gubbay *et al.* (1990) on this novel gene family in mice, one member of the family mapping to the sex-determining region of the Y chromosome satisfied various predictions made for a testis-determining gene. Over and above the molecular evidence that a portion of *Sry* is deleted from the Y chromosome in XY female mice shown genetically to have a mutation in the testis-determining gene (Lovell-Badge & Robertson, 1990), any candidate for *TDY* should be expressed at a time and in a tissue consistent with its presumed action. In mice, *Tdy* should therefore be expressed in the male urogenital ridge at approximately 11.5 days *post coitum* (Table 2.2), just before formation of the testis cords which is considered the first recognisable difference between male and female embryogenesis. Expression of *Sry* conforms with this prediction, even though the level of transcripts found in the 11.5 day *post coitum* urogenital ridge was low. Thus, as with the human gene homologue, *SRY* or *Sry* protein could be a nuclear protein that binds DNA and acts as a transcriptional switch, orchestrating the action of other genes that control development of the testis.

Three further articles of direct relevance were published simultaneously in the same November 1990 number of *Nature*. Koopman *et al.* (1990) expanded upon the involvement of their candidate for *Tdy* which, as noted above, maps to the minimum sex-determining region of the Y chromosome. Expression of this gene, *Sry*, was (1) found to be limited to the period in which the testes begin to form, (2) confined to gonadal tissue, and (3) did not require the presence of germ cells. This additional evidence enabled the authors to endorse a primary rôle for *Sry* in mouse sex determination. The gene is expressed in one of the somatic cell lineages present in the developing genital ridge, *Sry* being envisaged to initiate a cascade of gene expression to regulate formation of the gonad. Triggering of differentiation along the Sertoli cell pathway is a key event. This would be compatible with the observed rapid cessation of *Sry* transcription after testis cord formation, although this latter point may deserve reevaluation and certainly does not hold for sheep (C. Cotinot, personal communication; and see below).

Other supportive evidence was drawn from instances of sex reversal. This condition in XY females results from the failure of testis determination or testis differentiation pathways (see Chapters 6 and 7). Some XY females with gonadal dysgenesis have lost the sex-determining region from the Y chromosome by terminal exchange between the sex chromosomes (Fig. 2.9) or by other deletions. If *SRY* is the testis-determining factor, then some sex-reversed XY females without Y chromosome deletions could have undergone mutations in *SRY* (Berta *et al.*, 1990). A screening of XY human females and normal XY males for alterations in *SRY* using the single strand conformational polymorphism assay, a technique capable of detecting point mutations, followed by DNA sequencing, revealed a *de novo* mutation in the *SRY* gene of one XY female – not present in the patient's normal father and brother. Single base pair changes in the conserved region of *SRY* were detected. The *de novo* mutation associated with sex reversal provided strong evidence that *SRY* is directly involved in male sex determination (Berta *et al.*, 1990).

Jäger *et al* (1990) similarly identified a mutation in *SRY* in 1 of 12 sex-reversed XY human females with gonadal dysgenesis who do not lack large segments of the short arm of the Y chromosome. The four-nucleotide deletion occurred in a sequence of *SRY* encoding a conserved DNA-binding motif and resulted in a frame shift presumably leading to a non-functional protein. The mutation occurred *de novo* because the father of the 'sporadic' XY female who bore it had the normal sequence at the

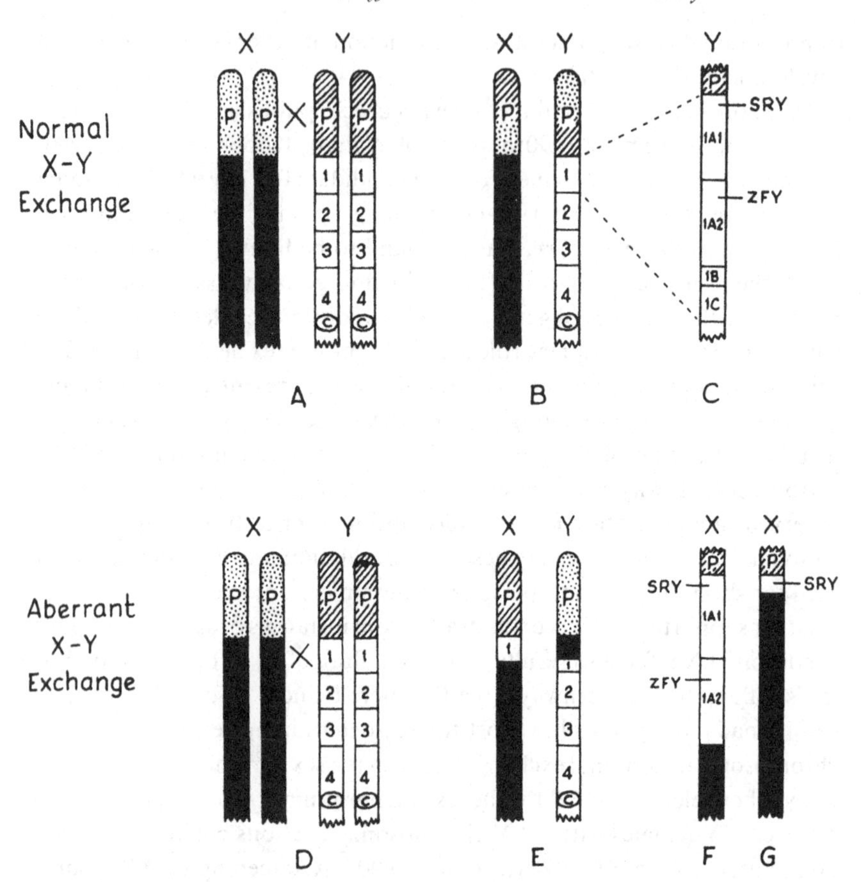

Fig. 2.9. A close-up view of the region of exchange between the X and Y chromosomes indicating the manner of aberrant exchange outside the so-called pairing region, enabling an X chromosome to carry *SRY*. Variation in sexual phenotype has been suggested to depend on the proximity of *SRY* to the X boundary. (From Burgoyne, 1992.)

corresponding position. The results again provided strong evidence for *SRY* being *TDF*, although another gene with the same embryonic expression pattern as *SRY* could have been present in the Y chromosome 35 kilobase region that is critical for human sex determination.

Most exciting and elegant were the subsequent experiments of the two London teams of Goodfellow and Lovell-Badge (1991). In order to demonstrate that the sex-determining sequence on the Y chromosome had indeed been isolated and correctly identified, then a convincing demonstration would be to introduce the candidate gene into a newly-fertilised egg of XX sex chromosome constitution and change the sex of the developing embryo. This has now been achieved in mice, switching would-be females into transgenic males (Koopman *et al.*, 1991). In essence, *Sry* on a 14 kilobase genomic DNA fragment microinjected into one of the pronuclei was able to induce testis differentiation and subsequent male development in chromosomally female transgenic mice. Thus, the 14 kilobase murine genomic fragment must have acted as a transgene and provided the correct regulatory sequences in at least a proportion of instances, thereby demonstrating many of the genetic and biological properties anticipated in a Y-located, testis-determining gene. However, the human *SRY* gene introduced in a similar manner on a 25 kilobase genomic fragment did not succeed in promoting sex reversal in mice.

The microinjection technique usually involves introducing the gene sequence into a male pronucleus, invariably the larger of the two, in the hope that it will become integrated into the genome of the embryo. This form of experimental approach has been known to be feasible since 1982 when Palmiter *et al.* demonstrated that a gene construct for growth hormone introduced into the pronuclei of mice could lead to the formation of transgenic giant mice, i.e. significantly larger than mature male mice of that strain (Fig. 1.5). One of the problems noted in 1982 was that the overall success of the technique, as indicated by the proportion of offspring exhibiting the introduced trait, was low, a fact highlighted in the review of Wilmut, Hooper & Simons (1991). To a lesser extent, this shortcoming is apparent in the report of Koopman *et al.* (1991), but perhaps it is not entirely unexpected. The 14 kilobase sequence introduced into the XX mouse zygotes would need to become integrated into the genome at an appropriate place and time, transcribed and expressed, and failure is possible at any of these stages. Nonetheless, introduction of the putative male-determining gene resulted in sex reversal in three of 11 XX mice transgenic for *Sry*; i.e. three were sex-reversed males but eight remained females. So *Sry* alone can induce changes leading to male development even when acting on an XX genetic backcloth, and the experiment thus demonstrates that female embryos possess all the other genes required for forming a phenotypic male, presumably upon activation by the *Sry* gene. Although demonstrably male with penis and testes, such transgenic animals would have been sterile due to a complete absence of spermatogenesis. At least two other genes on the Y chromosome are needed to promote formation of spermatozoa. In addition, it has been appreciated for some time that the presence of two X chromosomes in a male mouse always results in sterility, as germ cells are prevented from progressing beyond the stage of prospermatogonia.

There remains the important question as to why the majority of treated embryos were not sex-reversed. Because the microinjected pronuclei received multiple copies of the gene sequence, this quantitative aspect could have been relevant, for the three transgenic males may have carried more copies of *Sry* than the eight remaining females. Females mosaic for the transgene, with only a small proportion of the cells making up the somatic portion of the genital ridge carrying functional *Sry* gene copies, would have been an explanation. At the level of expression, it is not known from the report of Koopman *et al.* (1991) whether *Sry* was in fact expressed in all 11 of the transgenics, or if eight of the females failed to express it because it was inserted in an inappropriate region of the genome. In other words, the *Sry* gene may have been sensitive to positional effects. But expression in the embryonic gonad could be tested since one of the transgenic females has transmitted the transgene to her progeny.

The Koopman *et al.* (1991) paper also reported making mice transgenic for the human *SRY* gene. Although transgene expression in the developing gonads was demonstrated, no sex reversal of XX embryos could be detected. The question therefore arises as to whether the (human) *SRY* gene can be effective in mice, and indeed whether *SRY* alone could fully induce maleness. The masculinisation of four XX human cases with only 35 kilobases of *SRY*-containing Y chromosome DNA extended to development of testicular tissue, but none showed completely normal development of male genitalia (Palmer *et al.*, 1990). Further studies examining the potential for *SRY*-induced modifications in mice would be anticipated, not least to clarify whether the human transcript is stable and can be correctly processed. An appropriate human–mouse interaction would depend upon the amino acid sequence within the conserved DNA binding domain. Moreover, the mode of action of *SRY* could differ from that of the mouse gene, conceivably interacting with an alternative gene(s) in the human pathway (Affara, 1991). A recent report on *SRY* itself indicates that it is an intronless gene, encodes a protein of 204 amino acids, and contains a small GC-rich promoter at its 5′ sequence (Su & Lau, 1993). Indeed, evidence from the human Y chromosome suggests that *SRY* protein is encoded by a single exon (Behlke *et al.*, 1993).

In summary, the candidate gene *Sry* acts over a short period of time in mice to initiate testis development by promoting differentiation of Sertoli cells from the supporting cell lineage of the gonad. It could achieve this through interaction with other genes, some of which will be involved in the regulation of *Sry*, others of which will be downstream targets of *Sry* (see Haqq *et al.*, 1993). In man, one of the genes frequently considered to be a target of *SRY* is the gene *WT*, for Wilms' tumour, even though its specific involvement in gonad determination and differentiation is not clear, but it is expressed in Sertoli cells. However, a primary involvement of the *WT* gene is in a malignancy of childhood that may involve one or both kidneys (Armstrong *et al.*, 1992). Mutations that inactivate the gene confer susceptibility to Wilms' tumour, in some manner converting the protein from a transcriptional repressor of its target sequence to a transcriptional activator (Kreidberg *et al.*, 1993; Park, Schalling & Bernard, 1993). Other interacting genes must map elsewhere in the genome, either on the X chromosome or on autosomes as for the *AMH* gene (see Chapter 4), because *Sry* has been shown to be the only Y-linked gene required to bring about male development in mice. But although undoubtedly initiated by the Y chromosome, the process of testicular development in mice involves a complex interaction with at least three autosomally-located genes (Eicher & Washburn, 1986). Even so, in the eloquent phrasing of Wilkie (1991), the genetic essence of masculinity is seemingly a tiny strand of DNA, and *Sry* can therefore be equated with *Tdy*. Future experiments will doubtless focus on the manner whereby the *Sry* gene is activated at a critical moment or moments in the unfolding of the embryonic genome, and the extent to which *Sry* itself regulates a cascade of gene action by encoding a DNA binding protein (i.e. acts as a switch). Studies in marsupials should also be revealing, for a Y-linked *Sry* homologue has been identified (Foster *et al.*, 1992).

As a perspective on the evolution of *SRY*/*Sry* and its rôle as a critical regulator of development, its encoded protein sequence is well conserved between species in the central 'high mobility group' domain (HMG) of 78–80 amino acids (e.g. Payen & Cotinot, 1993). However, among the genes similar to *SRY*, termed *SRY* box genes (or *SOX* genes), none of those so far examined is homologous to *SRY* outside the HMG box region. Whitfield, Lovell-Badge & Goodfellow (1993) investigated the coding sequence of *SRY* in various primates, and found that evolution has been rapid in the regions flanking the conserved domain; similar evidence has been presented for rodents (Tucker & Lundrigan, 1993). Because reduced or inappropriate *SRY* activity can cause sex reversal, Whitfield *et al.* (1993) commented that rapid evolution leading to populations with different *SRY* sequences could be a significant cause of reproductive isolation.

Further molecular considerations

Despite the historical perspective presented above that brings the discussion to accept a sex-determining rôle for the gene termed *SRY* (man) or *Sry* (mice), it is by no means clear exactly how the DNA sequence would act to impose its tissue organising influence. The explanation invariably offered is that its protein product would serve as a transcription-regulating factor for other genes, and the precise focus here seems to be in terms of a binding interaction between DNA and a protein domain. Recombinant *SRY* protein has been shown to bind to DNA target sequences that are recognised by some of these transcription factors (Nasrin *et al.*, 1991; Harley *et al.*, 1992). The protein domain in question is referred to as the HMG domain (see above), and it is a motif common to a series of transcription factors, sex-determining proteins and other quite distinct proteins that associate with DNA.

Such high mobility group (HMG) proteins are a series of acid-soluble non-histone nuclear proteins. Two-thirds of the protein molecule is composed of two homologous repeats of an 80 amino acid sequence, this being termed the HMG box. Because it exists in diverse biological instructional systems, including the *SRY* gene and – as noted – in yeast proteins that are involved in mating type control, the question arises as to what the specific function of the domain actually is. The HMG domain is thought to have a major structural rôle in causing the bending, looping or supercoiling of DNA, such a rearrangement seemingly being essential to present the correct conformation for transcription and further classes of DNA rearrangement. Indeed, a single HMG domain may be able to bend DNA almost back on itself (Lilley, 1992). Moreover, this same reviewer added the potent remark that the actual structure of the DNA may be as or more important than sequence requirements. The speculation was added that the HMG in chromatin may be involved in the sealing of loops in a manner reminiscent of that proposed for another histone, H1. Hence, there is the probability of a functional connection between the HMG domain and the manipulation of DNA structure. In fact, Paull, Haykinson & Johnson (1993) have demonstrated that HMG1 and HMG2 can bend DNA extremely efficiently, forming circles as small as 66 base pairs, and even circles of 59 base pairs at high concentrations of HMG protein. These authors suggest that an important biological function of HMG1 and HMG2 is to facilitate cooperative interactions between *cis*-acting proteins by promoting DNA flexibility.

As to the DNA binding specificity (Fig. 2.10), this is determined not only by the recognition helix that sits in the major groove of the DNA, but also by a flexible segment at the amino-terminal end of the homeobox which appears to wrap around the DNA so as to contact the minor groove (see Short, 1993). The carboxyl-terminal portion of the homeobox domain has structural similarity to the helix–turn–helix DNA binding domain of some regulatory proteins in yeast and prokaryotes. Upon DNA binding of this particular motif, the protruding recognition helix becomes embedded in the

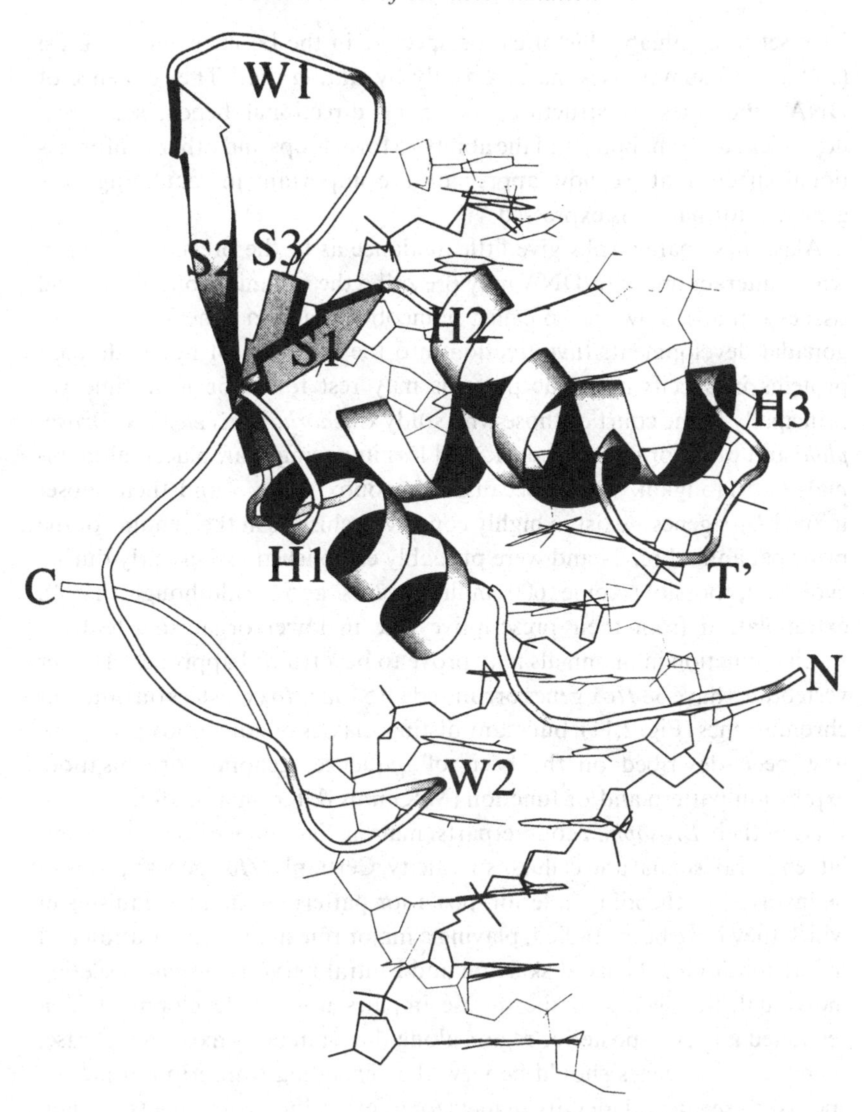

Fig. 2.10. An example of a DNA binding domain with its recognition helix, which interacts with DNA in the major groove. (Courtesy of Drs S. K. Burley, J. W. R. Schwabe and A. A. Travers.)

DNA major groove of its target site and is responsible for making many, but certainly not all, base-specific DNA contacts (Brennan, 1993). Once again, three-dimensional considerations are paramount, for interactions between protein sidechains and DNA base pairs depend in a highly complex way upon conformation. Such DNA–protein interactions have been set in a valuable historical perspective in the book review by Luisi (1993), who summarises matters neatly by stating that 'The sequence of DNA modulates its structure, imparting directional bends, sequence-dependent deformability and the ability to form loops and other conformational effects that we now appreciate are important in regulating how genetic information is expressed.'

Alas, these paragraphs give little guidance as to the precise manner in which interactions with DNA may prescribe the organisation of cells and tissues – that is how the so-called homeobox leads to genetic control of gonadal development. Investigation into the influence of homeodomain proteins in specifying tissue patterns may rest for some little time yet principally in the court of those who study *Caenorhabditis elegans*, *Drosophila* and other organisms considered less imposing than placental mammals (see Hodgkin, 1990). Because homeobox genes – and their subset termed *Hox* genes – exist in highly conserved clusters in the genome of the principal animal phyla and were probably established rather early during evolution, possibly some 600 million years ago (Riddihough, 1992), extrapolation from their prescriptive rôle in lower organisms to their specific function in mammals may prove to be a fruitful approach. Higher vertebrates have 38 *Hox* genes organised into four *Hox* clusters on different chromosomes (Fig. 2.11), but many distinct classes of homeobox gene have now been described on the basis of sequence, genomic organisation, expression pattern and/or function (McGinnis & Krumlauf, 1992).

As in their *Drosophila* counterparts, mammalian homeobox genes exhibit temporal, spatial and cellular specificity. Certainly, *Hox* genes appear to be involved in encoding anterior–posterior patterning in all organisms in which they have been studied, playing a major rôle in the organisation and differentiation of the axial skeleton and central nervous system. Deleting individual *Hox* genes in the mouse impairs normal development in a restricted anterior–posterior region along the main body axis. In a phrase, therefore, *Hox* genes should be viewed as encoding transcription factors specifying segmental identity in metazoans extending from insects to man; they are conveying positional information rather than being involved in specification of cell type (reviewed by Kim & Kessel, 1993).

Accelerated development of male embryos

Returning to the more immediate matter of sexual differentiation, there is now persuasive evidence that male embryos usually develop slightly faster than females in rodents, ruminants and primates. Indeed, a precocious

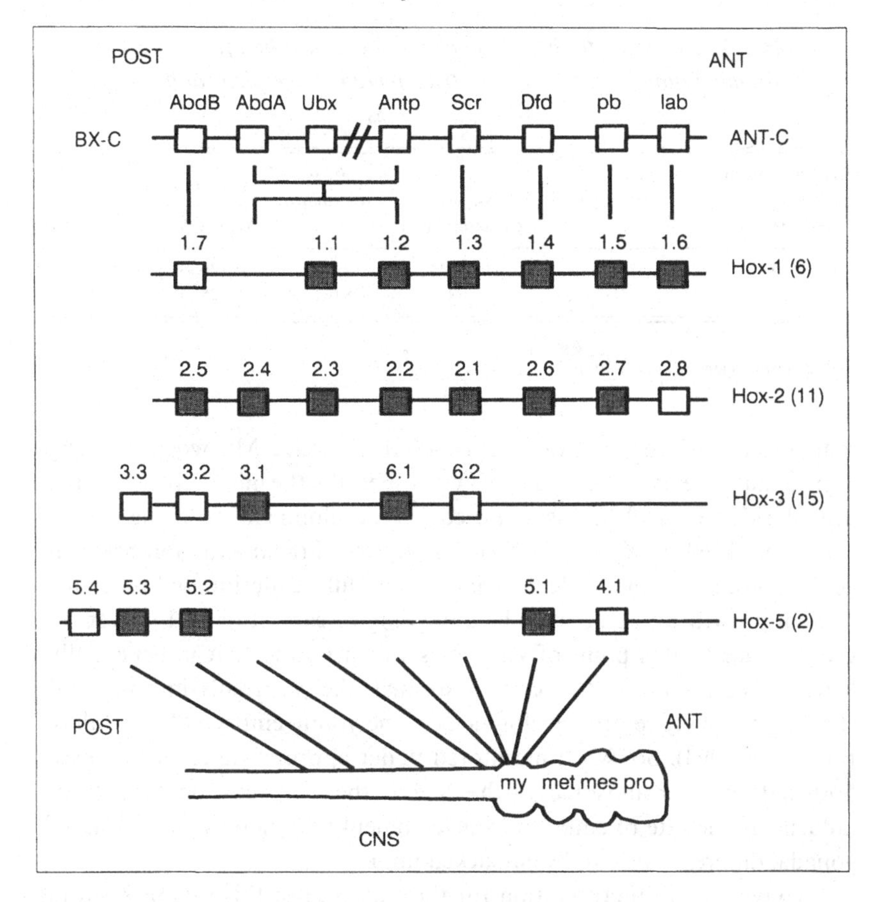

Fig. 2.11. Organisation and pattern of expression in the central nervous system (CNS) of four *Hox* gene clusters of mouse and the *Antennapedia* (ANT-C) and *Bithorax* (BX-C) complexes of *Drosophila*. There are strong similarities between these species in the organisation and spatial patterns of expression of these genes. (From Rossant & Joyner, 1989.)

development of male relative to female embryos is not only demonstrable by the morula–blastocyst stage in mice (Tsunoda, Tokunaga & Sugie, 1985; Burgoyne, 1993; Valdivia *et al.*, 1993; Table 2.3) and cattle (Itoh & Goto, 1986; Avery, Madison & Greve, 1991; Xu *et al.*, 1992), but may already exist by the time of the early cleavage divisions. In other words, sexual dimorphisms may occur before differentiation of the gonads. In some manner, the presence of a Y chromosome may act to programme an accelerated development of embryos, perhaps suggesting that the sex-determining sequence interacts with genes regulating the rate of growth or,

Table 2.3. *To illustrate the influence on sex ratio when fast- and slow-developing mouse embryos were transferred to recipient animals as separate groups*

Classified rate of embryonic development	No. of embryos transferred	No. of implantations (%)	No. of resorptions (%)	No. of foetuses	
				Males (%)	Females (%)
Fast	74	60 (81.1)	39 (52.7)	16 (76.2)	5 (23.8)
Slow	113	113 (100)	78 (69.0)	9 (25.7)	26 (74.3)

Note:
Adapted from Valdivia *et al.* (1993).

more specifically, the duration of individual cell cycles. Mittwoch had long argued that accelerated gonadal growth, especially the initial growth of the genital ridge (Fig. 2.12), is required for development as a male (e.g. Mittwoch, 1969, 1986, 1989, 1992). The nature of the interaction between gene programmes for sex determination and differentiation and those for rates of growth would seem to be a pressing research topic. However, to give balance to this point of view, it is worth noting that in genetically female mice (or at least in certain strains), the paternally-imprinted X chromosome may retard development of the young embryo (Thornhill & Burgoyne, 1993). So, when monitored under appropriate circumstances, both paternal chromosomes – the X and the Y – may be capable of influencing the rate of embryonic development and thereby the nature of gonadal differentiation to be embarked upon.

A conventional interpretation for the existence of this dimorphism in rates of development between the sexes might be that because the mean birth weight of males is invariably heavier, yet the duration of gestation within a species is closely comparable for males and females, then males would need to grow faster than females to achieve this difference in weight. But a more subtle aspect, and indeed an imperative, would seem to be that males must embark on the programme of events concerned with sexual differentiation, especially gonadal differentiation, sooner than females in order to avoid being compromised by exposure to the constitutive or default ovarian pathway (Eicher & Washburn, 1986). Only by achieving a relatively precocious development can genetic males consistently and unequivocally escape from this limitation, and thereby avoid situations of 'timing mismatch' that could lead to ambiguity in the cellular constitution and form of the gonads (see Burgoyne, 1988, 1989; Burgoyne & Palmer, 1991). Whether the temporal advantage achieved by males is primarily due

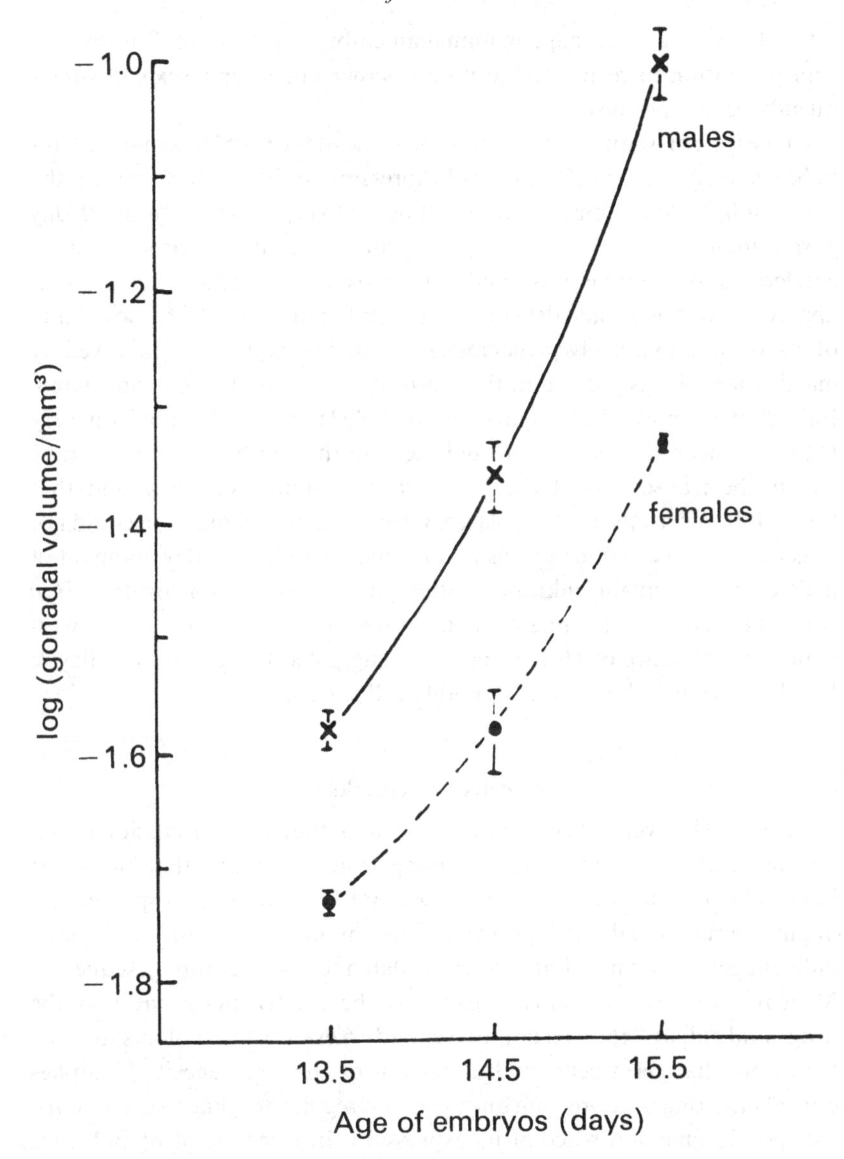

Fig. 2.12. To illustrate the accelerated growth of the differentiating gonadal tissues in male mammals (rats) when compared with females at a corresponding stage of development. (Taken from Mittwoch, 1986.)

to an accelerating influence promoted by *Sry*, or a retarding influence associated with the paternal X chromosome, or a combination of both, awaits further experimental observations in diverse mammalian species. It would certainly be of interest to attempt to enhance or retard the growth rates of early cleavage stage mammalian embryos in culture, followed by transplantation to recipients, to note whether phenotypic sex can subsequently be compromised.

At the time of writing these paragraphs, one of the problems that remains to be resolved concerns the timing of expression of *Sry*. Until very recently, and as indicated in Table 2.2, this was thought to be not before the tenth day *post coitum* in mice. So which gene products could be prompting an accelerated development of male embryos if the gene influence that apparently imposes male development is still in abeyance? The possibility of a short and extremely precocious burst of *Sry* expression, followed by inactivation of this gene until the tenth day, may need to be considered. Indeed, the report of Zwingman *et al.* (1993), although based on very limited evidence (Table 2.4), would indicate that both *Sry* and *Zfy* may already be transcribed at the 2-cell stage of mouse embryos and that transcription persists until the blastocyst stage; actual expression could not be certain. The gene interactions that promote accelerated development of male embryos remain unknown, although there is a suspicion that *Hya* could be involved in enhancing the rate of cell division. A growth-regulating influence of H-Y antigen was suggested by Heslop, Bradley & Baird (1989), possibly acting to modify cell surfaces.

Concluding remarks

A fact not to be overlooked in the excitement of the earlier discussion is that it is, in effect, only the testis-determining switch or trigger that has so far been isolated. Full determination of sex and formation of so splendid an organ as a functional testis (or ovary) must involve programming by many different genes, and not simply by the isolated testis-determining sequence. Moreover, diverse genes must already have been active in formation of the urogenital ridges – the presumptive gonads – even before colonisation by the primordial germ cells, and activation of the *Sry* gene itself implies critical targeting by more than one upstream regulatory gene to account for the specific time and place of its expression. Irrespective of its influence during early cleavage, the transient but potent switch towards maleness imposed by *Sry* during approximately 2 days in the mouse must be seen as a vital stage of confluence between the processes of sex determination and sexual differentiation. Subsequent organisation of the embryonic testis may depend on key products from the pre-Sertoli cells (Singh, Matsukuma & Jones, 1987; Lovell-Badge & Robertson, 1990).

Some of the most recent studies on *Sry* may lead to completely new

Table 2.4. *Observations concerning the possible transcription of the genes* Sry, Zfy *and* Amh *at successive stages of preimplantation development in mouse embryos*

Gene under examination	Method of detecting gene product	Positive reverse transcription–polymerase chain reactions according to stage of embryonic development			
		2 cells	4–8 cells	Morula	Blastocyst
Zfy	Ethidium bromide gel silver staining	2 (3)	4 (8)[a]	0 (3) 1 (1)	3 (9) 1 (2)
Sry	Ethidium bromide gel silver staining	1 (3)	0 (6)	0 (5) 1 (2)	5 (13) 2 (2)
Amh	Ethidium bromide gel	Not determined	Not determined	Not determined	0 (3)

Notes:
Adapted from the report of Zwingman *et al.* (1993).
[a] Except in the case of 2-cell embryos, the total number of determinations from more than one RNA preparation is shown in brackets.

interpretations concerning its expression. Conformational considerations could imply an active rôle preventing translation of a protein product. In brief, *Sry* expression has been detected in adult testis (Rossi *et al.*, 1993) and is germ-cell dependent (Koopman *et al.*, 1990), and the most abundant *Sry* transcripts in such testis tissue, as opposed to the embryonic genital ridge, are present as circular RNA molecules (Capel *et al.*, 1993). This is currently viewed as an unusual arrangement, but it may in fact be more widespread than hitherto appreciated due to difficulties in its detection. Thus, views on the scope and timing of *Sry* expression would certainly appear to be open to modification.

As noted earlier in this chapter, maleness tends to be given special prominence in discussions of mammalian sexual differentiation, with femaleness being treated essentially as the permissive, secondary or alternative condition (i.e. the default pathway) in the absence of the prior action of any male-determining sequences. Perhaps this general interpretation should be open to modification in the late twentieth century and rather more active consideration given to the positive steps involved in the formation of an ovary – which itself should be viewed as a specific and incisive process stemming from appropriate gene-regulated events, even though occurring slightly later than the male gonadal pathway.

Finally, since aberrant or anomalous development of the gonads is well documented and, as will be discussed later in the text, may involve unilateral development of an ovotestis or testis with contralateral development of a functional ovary in a genetic female, some serious thought must be given to seemingly independent gene programmes unfolding in the two distinct gonads coupled with different rates of development. Endeavouring to understand the regulation of such asymmetrical development should provide further insights into the differentiation of functional gonads, particularly if a full analysis of the genetic and molecular determinants of that disorder can be obtained.

References

Affara, N. A. (1991). Sex and the single Y. *BioEssays*, **13**, 475–8.

Affara, N. A., Ferguson-Smith, M. A., Magenis, R. E., Tolmie, J. L., Boyd, E., Cooke, A., Jamieson, D., Kwok, K., Mitchell, M. & Snadden, L. (1987). Mapping the testis determinants by an analysis of Y-specific sequences in males with apparent XX and XO karyotypes and females with XY karyotypes. *Nucleic Acids Research*, **15**, 7325–42.

Affara, N. A., Ferguson-Smith, M. A., Tolmie, J., Kwok, K., Mitchell, M., Jamieson, D., Cooke, A. & Florentin, L. (1986). Variable transfer of Y-specific sequences in XX males. *Nucleic Acids Research*, **14**, 5375–87.

Andersson, M., Page, D. C. & de la Chapelle, A. (1986). Chromosomal Y-specific DNA is transferred to the short arm of X chromosome in human XX males. *Science*, **233**, 786–8.

Armstrong, J. F., Pritchard-Jones, K., Bickmore, W. A., Hastie, N. D. & Bard, J. B. L. (1992). The expression of the Wilms' tumour gene, *WTI*, in the developing mammalian embryo. *Mechanisms of Development*, **40**, 85–97.

Avery, B., Madison, V. & Greve, R. (1991). Sex and development in bovine *in vitro* fertilised embryos. *Theriogenology*, **35**, 953–63.

Beatty, R. A. (1960). Chromosomal determination of sex in mammals. *Memoirs. Society for Endocrinology*, **7**, 45–8.

Beatty, R. A. (1970). Genetic basis for the determination of sex. *Philosophical Transactions of the Royal Society of London, Series B*, **259**, 3–13.

Behlke, M., Bogan, J. S., Beer-Romero, P. & Page, D. C. (1993). Evidence that the *SRY* protein is encoded by a single exon on the human Y chromosome. *Genomics*, **17**, 736–9.

Berta, P., Hawkins, J. R., Sinclair, A. H., Taylor, A., Griffiths, B. L., Goodfellow, P. N. & Fellous, M. (1990). Genetic evidence equating *SRY* and the testis-determining factor. *Nature (London)*, **348**, 448–50.

Billingham, R. E. & Silvers, W. K. (1960). Studies on tolerance of the Y chromosome antigen in mice. *Journal of Immunology*, **85**, 14–26.

Bird, A. P. (1986). CpG-rich islands and the function of DNA methylation. *Nature (London)*, **321**, 209–13.

Bishop, C. E., Weith, A., Mattei, M.-G. & Roberts, C. (1988). Molecular aspects of sex determination in mice: an alternative model for the origin of the *Sxr* region. *Philosophical Transactions of the Royal Society of London, Series B*, **322**, 119–24.

Brandeis, M., Ariel, M. & Cedar, H. (1993). Dynamics of DNA methylation during development. *BioEssays*, **15**, 709–13.

Brennan, R. G. (1993). The winged-helix DNA-binding motif: another helix-turn-helix takeoff. *Cell*, **74**, 773–6.

Brockdorff, N., Ashworth, A., Kay, G., McCabe, V. M., Norris, D. P., Cooper, P. J., Swift, S. & Rastan, S. (1992). The product of the mouse *Xist* gene is a 15 kb inactive X-specific transcript containing no conserved ORF and located in the nucleus. *Cell*, **71**, 515–26.

Brown, C. J., Ballabio, A., Rupert, J. L., Lafrenière, R. G., Grompe, M., Tonlorenzi, R. & Willard, H. F. (1991). A gene from the region of the human X inactivation centre is expressed exclusively from the inactive X chromosome. *Nature (London)*, **349**, 38–44.

Brown, C. J., Hendrich, B. D., Rupert, J. L., Lafrenière, R. G., Xing, Y., Lawrence, J. & Willard, H. F. (1992). The human *XIST* gene: analysis of a 17 kb inactive X-specific RNA that contains conserved repeats and is highly localised within the nucleus. *Cell*, **71**, 527–42.

Bull, J. J. (1980). Sex determination in reptiles. *Quarterly Review of Biology*, **55**, 3–20.

Bull. J. J. (1983). *Evolution of Sex Determining Mechanisms*. Menio Park, California: Benjamin Cummings.

Bull, J. J. (1987). Temperature-sensitive periods of sex determination in a lizard: similarities with turtles and crocodilians. *Journal of Experimental Zoology*, **241**, 143–8.

Burgoyne, P. S. (1982). Genetic homology and crossing over in the X and Y chromosomes of mammals. *Human Genetics*, **61**, 85–90.

Burgoyne, P. S. (1986). Mammalian X and Y crossover. *Nature (London)*, **319**, 258–9.

Burgoyne, P. S. (1988). Role of mammalian Y chromosome in sex determination. *Philosophical Transactions of the Royal Society of London, Series B*, **322**, 63–72.
Burgoyne, P. S. (1989). Thumbs down for zinc finger. *Nature (London)*, **342**, 860–2.
Burgoyne, P. S. (1992). Y chromosome function in mammalian development. *Advances in Developmental Biology*, **1**, 1–29.
Burgoyne, P. S. (1993). A Y-chromosomal effect on blastocyst cell number in mice. *Development*, **117**, 341–5.
Burgoyne, P. S., Ansell, J. D. & Tournay, A. (1986). Can the indifferent mammalian XX gonad be sex-reversed by interaction with testicular tissue. In *Development and Function of the Reproductive Organs*, ed. A. Eshkol, B. Eckstein *et al.*, Serono Symposia Review No. 11, pp. 23–39. New York: Raven Press.
Burgoyne, P. S., Levy, E. R. & McLaren, A. (1986). Spermatogenic failure in male mice lacking H-Y antigen. *Nature (London)*, **320**, 170–2.
Burgoyne, P. S. & Palmer, S. J. (1991). The genetics of XY sex reversal in the mouse and other mammals. *Seminars in Developmental Biology*, **2**, 277–84.
Byskov, A. G. & Hoyer, P. E. (1988). Embryology of mammalian gonads and ducts. In *The Physiology of Reproduction*, ed. E. Knobil, J. Neill *et al.*, chapter 8, pp. 265–302. New York: Raven Press.
Capel, B., Swain, A., Nicolis, S., Hacker, A., Walter, M., Koopman, P., Goodfellow, P. & Lovell-Badge, R. (1993). Circular transcripts of the testis-determining gene *Sry* in adult mouse testis. *Cell*, **73**, 1019–30.
Cattanach, B. M. (1975). Control of chromosome inactivation. *Annual Review of Genetics*, **9**, 1–18.
Cattanach, B. M., Pollard, C. E. & Hawkes, S. G. (1971). Sex-reversed mice: XX and XO males. *Cytogenetics*, **10**, 318–37.
Chalmers, C., Wiberg, U. & Hunter, R. H. F. (1988). H-Y transplantation antigen status of two intersex pigs. *Proceedings of the Society for the Study of Fertility*, No. **62**, 36.
Chandra, H. S. (1985). Is human X chromosome inactivation a sex-determining device? *Proceedings of the National Academy of Sciences, USA*, **82**, 6947–9.
Chandra, H. S. (1986). X chromosomes and dosage compensation. *Nature (London)*, **319**, 18.
Charnier, M. (1966). Action de la température sur la sex-ratio chez l'embryon *d'Agama agama. Société de Biologie Ouest Africa*, **160**, 620–2.
Cotinot, C., McElreavey, K. & Fellous, M. (1993). Sex determination: genetic control. *In Reproduction in Mammals and Man*, ed. C. Thibault, M. C. Levasseur & R. H. F. Hunter, pp. 213–26. Paris: Ellipses.
Crew, F. A. E. (1965). *Sex-Determination*, 4th edn. London: Methuen.
Davis, R. M. (1981). Localisation of male determining factors in man: a thorough review of structural anomalies of the Y chromosome. *Journal of Medical Genetics*, **18**, 161–95.
Deeming, D. C. & Ferguson, M. W. J. (1988). Environmental regulation of sex determination in reptiles. *Philosophical Transactions of the Royal Society of London, Series B*, **322**, 19–39.
Durbin, E. J., Stalvey, J. R. D. & Erickson, R. P. (1989). Northern analyses using single-stranded probes do not support a role for GATA/GACA repeats in sex determination in mice and men. *Molecular Reproduction and Development*, **1**, 116–21.
Eicher, E. M. & Washburn, L. L. (1983). Inherited sex reversal in mice:

Identification of a new primary sex-determining gene. *Journal of Experimental Zoology*, **228**, 297–304.

Eicher, E. M. & Washburn, L. L. (1986). Genetic control of primary sex determination in mice. *Annual Review of Genetics*, **20**, 327–60.

Eichwald, E. J. & Silmser, C. R. (1955). Untitled communication. *Transplantation Bulletin*, **2**, 148–149.

Epplen, J. T., McCarrey, J. R., Sutou, S. & Ohno, S. (1982). Base sequence of a cloned snake W-chromosome DNA fragment and identification of a male-specific putative mRNA in the mouse. *Proceedings of the National Academy of Sciences, USA*, **79**, 3798–802.

Erickson, R. P. & Verga, V. (1989). Minireview: Is zinc-finger Y the sex-determining gene? *American Journal of Human Genetics*, **45**, 671–4.

Evans, H. J., Buckton, K. E., Spowart, G. & Carothers, A. D. (1979). Heteromorphic X chromosomes in 46 XX males: evidence for the involvement of X–Y interchange. *Human Genetics*, **49**, 11–31.

Evans, E. P., Burtenshaw, M. D. & Cattanach, B. M. (1982). Meiotic crossing-over between the X and Y chromosomes of male mice carrying the sex-reversing (*Sxr*) factor. *Nature (London)*, **300**, 443–5.

Ewert, M. A. & Nelson, C. E. (1991). Sex determination in turtles: diverse patterns and some possible adaptive values. *Copeia*, No. 1, 50–69.

Ferguson-Smith, M. A. (1966). X–Y chromosomal interchange in the aetiology of true hermaphroditism and of the XX Klinefelter's syndrome. *Lancet*, **ii**, 475–6.

Ferguson-Smith, M. A. (1991). Genotype-phenotype correlations in individuals with disorders of sex determination and development including Turner's syndrome. *Seminars in Developmental Biology*, **2**, 265–76.

Ferguson-Smith, M. A. & Affara, N. A. (1988). Accidental X–Y recombination and the aetiology of XX males and true hermaphrodites. *Philosophical Transactions of the Royal Society of London, Series B*, **322**, 133–44.

Fisher, R. A. (1930). *The Genetical Theory of Natural Selection*. Oxford: Clarendon Press.

Ford, C. E., Jones, K. W., Polani, P. E., de Almeida, J. C. & Briggs, J. H. (1959). A sex-chromosome anomaly in a case of gonadal dysgenesis (Turner's syndrome). *Lancet*, **i**, 711–13.

Foster, J. W., Brennan, F. E., Hampikian, G. K., Goodfellow, P. N., Sinclair, A. H., Lovell-Badge, R., Selwood, L., Renfree, M. B., Cooper, D. W. & Marshall Graves, J. A. (1992). Evolution of sex determination and the Y chromosome: *SRY*-related sequences in marsupials. *Nature (London)*, **359**, 531–3.

Gallien, L. (1965). Genetic control of sexual differentiation in vertebrates. In *International Conference on Organogenesis*, vol. 23, ed. R. L. de Haan & H. Ursprung, PP. 583–610. New York: Holt.

Gardner, R. L., Lyon, M. F., Evans, E. P. & Burtenshaw, M. D. (1985). Clonal analysis of X-chromosome inactivation and the origin of the germ line in the mouse embryo. *Journal of Embryology and Experimental Morphology*, **88**, 349–63.

Gartler, S. M., Rivest, M. & Cole, R. E. (1980). Cytological evidence for an inactive X chromosome in murine oogonia. *Cytogenetics and Cell Genetics*, **28**, 203–7.

George, F. W. & Wilson, J. D. (1988). Sex determination and differentiation. In *The Physiology of Reproduction*, ed. E. Knobil, J. Neill *et al.*, chapter 1, pp. 3–26. New York, Raven Press.

Gubbay, J., Collignon, J., Koopman, P., Capel, B., Economou, A., Münsterberg, A., Vivian, N., Goodfellow, P. & Lovell–Badge, R. (1990). A gene mapping to the sex-determining region of the mouse Y chromosome is a member of a novel family of embryonically expressed genes. *Nature (London)*, **346**, 245–50.

Haqq, C. M., King, C. Y., Donahoe, P. K. & Weiss, M. A. (1993). *SRY* recognizes conserved DNA sites in sex-specific promoters. *Proceedings of the National Academy of Sciences, USA*, **90**, 1097–101.

Harley, V. R., Jackson, D. I., Hextall, P. J., Hawkins, J. R., Berkovitz, G. D., Sockanathan, S., Lovell-Badge, R. & Goodfellow, P. N. (1992). DNA binding activity of recombinant *SRY* from normal males and XY females. *Science*, **255**, 453–6.

Head, G., May, R. M. & Pendleton, L. (1987). Environmental determination of sex in the reptiles. *Nature (London)*, **329**, 198–9.

Heslop, B. F., Bradley, M. P. & Baird, M. A. (1989). A proposed growth regulatory function for the serologically detectable sex-specific antigen H-Y. *Human Genetics*, **81**, 99–104.

Hodgkin, J. (1990). Sex determination compared in *Drosophila* and *Caenorhabditis*. *Nature (London)*, **344**, 721–8.

Hunter, R. H. F., Baker, T. G. & Cook, B. (1982). Morphology, histology and steroid hormones of the gonads in intersex pigs. *Journal of Reproduction and Fertility*, **64**, 217–22.

Hunter, R. H. F., Chalmers, C. & Cavazos, F. (1988). Intersexuality in domestic pigs: a guide to mechanisms of gonadal differentiation? *Animal Breeding Abstracts*, **56**, 785–91.

Hunter, R. H. F., Cook, B. & Baker, T. G. (1985). Intersexuality in five pigs, with particular reference to oestrous cycles, the ovotestis, steroid hormone secretion and potential fertility. *Journal of Endocrinology*, **106**, 233–42.

Itoh, S. & Goto, T. (1986). Sex frequency of offspring from different developmental stage of cattle embryos. *Japanese Journal of Animal Artificial Insemination Research*, **8**, 95–9.

Jacobs, P. A. & Ross, A. (1966). Structural abnormalities of the Y chromosome in man. *Nature (London)*, **210**, 352–4.

Jacobs, P. A. & Strong, J. A. (1959). A case of human intersexuality having a possible XXY sex-determining mechanism. *Nature (London)*, **183**, 302–3.

Jäger, R. J., Anvret, M., Hall, K. & Scherer, G. (1990). A human XY female with a frame shift mutation in the candidate testis-determining gene *SRY*. *Nature (London)*, **348**, 452–4.

Jagiello, G. M., Tantravahi, U., Ducayen, M. B. & Erlanger, B. F. (1987). Immunocytochemical evidence for methylation of the inactive X chromosome in human foetal oogonia. *Proceedings of the Society for Experimental Biology and Medicine*, **186**, 223–8.

Jones, K. W. (1989). Inactivation phenomena in the evolution and function of sex chromosomes. In *Evolutionary Mechanisms in Sex Determination*, ed. S. S. Wachtel, pp. 69–78. Boca Raton, Florida: CRC Press.

Jones, K. W. & Singh, L. (1981). Conserved repeated DNA sequences in vertebrate sex chromosomes. *Human Genetics*, **58**, 46–53.

Jost, A. (1953). Problems of fetal endocrinology: the gonadal and hypophyseal hormones. *Recent Progress in Hormone Research*, **8**, 379–413.

Kafri, T., Ariel, M., Brandeis, M., Shemer, R., Urven, L., McCarrey, J., Cedar, H. & Razin, A. (1992). Developmental pattern of gene specific DNA methylation in the mouse embryo and germ line. *Genes and Development*, **6**, 705–14.

Kallikin, L. M., Fujimoto, H., Witt, M. P., Verga, V. & Erickson, R. P. (1989). A genomic clone of *Zfy-1* from a Y^{DOM} mouse strain detects post-meiotic gene expression of *Zfy* in testes. *Biochemical and Biophysical Research Communications*, **165**, 1286–91.

Kay, G. F., Penny, G. D., Patel, D., Ashworth, A., Brockdorff, N. & Rastan, S. (1993). Expression of *Xist* during mouse development suggests a role in the initiation of X chromosome inactivation. *Cell*, **72**, 171–83.

Kim, M. H. & Kessel, M. (1993). Homeobox genes as regulators of vertebrate development. *Agbiotech News and Information*, **5**, 189N–94N.

Koller, P. C. & Darlington, C. D. (1934). The genetical and mechanical properties of the sex-chromosomes. I. *Rattus norvegicus*. *Journal of Genetics*, **29**, 159–73.

Koopman, P., Gubbay, J., Collignon, J. & Lovell-Badge, R. (1989). *Zfy* gene expression patterns are not compatible with a primary rôle in mouse sex determination. *Nature (London)*, **342**, 940–2.

Koopman, P., Gubbay, J., Vivian, N., Goodfellow, P. & Lovell-Badge, R. (1991). Male development of chromosomally female mice transgenic for *Sry*. *Nature (London)*, **351**, 117–21.

Koopman, P., Münsterberg, A., Capel, B., Vivian, N. & Lovel-Badge, R. (1990). Expression of a candidate sex-determining gene during mouse testis differentiation. *Nature (London)*, **348**, 450–2.

Kratzer, P. G. & Chapman, V. M. (1981). X chromosome reactivation in oocytes of *Mus caroli*. *Proceedings of the National Academy of Sciences, USA*, **78**, 3093–7.

Kreidberg, J. A., Sariola, H., Loring, J. M., Maeda, M., Pelletiér, J., Housman, D. & Jaenisch, R. (1993). *WT-1* is required for early kidney development. *Cell*, **74**, 679–91.

LeBarr, D. K., Blecher, S. R. & Moger, W. H. (1991). Development of the normal XY male and sex-reversed XX Sxr pseudomale mouse epididymis. *Molecular Reproduction and Development*, **28**, 9–17.

Levinson, G., Marsh, J. L., Epplen, J. T. & Gutman, G. A. (1985). Cross-hybridizing snake satellite, *Drosophila*, and mouse DNA sequences may have arisen independently. *Molecular Biology and Evolution*, **2**, 494–504.

Lilley, D. M. J. (1992). HMG has DNA wrapped up. *Nature (London)*, **357**, 282–3.

Lovell-Badge, R. & Robertson, E. (1990). XY female mice resulting from a heritable mutation in the primary testis-determining gene, *Tdy*. *Development*, **109**, 635–46.

Luisi, B. (1993). The crystal maze. *Trends in Genetics*, **9**, 401.

Lyon, M. F. (1961). Gene action in the X-chromosome of the mouse (*Mus musculus L.*). *Nature (London)*, **190**, 372–3.

Lyon, M. F. (1962). Sex chromatin and gene action in the mammalian X-chromosome. *American Journal of Human Genetics*, **14**, 135–48.

Lyon, M. F. (1974). Mechanisms and evolutionary origins of variable X-chromosome activity in mammals. *Proceedings of the Royal Society of London, Series B*, **187**, 243–68.

Lyon, M. F. (1986). X chromosomes and dosage compensation. *Nature (London)*, **320**, 313.

McGinnis, W. & Krumlauf, R. (1992). Homeobox genes and axial patterning. *Cell*, **68**, 283–302.

McLaren, A. (1981). *Germ Cells and Soma. A New Look at an Old Problem.* New Haven: Yale University Press.

McLaren, A. (1988*a*). Sex determination in mammals. *Trends in Genetics*, **4**, 153–7.
McLaren, A. (1988*b*). The developmental history of female germ cells in mammals. In *Oxford Reviews of Reproductive Biology*, vol. 10, ed. J. R. Clarke, pp. 162–79. Oxford: Oxford University Press.
McLaren, A. (1990). What makes a man a man? *Nature (London)*, **346**, 216–17.
McLaren, A. (1991*a*). Sex determination in mammals. In *Oxford Reviews of Reproductive Biology*, vol. 13, ed. S. R. Milligan, pp. 1–33. Oxford: Oxford University Press.
McLaren, A. (1991*b*). The making of male mice. *Nature (London)*, **351**, 96–7.
McLaren, A. & Monk, M. (1981). X chromosome activity in the germ cells of sex-reversed mouse embryos. *Journal of Reproduction and Fertility*, **63**, 533–7.
McLaren, A., Simpson, E., Tomonari, K., Chandler, P. & Hogg, H. (1984). Male sexual differentiation in mice lacking H-Y antigen. *Nature (London)*, **312**, 552–5.
Magenis, R. E., Webb, M. J., McKean, R. S., Tomar, D., Allen, L.J., Kammer, H., Van Dyke, D. L. & Lovrien, E. (1982). Translocation (X;Y) (p.22.33;p.11.2) in XX males; etiology of male phenotype. *Human Genetics*, **62**, 271–6.
Mardon, G. & Page, D. C. (1989). The sex-determining region of the mouse Y chromosome encodes a protein with a highly acidic domain and 13 zinc fingers. *Cell*, **56**, 765–70.
Mardon, G., Mosher, R., Disteche, C. M., Nishioka, Y., McLaren, A. & Page, D. C. (1989). Duplication, deletion and polymorphism in the sex-determining region of the mouse Y chromosome. *Science*, **243**, 78–80.
Maynard Smith, J. (1978). *The Evolution of Sex*. Cambridge: Cambridge University Press.
Merchant, H. (1975). Rat gonadal and ovarian organogenesis with and without germ cells. An ultrastructural study. *Developmental Biology*, **44**, 1–21.
Migeon, B. R. & Jelalian, K. (1977). Evidence for two active X chromosomes in germ cells of female before meiotic entry. *Nature (London)*, **269**, 242–3.
Migeon, B. R., Luo, S., Stasiowski, B. A., Jani, M., Axelman, J., van Dyke, D. L., Weiss, L., Jacobs, P. A., Yang-Feng, T. L. & Wiley, J. E. (1993). Deficient transcription of *XIST* from tiny ring X chromosomes in females with severe phenotypes. *Proceedings of the National Academy of Sciences, USA*, **90**, 12025–9.
Mintz, B. (1968). Hermaphroditism, sex chromosomal mosaicism and germ cell selection in allophenic mice. *Journal of Animal Science*, **27**, Suppl. 1, 51–60.
Mittwoch, U. (1967). *Sex Chromosomes*. New York & London: Academic Press.
Mittwoch, U. (1969). Do genes determine sex? *Nature (London)*, **221**, 446–8.
Mittwoch, U. (1977). H-Y antigen and the growth of the dominant gonad. *Journal of Medical Genetics*, **14**, 335–8.
Mittwoch, U. (1986). Males, females and hermaphrodites. *Annals of Human Genetics*, **50**, 103–21.
Mittwoch, U. (1989). Sex differentiation in mammals and tempo of growth: probabilities *vs*. switches. *Journal of Theoretical Biology*, **137**, 445–55.
Mittwoch, U. (1992). Sex determination and sex reversal: genotype, phenotype, dogma and semantics. *Human Genetics*, **89**, 467–79.
Monk, M. & Harper, M. I. (1979). Sequential X-chromosome inactivation coupled with cellular differentiation in early mouse embryos. *Nature (London)*, **281**, 311–13.

Monk, M. & McLaren, A. (1981). X-chromosome activity in foetal germ cells of the mouse. *Journal of Embryology and Experimental Morphology*, **63**, 75–84.
Monk, M., Boubelik, M. & Lehnert, S. (1987). Temporal and regional changes in DNA methylation in the embryonic, extraembryonic and germ cell lineage during mouse embryo development. *Development*, **99**, 371–82.
Müller, U. & Wachtel, S. S. (1991). Are testis-secreted H-Y serological antigen and Müllerian inhibiting substance the same? *American Journal of Human Genetics*, **49**, S22.
Müller, U., Aschmoneit, I., Zenzes, M. T. & Wolf, U. (1978). Binding studies of H-Y antigen in rat tissues. Indications for a gonad-specific receptor. *Human Genetics*, **43**, 151–7.
Nasrin, N., Buggs, C., Kong, X.F., Carnazza, J., Goebl, M. & Alexander-Bridges, M. (1991). DNA-binding properties of the product of the testis-determining gene and a related protein. *Nature (London)*, **354**, 317–20.
O,W.-S., Short, R. V., Renfree, M. B. & Shaw, G. (1988). Primary genetic control of somatic sexual differentiation in a mammal. *Nature (London)*, **331**, 716–17.
Ohno, S. (1967). *Sex Chromosomes and Sex-Linked Genes*. Berlin, Heidelberg, New York: Springer-Verlag.
Ohno, S. (1976). Major regulatory genes for mammalian sexual development. *Cell*, **7**, 315–21.
Ohno, S. (1977). The original function of MHC antigens as the general plasma membrane anchorage site of organogenesis-directing proteins. *Immunological Reviews*, **33**, 59–69.
Ohno, S. (1978). The role of H-Y antigen in primary sex determination. *Journal of the American Medical Association*, **239**, 217–20.
Ohno, S. (1989). From GATA-GACA repeats to *Dictyostelium* cell adhesion protein to C-CAM, N-CAM, and gonad-organizing proteins. In *Evolutionary Mechanisms in Sex Determination*, ed. S. S. Wachtel, pp. 15–23. Boca Raton, Florida: CRC Press.
Ohno, S., Klinger, H. P. & Atkin, N. B. (1962). Human oogenesis. *Cytogenetics*, **1**, 42–51.
Ohno, S., Nagai, Y., Ciccarese, S. & Iwata, H. (1979). Testis-organizing H-Y antigen and the primary sex-determining mechanism of mammals. *Recent Progress in Hormone Research*, **35**, 449–76.
Page, D. C. (1986). Sex reversal: deletion mapping the male-determining function of the human Y chromosome. *Cold Spring Harbor Symposia on Quantitative Biology*, **51**, 229–35.
Page, D. C., Brown, L. G. & de la Chapelle, A. (1987). Exchange of terminal portions of the X- and Y-chromosomal short arms in human XX males. *Nature (London)*, **328**, 437–40.
Page, D. C. & de la Chapelle, A. (1984). The parental origin of X chromosomes in XX males determined using restriction fragment length polymorphisms. *American Journal of Human Genetics*, **36**, 565–75.
Page, D. C., Fisher, E. M. C., McGillivray, B. & Brown, L. G. (1990). Additional deletion in sex-determining region of human Y chromosome resolves paradox of X,t (Y;22) female. *Nature (London)*, **346**, 279–81.
Page, D. C., Mosher, R., Simpson, E. M., Fisher, E. M. C., Mardon, G., Pollack, J., McGillivray, B., de la Chapelle, A. & Brown, L. G. (1987). The sex-determining region of the human Y chromosome encodes a finger protein. *Cell*, **51**, 1091–104.
Palmer, M. S., Berta, P., Sinclair, A. H., Pym, B. & Goodfellow, P. N. (1990).

Comparison of human *ZFY* and *ZFX* transcripts. *Proceedings of the National Academy of Sciences, USA*, **87**, 1681–5.

Palmer, M. S., Sinclair, A. H., Berta, P., Ellis, N. A., Goodfellow, P. N., Abbas, N. E. & Fellous, M. (1989). Genetic evidence that *ZFY* is not the testis-determining factor. *Nature (London)*, **342**, 937–9.

Palmiter, R. D., Brinster, R. L., Hammer, R. E., Trumbauer, M. E., Rosenfeld, M. G., Birnberg, N. C. & Evans, R. M. (1982). Dramatic growth of mice that develop from eggs microinjected with metallothionein-growth hormone fusion genes. *Nature (London)*, **300**, 611–15.

Park, S., Schalling, M., Bernard, A. *et al.* (1993). The Wilms tumour gene *WT1* is expressed in murine mesoderm-derived tissues and mutated in a human mesothelioma. *Nature Genetics*, **4**, 415–20.

Paull, T. T., Haykinson, M. J. & Johnson, R. C. (1993). The nonspecific DNA-binding and -bending proteins HMG1 and HMG2 promote the assembly of complex nucleoprotein structures. *Genes and Development*, **7**, 1521–34.

Payen, E. J. & Cotinot, C. Y. (1993). Comparative HMG-box sequences of the *SRY* gene between sheep, cattle and goats. *Nucleic Acids Research*, **21**, 2772.

Renfree, M. B., Wilson, J. D., Short, R. V., Shaw, G. & George, F. W. (1992). Steroid hormone content of the gonads of the tammar wallaby during sexual differentiation. *Biology of Reproduction*, **47**, 644–7.

Riddihough, G. (1992). Homing in on the homeobox. *Nature (London)*, **357**, 643–4.

Rossant, J. & Joyner, A. L. (1989). Towards a molecular-genetic analysis of mammalian development. *Trends in Genetics*, **5**, 277–83.

Rossi, P., Dolci, S., Albanesi, C., Grimaldi, P. & Geremia, R. (1993). Direct evidence that the mouse sex-determining gene *Sry* is expressed in the somatic cells of male fetal gonads and in the germ cell line in the adult testis. *Molecular Reproduction and Development*, **34**, 369–73.

Schäfer, R., Ali, S. & Epplen, J. T. (1986). The organisation of the evolutionarily conserved GATA/GACA repeats in the mouse genome. *Chromosoma*, **93**, 502–10.

Schneider-Gadicke, A., Beer-Romero, P., Brown, L. G., Nussbaum, R. & Page, D. C. (1989). *ZFX* has a gene structure similar to *ZFY*, the putative human sex determinant, and escapes X-inactivation. *Cell*, **57**, 1247–58.

Schwabe, J. W. R. & Travers, A. A. (1993). What is evolution playing at? *Current Biology*, **3**, 628–30.

Selker, E. U. (1990). DNA methylation and chromatin structure: a view from below. *TIBS*, **15**, 103–7.

Shaw, G., Renfree, M. B. & Short, R. V. (1990). Primary genetic control of sexual differentiation in marsupials. *Australian Journal of Zoology*, **37**, 443–50.

Short, N. (1993). Forty years of molecular information. *Nature (London)*, **362**, 783–4.

Short, R. V. (1982). Sex determination and differentiation. In *Reproduction in Mammals*, 2nd edn, vol. 2, ed. C. R. Austin & R. V. Short, pp. 70–113. Cambridge & London: Cambridge University Press.

Simpson, E., Chandler, P., Hunt, R., Hogg, H., Tomonari, K. & McLaren, A. (1986). H-Y status of X/XSxr′ male mice: *in vivo* tests. *Immunology*, **57**, 345–9.

Simpson, E., Chandler, P., Goulmy, E., Disteche, C. M., Ferguson-Smith, M. A. & Page, D. C. (1987). Separation of the genetic loci for H-Y antigen and for testis determination on human Y chromosome. *Nature (London)*, **326**, 876–8.

Sinclair, A. H., Berta, P., Palmer, M. S., Hawkins, J. R., Griffiths, B. L., Smith, M. J. , Foster, J. W., Frischauf, A.-M., Lovell-Badge, R., Goodfellow, P. N. (1990). A gene from the human sex-determining region encodes a protein with homology to a conserved DNA binding motif. *Nature (London)*, **346**, 240–4.

Sinclair, A. H., Foster, J. W., Spencer, J. A., Page, D. C., Palmer, M., Goodfellow, P. N. & Marshall-Graves, J. A. (1988). Sequences homologous to *ZFY*, a candidate human sex-determining gene, are autosomal in marsupials. *Nature (London)*, **336**, 780–3.

Singh, L. & Jones, K. W. (1982). Sex reversal in the mouse (*Mus musculus)* is caused by a recurrent non-reciprocal crossover involving the X and an aberrant Y chromosome. *Cell*, **28**, 205–16.

Singh, L., Matsukuma, S. & Jones, K. W. (1987). Testis development in a mouse with 10% of XY cells. *Developmental Biology*, **122**, 287–90.

Singh, L., Phillips, C. & Jones, K. W. (1984). The conserved nucleotide sequences of *Bkm*, which define *Sxr* in the mouse, are transcribed. *Cell*, **36**, 111–20.

Singh, L., Purdom, I. F. & Jones, K. W. (1976). Satellite DNA and evolution of sex chromosomes. *Chromosoma*, **59**, 43–62.

Singh, L., Purdom, I. F. & Jones, K. W. (1980). Sex chromosome-associated satellite DNA: evolution and conservation. *Chromosoma*, **79**, 137–57.

Su, H. & Lau, Y.-F. C. (1993). Identification of the transcriptional unit, structural organisation, and promoter sequence of the human sex-determining region Y (*SRY*) gene, using a reverse genetic approach. *American Journal of Human Genetics*, **52**, 24–38.

Takagi, N. & Abe, K. (1990). Detrimental effects of two active X chromosomes on early mouse development. *Development*, **109**, 189–201.

Takagi, N. & Sasaki, M. (1975). Preferential inactivation of the paternally derived X chromosome in the extraembryonic membranes of the mouse. *Nature (London)*, **256**, 640–2.

Teplitz, R. & Ohno, S. (1963). Postnatal induction of oogenesis in the rabbit (*Oryctolagus cuniculus*). *Experimental Cell Research*, **31**, 183–9.

Thornhill, A. R. & Burgoyne, P. S. (1993). A paternally imprinted X chromosome retards the development of the early mouse embryo. *Development*, **118**, 171–4.

Tsunoda, Y., Tokunaga, T. & Sugie, T. (1985). Altered sex ratio of live young after transfer of fast and slow developing mouse embryos. *Gamete Research*, **12**, 301–4.

Tucker, P. K. & Lundrigan, B. L. (1993). Rapid evolution of the sex determining locus in Old World mice and rats. *Nature (London)*, **364**, 715–17.

Valdivia, R. P. A., Kunieda, T., Azuma, S. & Toyoda, Y. (1993). PCR sexing and developmental rate differences in preimplantation mouse embryos fertilised and cultured *in vitro*. *Molecular Reproduction and Development*, **35**, 121–6.

Vergnaud, G., Page, D. C., Simmler, M.-C., Brown, L., Rouyer, F., Noel, B., Botstein, D., de la Chapelle, A. & Weissenbach, J. (1986). A deletion map of the human Y chromosome based on DNA hybridization. *American Journal of Human Genetics*, **38**, 109–24.

Watchel, S. S. (1983). *H-Y Antigen and the Biology of Sex Determination*. New York: Grune & Stratton.

Wachtel, S. S., Ohno, S., Koo, G. C. & Boyse, E. A. (1975). Possible role for H-Y antigen in the primary determination of sex. *Nature (London)*, **257**, 235–6.
Waibel, F., Scherer, G., Fraccaro, M., Hustinx, T. W. J., Weissenbach, J., Wieland, J., Mayerova, A., Back, E. & Wolf, U. (1987). Absence of Y-specific DNA sequences in human 46,XX true hermaphrodites and in 45,X mixed gonadal dysgenesis. *Human Genetics*, **76**, 332–6.
Welshons, W. J. & Russell, L. B. (1959). The Y-chromosome as the bearer of male determining factors in the mouse. *Proceedings of the National Academy of Sciences, USA*, **45**, 560–6.
West, J. D. (1982). X-chromosome expression during mouse embryogenesis. In *Genetic Control of Gamete Production and Function*, ed. P. G. Crosignani, B. L. Rubin & M. Fraccaro, pp. 49–91. New York: Academic Press.
West, J. D., Frels, W. I., Chapman, V. M. & Papaioannou, V. E. (1977). Preferential expression of the maternally derived X-chromosome in the mouse yolk sac. *Cell*, **12**, 873–82.
Whitfield, L. S., Lovell-Badge, R. & Goodfellow, P. N. (1993). Rapid sequence evolution of the mammalian sex-determining gene *SRY*. *Nature (London)*, **364**, 713–15.
Wiberg, U.H. (1985). H-Y transplantation antigen in human XO females. *Human Genetics*, **69**, 15–18.
Wiberg, U.H. (1987). Facts and considerations about sex-specific antigens. *Human Genetics*, **76**, 207–19.
Wiberg, U.H. & Gunther, E. (1985). Female wood lemmings with the mutant X*-chromosome carry the H-Y transplantation antigen. *Immunogenetics*, **21**, 91–6.
Wilkie, T. (1991). Mice embryos' sex changed. *The Independent*, 9 May.
Wilmut, I., Hooper, M. L. & Simons, J. P. (1991). Genetic manipulation of mammals and its application in reproductive biology. *Journal of Reproduction and Fertility*, **92**, 245–79.
Wolf, U., Fraccaro, M., Mayerová, A., Hecht, T., Maraschio, P. & Hameister, H. (1980). A gene controlling H-Y antigen on the X chromosome: tentative assignment by deletion mapping to Xp.22.3. *Human Genetics*, **54**, 149–54.
Wolf, U., Schempp, W. & Scherer, G. (1992). Molecular biology of the human Y chromosome. *Reviews in Physiology, Biochemistry and Pharmacology*, **121**, 145–213.
Xu, K. P., Yadav, B. R., King, W. A. & Betteridge, K. J. (1992). Sex-related differences in developmental rates of bovine embryos produced and cultured *in vitro*. *Molecular Reproduction and Development*, **31**, 249–52.
Yen, P. H., Patel, P., Chinault, A. C., Mohandas, T. & Shapiro, L. J. (1984). Differential methylation of hypoxanthine phosphoribosyltransferase genes on active and inactive human X chromosomes. *Proceedings of the National Academy of Sciences, USA*, **81**, 1759–63.
Zenzes, M. T., Müller, U., Aschmoneit, I. & Wolf, U. (1978). Studies on H-Y antigen in different cell fractions of the testis during pubescence. *Human Genetics*, **45**, 297–303.
Zwingman, T., Erickson, R. P., Boyer, T. & Ao, A. (1993). Transcription of the sex-determining region genes *Sry* and *Zfy* in the mouse preimplantation embryo. *Proceedings of the National Academy of Sciences, USA*, **90**, 814–17.

3

Differentiation of the gonads

Introduction

In a relatively condensed manner, this chapter attempts to portray some of the highlights during formation of the eutherian gonads. These organs are uniquely derived from migration and movement of germinal and somatic cells, and give the first clear evidence of sexual dimorphism in specific tissues. The treatment that follows focuses on the question of their origins during early embryogenesis and then on the nature of the interactions as the cell lines of the gonad differentiate and develop. Although discussion at the level of cellular organisation is both stimulating and illuminating as a teaching approach, and in the present instance emphasises the proximity between embryonic renal, gonadal and adrenal tissues, the ultimate and incisive problem for the 1990s must concern the regulation of gonadal formation at the molecular level. Precisely how does gene action programme cellular differentiation to establish the respective lineages? At the time of writing, an appropriate body of information is not available on this topic, although it is assumed that specific proteins are induced to regulate transcription.

Numerous reviews exist on the processes involved in formation of the gonads in eutherian mammals. Extensive treatments include those of Gillman (1948), Witschi (1951), Burns (1961), Clermont & Huckins (1961), Clark & Eddy (1975), Zuckerman & Weir (1977), Jones (1978), McLaren (1981*a*), Baker (1982), Byskov (1981, 1982, 1986), Byskov & Høyer (1988), Wartenberg (1989) and Jost & Magre (1993). The essays of Mintz (1959), Jost (1972*a*), Wilson (1978), Eddy *et al.* (1981), Jost, Magre & Agelopoulou (1981) and De Felici, Dolci & Pesce (1992) should also prove helpful, and likewise that of Maitland & Ullmann (1993) for marsupials.

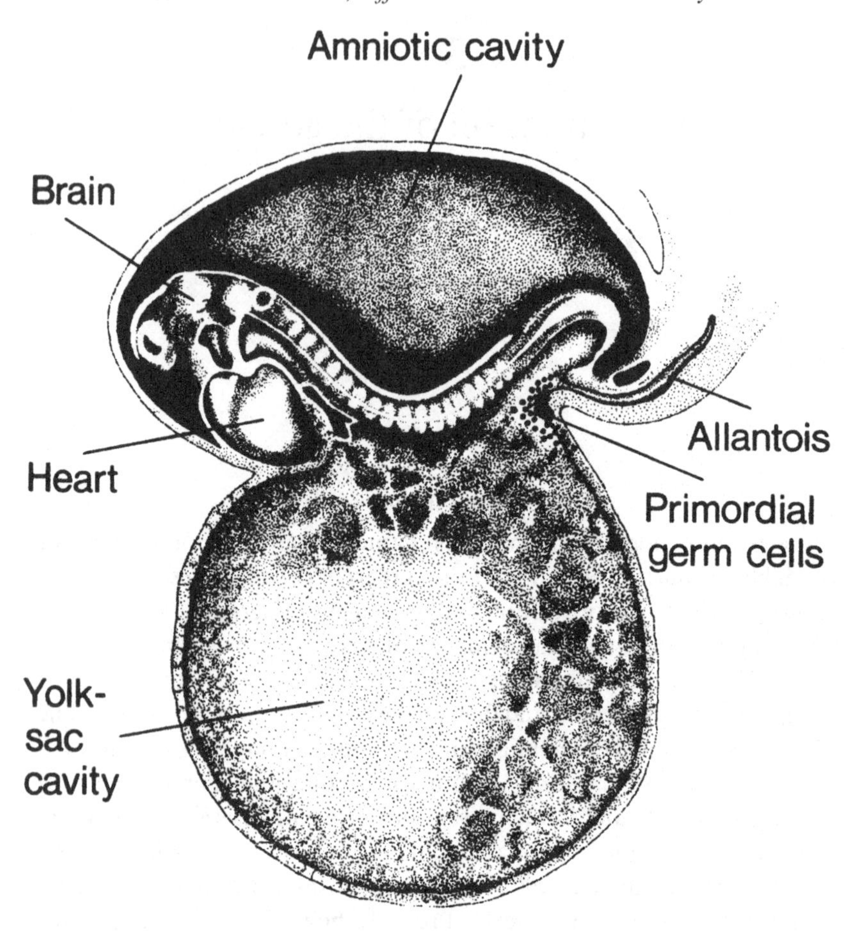

Fig. 3.1. The classical illustration from Witschi (1948) showing the later stages of migration of the primordial germ cells as they become concentrated in the region of the hind gut. (Redrawn courtesy of Professor T. G. Baker.)

Origin of primordial germ cells

Following formation of a male or female zygote, as determined by the sex chromosome constitution of the fertilising spermatozoon, the story of relevant steps in sexual differentiation as told in many texts usually jumps to the stage of the young embryo (Fig. 3.1). This is when primordial germ cells – progenitors of the sex cells – can be identified with confidence in a region of the yolk sac membranes (Witschi, 1948), and bilaterally arranged ridges of mesoderm (ventral to the mesonephric rudiments) are recognisable in the embryo proper as the gonadal primordia. Under the influence of

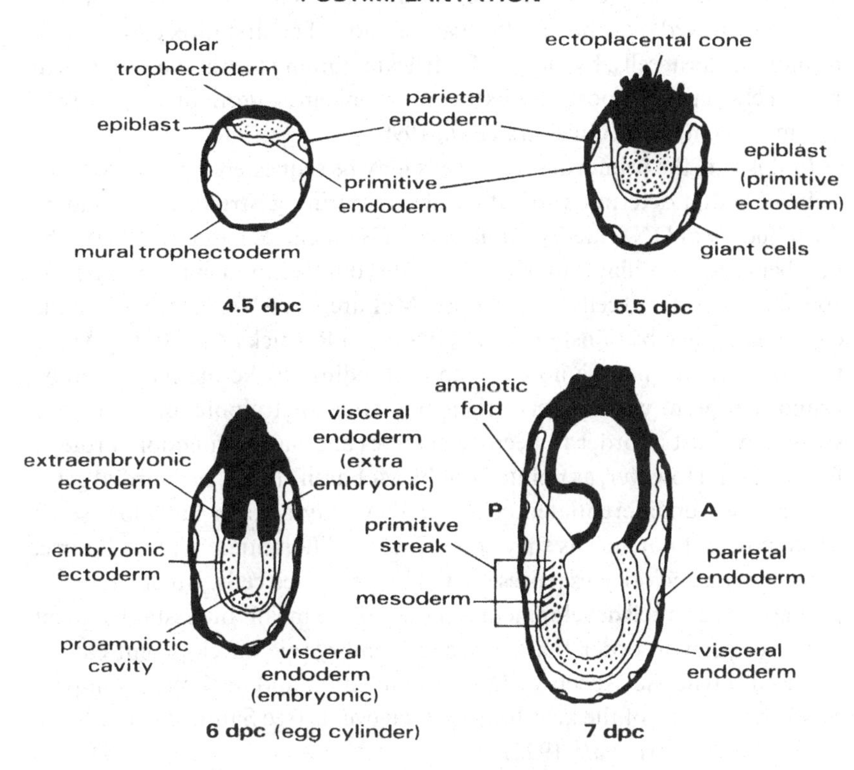

Fig. 3.2. Diagrammatic representation of early stages of development of the mouse embryo to illustrate the region of primitive ectodermal lineage termed the epiblast, wherein the primordial germ cells are thought to arise. (Adapted from Glover & Hames, 1989.)

their diploid genetic complement, primordial germ cells commence to synthesise glycogen and to express the enzyme alkaline phosphatase which can be demonstrated by specific staining reactions, thereby enabling a pin-pointing of these unique cells. Primordial germ cells may also be distinguished from somatic cells because of their significantly greater size, their large rounded nuclei and clear cytoplasm. Their precise initial location and indeed identity remain under a question mark in most mammals, although they are thought to be derived from primitive ectodermal cells of the inner cell mass (Gardner & Rossant, 1979; Gardner *et al.*, 1985), as inferred by the study of Ożdżenski (1967), in other words from the epiblast (McLaren, 1983; Fig. 3.2). Microsurgical experiments involving cultured portions of very young mouse embryos also indicate that the precursor cells for the primordial germ cells arise in the epiblast (Snow & Monk, 1983;

Ginsburg, Snow & McLaren, 1990), from which position they have moved – or been moved – at the onset of gastrulation. The first direct sighting of primordial germ cells, by means of a delicate staining technique, appears to be as a cluster in the posterior extra-embryonic mesoderm of very early 7-day mouse embryos (Ginsburg *et al.*, 1990).

In this site, the primordial germ cells may be sequestered away from the influences that determine somatic cell fate during gastrulation, including the influence of DNA methylation (Monk, Boubelik & Lehnert, 1987). The numbers first identifiable in this extra-embryonic location may be some 50–100 tightly-clustered cells in the mouse (McLaren, 1992) rather than just the eight highlighted by Ginsburg *et al.* (1990) and Resnick *et al.* (1992). At the time of writing, there is no clear understanding of the manner whereby primordial germ cells preserve their potential for totipotency, as in one sense they must in order to become gametes and have a functional rôle at fertilisation. However, as pointed out by McLaren (1992), primordial germ cells are not undifferentiated cells, for they may already have lost some developmental potency as they commence to differentiate and, of course, spermatozoa and oocytes represent highly specialised cells. An outstanding problem in germ cell development is therefore the means of reestablishment of full potency in order that subsequent embryonic development can be achieved. Dynamic aspects of DNA methylation and demethylation processes may be one of the keys to these vital events (see Sanford *et al.*, 1987; Monk, 1990; Kafri *et al.*, 1992).

The gene cascade involved in regulating this vital phase of cellular differentiation is not understood. Nonetheless, it is fully accepted that the germ cell line is established and can be identified relatively early in embryogenesis, although – as judged by X chromosome inactivation (see Chapter 2) – somatic cell lines may be allocated before the germ cell line, at least in mice. A contrary view, derived from the use of retroviruses as developmental probes, suggests that the germ cell line is set aside as early as the 4- to 8-cell stage, and perhaps before allocation of cells to somatic tissues (see Byskov & Hoyer, 1988).

Migration and multiplication of primordial germ cells

Passing over intervening stages that have still to be fully unravelled in detail, an initially small population of primordial germ cells will commence migration from the dorsal endoderm of the yolk sac via the hind-gut epithelium and mesenteries and then the adjoining ventral mesonephric tissues into the gonadal anlagen (Witschi, 1948). Depending on chromosomal constitution, these latter regions will give rise to an ovary or a testis or, exceptionally, to an ovotestis (see Chapters 5, 6 and 7). Two of the many outstanding questions on this topic are (1) precisely what initiates the phase of migration of the germ cells from an extra gonadal site, and (2) the informational cues required for successful completion of the process. A long-standing response to this second question is that of putative chemotactic substances emitted from the genital ridges acting to guide the primordial germ cells (see Witschi, 1948; Rogulska, 1969). Cell surface antigens are thought to play a rôle in the events of recognition and interaction between cells during the migration of germ cells (Eddy *et al.*, 1981), which may follow pre-existing pathways on adhesive substrates. As a component of the extra cellular matrix, glycoproteins such as fibronectin may 'illuminate' a migratory pathway and indeed serve to stimulate translocation of the migrating cells (Fujimoto, Yoshinaga & Kono, 1985; Alvarez-Buylla & Merchant-Larios, 1986; De Felici, Dolci & Pesce, 1992). Primordial germ cells apparently show locomotion *in vitro* only when fibronectin is present in the substrate and this glycoprotein is found along the germ cell trajectory *in vivo* (Merchant-Larios & Alvarez-Buylla, 1986). Even so, if fibronectin is present in significant amounts elsewhere in the developing embryo at this time, then it would seem difficult to assign the protein a specific guiding rôle in the migratory phenomenon. Multiple signals are almost certainly involved (De Felici *et al.*, 1992). The fact that fibronectin is abundant around the basal lamina of the hind-gut and blood vessels, especially the dorsal aorta, is relevant and yet only rarely were primordial germ cells found attached to these fibronectin-rich structures. Hence there is the need to postulate the influence of other active factors to explain the overall pattern of an orientated locomotion of primordial germ cells (Merchant-Larios & Alvarez-Buylla, 1986). In terms of specificity of the guiding stimulus, mouse germ cells have been demonstrated to colonise chick gonadal tissues in transplantation experiments (i.e. grafts of mouse hind-gut in the coelomic cavity of young chick embryos), so there would appear to be rather powerful informational programmes in operation that cross species boundaries (Rogulska, Oždženski & Komar, 1971).

Another approach used to demonstrate an attraction for mouse primordial germ cells by the genital ridges is the *in vitro* culture system reported by Godin, Wylie & Heasman (1990). In such a system, primordial germ cells migrate towards 10.5 day *post coitum* genital ridges in preference to other explanted organs. And, using primordial germ cells isolated from 8.5 day *post coitum* embryos, at the beginning of their migratory phase, culture medium conditioned by isolated 10.5 day *post coitum* genital ridges caused an increase in the number of primordial germ cells in the cultures. Taken together, these observations strongly suggest that the secretion of attractive and trophic factors by 10.5 day *post coitum* mouse genital ridges plays a key rôle in guiding the primordial germ cells. Moreover, they infer that the genital ridges are in some manner able to exert long-range effects on the population of primordial germ cells. In other words, the differentiating genital ridges release factors that influence both the numbers and the direction of migration of primordial germ cells *in vitro*, an influence that is presumed also to occur in the physiological situation. The attractant is thought to be transforming growth factor- $beta_1$ ($TGF\text{-}\beta_1$) or a closely related molecule (Godin & Wylie, 1991), but the involvement of different integrins in the events of primordial germ cell migration awaits further study (De Felici *et al.*, 1992).

As to the more specific question of what factor(s) might be prompting proliferation of primordial germ cells, *in vitro* studies suggest a cAMP-dependent mechanism and indicate that cAMP may be acting directly on mouse primordial germ cells (De Felici, Dolci & Pesce, 1993). This study made the additional observation that somatic cell support is required for optimal growth of primordial germ cells. In a separate report from the same laboratory, a synergism between cAMP, stem cell factor and/or leukaemia inhibitory factor was suggested to influence proliferation of mouse primordial germ cells *in vitro* (Dolci, Pesce & De Felici, 1993). A stem cell factor influence on survival and/or proliferation of primordial germ cells had previously been noted by Dolci *et al.* (1991), Godin *et al.* (1991) and Matsui *et al.* (1991), and an influence of leukaemia inhibiting factor by De Felici & Dolci (1991). These and other related findings were set in perspective by McLaren (1992) and De Felici *et al.* (1992). And, most recently, both stem cell factor and leukaemia inhibitory factor were noted to suppress programmed cell death (apoptosis) among mouse primordial germ cells in culture (Pesce *et al.*, 1993).

Despite the preceding suggestion (p. 72), the question remains as to why the primordial germ cells commence to proliferate as a discrete group so far from the presumptive gonads and are thereby subjected to a potentially hazardous phase of migration. No matter what the response to this question turns out to be, migration of the diploid germ cells into the genital tissues may provide or impose subtle organisational programmes upon the gonadal primordia in a way that would not have been possible if the germ cells had arisen *de novo* within the differentiating genital tissues. Migration of the germ cells from an extra-gonadal site may also permit a degree of flexibility in gonadal organisation that would not have existed if the germ cells had been elaborated in the primordia themselves.

Active amoeboid movements are thought to underlie passage of primordial germ cells through the differentiating tissues (Witschi, 1948, 1951; Zamboni & Merchant, 1973), pseudopodia-like processes being prominent on such cells and an amoeboid form of migration being demonstrable *in vitro* (Blandau, White & Rumery, 1963). However, Merchant-Larios & Alvarez-Buylla (1986) have emphasised that a mesonephric cell monolayer is necessary *in vitro* as a substrate and that even with their pseudopodia-like processes, primordial germ cells do not show locomotor activity directly on glass or plastic – an observation suggestive of specific adhesive properties. Passive displacement within the proliferating vascular network should also be considered, and would afford one explanation for germ cells becoming lodged in ectopic sites. Morphogenetic rearrangement by means of differential growth of tissues during organogenesis would contribute to relocation of the primordial germ cells, and could act to draw the cells passively from an extra to an intra-embryonic position (Zuckerman & Baker, 1977).

Errors during the phase of migration, that is failure of primordial germ cells to reach the genital ridges, are demonstrable by histochemical and microscopical techniques. Germ cells may be lost – or die – in various ways during migration to the genital ridges, but a now classical finding in the mouse is arrest in or close to the neighbouring mesonephric tissues or in the adrenal primordium (Upadhyay & Zamboni, 1982), that is among those cells that will eventually give rise to kidney or adrenal tissues. In brief, germ cells moving towards the genital ridges may be engulfed by mesonephric tissues moving towards the rudiments of the adrenal glands and thereby carried into such glands (Zamboni & Upadhyay, 1983). Primordial germ cells had earlier been noted in the adrenal glands of a 13 mm, 36-day-old male human foetus (McKay *et al.*, 1953), a 2-month-old male human foetus and a 5 to 6-week-old female foetus (Falin, 1969), so the phenomenon may be more frequent and extensive than generally realised.

Germ cells are capable of differentiation in such ectopic sites but, remarkably, they always differentiate as oocytes irrespective of their genetic sex, that is of the sex of the embryo. Thus, extra-gonadal XY germ cells behave and differentiate as XX germ cells normally would in the ovary, and such ectopic germ cells in foetal male mice enter meiosis at the same stage as oocytes do (Upadhyay & Zamboni, 1982; Zamboni & Upadhyay, 1983), doubtless due in part to the absence of Sertoli cell suppression. Hence, the suggestion remains that all germ cells should be viewed as potentially female, irrespective of their genetic sex, and will differentiate as such unless they become enclosed within the seminiferous cords of the testis or the neighbouring testicular territory (Upadhyay & Zamboni, 1982; Francavilla

& Zamboni, 1985). However, upon reaching a certain degree of differentiation in the ectopic site, they will degenerate and disappear without trace, perhaps primarily due to an absence of surrounding follicular cells. Germinal elements were no longer seen in the adrenal glands of mice older than 2–3 weeks of postnatal life. On the basis of these observations, Francavilla & Zamboni (1985) offered the viewpoint that 'the differentiation of XX germinal cells is an autonomous and ubiquitous process'. Indeed, the absence of any sign of sexual dimorphism in extra-gonadal germinal cells corresponds with Tarkowski's (1969) view that XY germ cells are capable of entering meiosis during foetal life, but are normally prevented from so doing by the somatic tissue (Sertoli cells) of the testis. The nature of the putative inhibitor is not yet known with certainty. These observations lead to the conclusion that whether primordial germ cells develop into oocytes or spermatogenic cells depends not on their own chromosome constitution but rather on the tissue environment in which they find themselves. Thus, gametogenesis follows the somatic sex of the gonadal tissue and not the genetic sex of the germ cells themselves. But although the primordial germ cells' chromosome constitution does not determine its phenotypic sex, it greatly influences its subsequent fate during oogenesis or spermatogenesis (McLaren, 1991*b*).

Primordial germ cells, first seen in the extra-embryonic mesoderm posterior to the primitive streak at 7–7½ days *post coitum* in mice, migrate in the wall of the hind-gut and up the dorsal mesentery (Ginsburg *et al.*, 1990). As to the number of primordial germ cells initially identifiable and embarking upon this phase of migration (and bearing in mind the preceding remarks concerning the germ cell line being set aside at the 4 to 8-cell stage), the histochemical and immunological techniques employed to date indicate a relatively small cohort – of the order of 10–100 in mice (Table 3.1). Indeed, Mintz (1968) has suggested that the germ cell line in mice arises from 2 to 9 stem cells. By means of alkaline phosphatase staining, Resnick *et al.* (1992) also suggested a population of about 8 positive-staining cells in the 7.0 days *post coitum* mouse embryo, a figure considered conservative by McLaren (1992). However, some 15–76 primordial germ cells were identifiable in the yolk sac and allantois of 8-day mouse embryos (Chiquoine, 1954), whilst in 9-day embryos some 170–350 germ cells were found in or close to the hind-gut epithelium (Bennett, 1956; Mintz & Russell, 1957). By the time they have entered the genital ridges, the number has risen to 2600–5700, indicating four or five divisions during the 4-day period of migration (Table 3.1 and 3.2). In due course, the number of germ cells increases massively by mitotic division so that 13-day mouse foetal gonads may contain $>$ 10 000

Table 3.1. *Location of mouse primordial germ cells identified by alkaline phosphatase staining during their migration to the genital ridges*[a]

Interval *post coitum* (days)	Embryonic location of primordial germ cells	No. of primordial germ cells (approx.)
8.5	At the base of the developing allantois	< 100
9.5	In the wall of the hind-gut	350
9.6–12.4	Migrating along the hind-gut mesentery to the dorsal body wall	1000 at 10.5 days
12.5	Migrating laterally to the genital ridges	4000
14.5	Male germ cells enter meiotic arrest; but female germ cells enter meiosis	20000–25000 maximum

Note:
[a] Data derived from various sources, most recently Godin *et al.* (1990) but also earlier studies such as those of Mintz (1957), Mintz & Russell (1957) and Tam & Snow (1981).

Table 3.2. *Observations on the entry of primordial germ cells into the genital ridges of mouse embryos*

Age *post coitum* (days)	Sex	No. of embryos	Distribution of primordial germ cells in: extra-gonadal sites Mean no.	(%)	genital ridge Mean no.	(%)	Total no.
10.5	♂	4	1030	(95)	44	(5)	1074
	♀	8	900	(92)	77	(8)	977
11.5	♂	6	221	(7)	3096	(93)	3317
	♀	2	193	(7)	2662	(93)	2855
13.5	♂	10	337	(1)	26785	(99)	27122
	♀	4	332	(1)	22130	(99)	22462

Notes:
Adapted from Tam & Snow (1981).
Note that, at any given stage, the number of primordial germ cells counted in male embryos always exceeded the number in female embryos, an observation previously commented on by Hardisty (1967).

germ cells (Tam & Snow, 1981). To some extent, these increases in numbers occur during the phase of migration (Witschi, 1948; Hardisty, 1967), but most conspicuously once the germ cells have reached the presumptive gonads and lost their pseudopodia. The consequences of this colonisation and vigorous multiplication of diploid cells may be of considerable significance to organisation of the gonad, even though one still finds statements to the contrary. This, of course, is not to deny the fact that 'gonads' can form in the absence of germ cells.

As a separate example of the dynamic status of germ cell numbers,

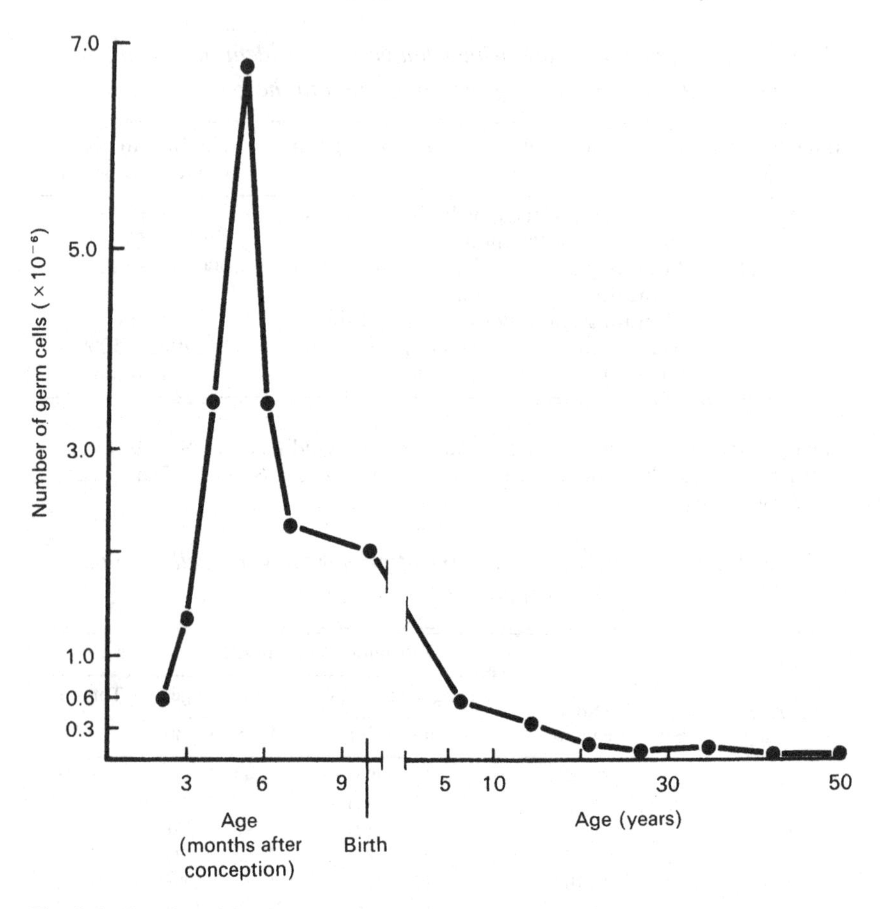

Fig. 3.3. Graph to illustrate changes in the total population of germ cells in the human ovary from the stage of early embryonic development through that of the foetal peak and so to the slow decline during the reproductive lifespan. (From Baker, 1982.)

human embryos contain only a few hundred of these migrating cells at 3–5 weeks of age (Witschi, 1948) but the number has reached some 600 000 by the eighth week and 7 million by the fifth month of pregnancy (Fig. 3.3), a peak that is followed by a very substantial wave of atresia (Baker, 1963). During the phase of germ cell multiplication in the gonad, a proportion of the oogonia will already have embarked upon the first meiotic division to become primary oocytes. Soon after the time of birth and the substantial wave of atresia, all the germ cells should have become primary oocytes with the chromosomes arranged at first meiotic prophase. In mice, a massive degeneration of oocytes in the pachytene stage occurs from the 16th day of pregnancy until 3–4 days after birth (Borum, 1961).

Although the above remarks have focused upon the ovaries, a comparable phase of multiplication of germ cells proceeds in the foetal testis. However, whereas the germ cells enter meiosis in females during foetal or, at the latest, neonatal life, initiation of meiosis in males is delayed until the time of puberty is imminent. Thus, in normal mouse development, XY germ cells in the testis enter a state of mitotic arrest as prospermatogonia when the female germ cells are entering meiosis, 13–15 days *post coitum*, and they resume mitotic proliferation in the immediate postnatal period. The first meiotic stages in the normal mouse testis are seen no earlier than 8–10 days after birth (McLaren, 1981*a*, 1983).

Formation and differentiation of gonads

Gonadal differentiation is held to be masterminded by the sex chromosomes, and the presence of a Y chromosome acts to programme maleness – that is formation of a testis – by virtue of one or more genes located on its short arm in mice and man (see Chapter 2). An influence of the Y chromosome pre-empts development of the gonads as ovaries. The precise manner whereby gene action is imposed to achieve the necessary cellular organisation typical of a gonad has yet to be unravelled. Various models have been proposed, including the one of Burgoyne (1988) in which the Y chromosome or rather *Tdy* is suggested to act autonomously in the supporting cell lineage to bring about Sertoli cell differentiation. All other aspects of foetal testicular development would then be triggered and directed by the Sertoli cells without further Y chromosome involvement. Sex determination may thus be equated with Sertoli cell determination (Burgoyne, 1988), and much of the differentiation that follows is controlled by hormones rather than by direct gene action (Jost *et al.*, 1973). Even so, and as a broader perspective, normal differentiation of the testis and ovary is controlled by specific autosomal genes as well as by those on the sex chromosomes, and development of a normal testis requires genes on the X chromosome as well as those on autosomes and on the Y. Hence, gonadal sex cannot be explained by the composition of the sex chromosomes alone (Wilson, 1978).

Since early in the twentieth century, it has been appreciated that the gonad anlage in vertebrates is bipotential, that is it is capable of developing into either an ovary or a testis. Working mainly with lower vertebrates, Witschi (1932) had stressed that two regions could be distinguished in the indifferent gonad (Fig. 3.4). The superficial region of the anlage, the cortex, would differentiate into an ovary whereas the medulla would differentiate into a testis. Hence, Witschi noted the need for proliferation of one region

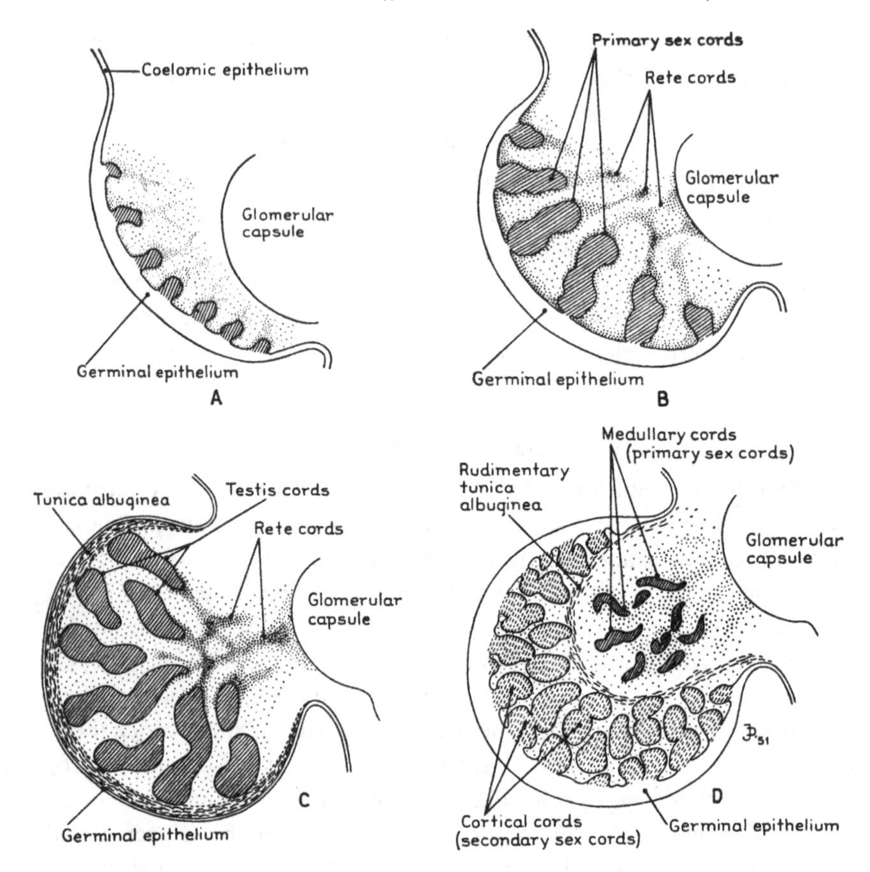

Fig. 3.4. The portrayal by Burns (1961) of the main features of gonadal differentiation in lower vertebrate (amniote) embryos. These are the views of Witschi that highlight the origin of the medullary and cortical components of the organ and some principal steps in the organisation of a testis or an ovary.

and regression of the other if a distinct ovary or testis is to form, and thus suggested antagonism between cortex and medulla (Jost, 1972*a*). But regression of the medulla in the incipient ovary is not immediate, and an ovotestis or even testis may form in a genetic female (Chapters 5, 6 and 7). Indeed, the longer the period during which the gonad remains indifferent, the more susceptible it becomes to external transforming influences. In mammals, development of the homogametic gonad – the ovary – can be viewed as a constitutive or default pathway (Jost *et al.*, 1973) although of course requiring a specific gene programme. Conversely, development of the heterogametic gonad – the testis – is said to require activation of a new circuit and thus diversion from the default circuit. Hence, factors that

Table 3.3. *The relative rate of appearance of key steps during differentiation and development of a male gonad in three domestic species. Testicular differentiation in mammals always commences before ovarian differentiation*[a]

Stage during male gonadal differentiation	Days *post coitum* in		
	mouse	pig	sheep
Migration of primordial germ cells	8.5–10.0	< 18	18
Colonisation of genital ridge by primordial germ cells	12.0–12.5	24–26	22
Gonadal differentiation occurring under influence of *Sry*	11.5–12.5	24–25	26
Appearance of stromal tissue characteristic of male gonad	12.0	26	29
Sertoli cells beginning to align into cords	12.5	28	31

Note:
[a] Data derived from various sources for mouse and sheep, whereas principally from Black & Erickson (1965), Moon & Hardy (1973) and Pelliniemi (1975) for pig.

reverse sex may serve as inducers or regressors of genes that comprise a pathway (Ohno, 1979).

Both somatic and germ cells must differentiate correctly if a functional gonad is to be formed. If the embryo is genetically male, that is of XY sex chromosome constitution, the primordial germ cells preferentially colonise the medullary region of the presumptive gonads whereas, if genetically female (XX), the germ cells become concentrated in the cortical region. Precisely why the germ cells assume these locations has not been determined, nor is it a fact that they always do so; loss to the adrenal glands has already been noted. Presumably, the primordial germ cells are usually guided, in part at least, by specific cues or attractant substances – chemotactants – released by appropriate portions of the developing gonad (Witschi, 1948; Godin *et al.*, 1990). Synthesis and release of such putative substances would likewise be programmed by gene action. Moreover, if distinct chemicals or different concentrations of the same substance are released differentially by the cortex and medulla of the developing gonads, then this would indicate an early rôle for sex-chromosome-associated gene expression in the somatic cells, perhaps even before dimorphism can be demonstrated microscopically in the gonads. Indeed, although gonadal differentiation is said to occur at 11.5–12.5 days *post coitum* in mice under the influence of *Sry* (Table 3.3), primordial germ cells have been migrating towards the genital ridges (presumptive gonads) since 9.5 days *post coitum*. And if primordial germ cells are responding to chemotactic signals or gradients to find their way to appropriate regions of the gonad (i.e. cortex or medulla), then conceivably genes involved in sex determination other than *Sry* could already be functioning between 9.5 and 11.5 days *post coitum*.

Despite the above inferences concerning differentiation of gonadal tissues, a critical question remains as to the extent to which the presumptive genital tissues have irreversibly embarked upon their sexual destiny before initial colonisation with germ cells. Zuckerman & Baker (1977) endorsed the earlier claim that the indifferent gonad contains all the cell types needed to render it capable of developing into either a testis or an ovary (Gillman, 1948). Perhaps this view could now be questioned afresh, even though experimental evidence suggests that the somatic cells themselves are alone sufficient for the preliminary organisation of a testis or an ovary. If not essential for gonadal differentiation, the primordial germ cells may nonetheless contribute organisational information of significance to the morphological integrity and subsequent physiological functions of the gonad.

A theoretically illuminating if technically difficult experiment in mammals would be to remove primordial germ cells from the yolk sac membranes of an individual of one sex and transplant them to the region of the genital ridges of an embryo of the opposite sex, having first removed its own germ cells from the yolk sac or destroyed them *in situ*. Performance of this experiment in mammals would be an advance on the impressive series of experiments in chick embryos reviewed by Haffen (1977). This elegant form of microsurgery might clarify to what extent – if at all – the primordial germ cells act to influence the formation and distribution of gonadal tissues (see Mintz & Russell, 1957). The study of O & Baker (1978) approached these questions *in vitro* by culturing XX germ cells mixed as a 'gonad' with XY somatic cells and vice versa. As noted above, once established within the appropriate region of the gonadal tissue, the relatively small cohort of germ cells multiplies massively by mitotic division to give a population of stem cells (oogonia or spermatogonia) and the gonad rudiment becomes organised progressively as an ovary or testis, in part due to rearrangement of tissues. Possibly of relevance here is the intriguing observation that in both the mouse and rat (Fig. 3.5), genital ridges are larger in XY than in XX embryos even before Sertoli cells can be recognised histologically (Mittwoch, 1985). This would suggest that an early step in the differentiation of Sertoli cells involves an increased proliferation rate.

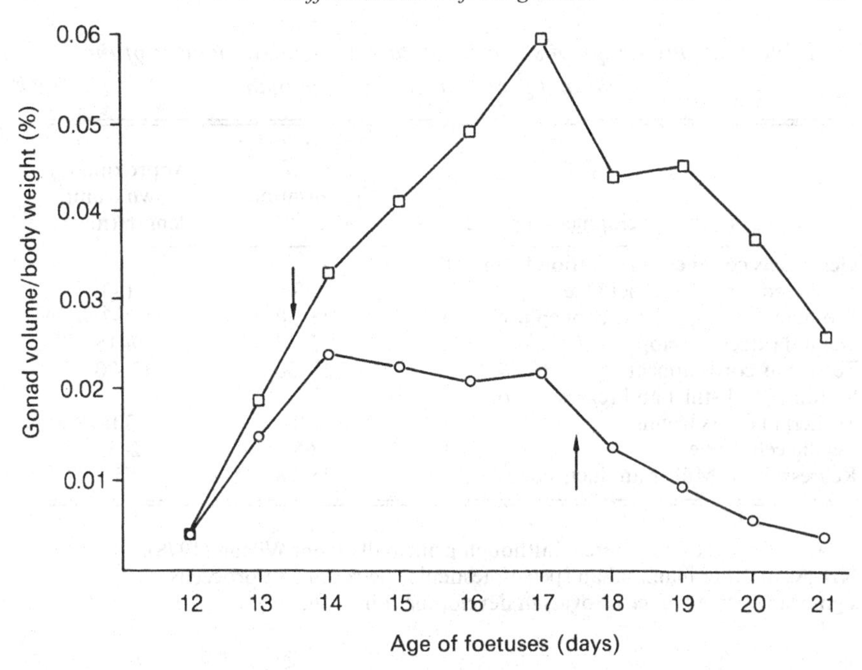

Fig. 3.5. An illustration from the review of Mittwoch (1992) to show the dichotomy in gonadal development between male (upper graph) and female rats (lower graph). The arrow in the former indicates the beginning of Sertoli cell formation, that in the latter the beginning of oocyte formation.

Cellular interactions and morphogenetic movements

Histological interpretations of the mammalian gonad have a distinguished history in the sense that, as long ago as 1878, Balfour suggested that the bulk (central mass) of the gonad originated in the mesonephros, a proposition endorsed by Witschi (1951). Indeed, the mesonephros and presumptive gonads interact intimately, and the mesonephros is also crucially involved in formation of the adrenal cortex.

The cellular basis for eutherian gonad formation may involve contributions from four distinct lineages: the coelomic epithelium, mesenchymal cells, the mesonephros and the germ cells themselves. However, there is no overall agreement as to whether the somatic cells of the gonad are derived from the coelomic epithelium by ingrowth or the mesonephros by outgrowth, or the extent to which they are of dual origin (McLaren, 1984). Discussion of this topic can be found in Jost *et al.* (1973), Merchant-Larios (1979), Upadhyay, Luciani & Zamboni (1979) and Wartenberg (1978, 1981). Migrating primordial germ cells that achieve the presumptive

Table 3.4. *Chronology of some key events during development of the gonads and genital duct system in man*

Stage of the developmental process	Approximate time after fertilisation (days)	Approximate crown–rump length (mm)
Germ cells commence migration from yolk sac endoderm to mesenchyme	19	1–3
Wolffian ducts appear as pronephric ducts	25–30	2–3
Genital ridges develop	37–45	7–15
Testicular cords appear	43–50	13–20
Sertoli cells distinct and regression of Müllerian ducts begins	60	30
Leydig cells appear	65	32–35
Regression of Müllerian ducts complete	75–78	55

Note:
Adapted from several sources, although principally from Wilson (1978).
NB: As in other mammalian species, testicular development proceeds significantly in advance of ovarian development in man.

gonads may be arrested within the coelomic epithelium (their presence there being misinterpreted by Waldeyer (1870) – hence the term germinal epithelium), or between the coelomic epithelium and the mesonephros in mesenchymal tissues (Everett, 1945). Closely associated somatic cells from both the mesonephros and the coelomic epithelium then advance to mix with germ cells in the mesenchymal tissue. An intrusion of connecting mesonephric cells forming cell cords or bodies, accompanied by mitotic multiplication of all cell types, gives rise to the gonadal ridge, soon to be recognisable macroscopically on the ventral side of the mesonephros. This gonadal organisation can be seen by the 10th day in mice (Upadhyay *et al.*, 1979; Byskov, 1982), by about 20 days in the pig (Patten, 1948), by the 22nd day in sheep (Zamboni, Bézard & Mauléon, 1979) and by about the 6th week of foetal life in man (Table 3.4; Wartenberg, 1978; Byskov, 1982).

As the gonads differentiate, organisation of a testis or an ovary assumes a characteristic form in most instances, invasion of mesenchymal cells being prominent in the male (Merchant-Larios & Taketo, 1991). But before gonadal sex differentiation is conspicuous, there is not a complete separation between the mesothelium and the underlying tissue, since a basal lamina is not yet fully formed. The earliest histological indication of dimorphism in the patterns of gonadal differentiation is elaboration of primordial Sertoli cells and their organisation as presumptive seminiferous cords in genetic males (Jost, 1972*b*; Jost & Magre, 1984), separated from the true epithelium by a basal lamina. Moreover, as suggested above, development of a testis is reflected in rapid growth of that organ (Mittwoch, 1970), and differentiation of testicular cell types occurs earlier than differentiation of ovarian cell types (Jost *et al.*, 1973; Jost & Magre, 1988). In other words, male gonads grow sooner and faster, not least since they must act to suppress femaleness. Hence, the undifferentiated phase before an ovary is distinguishable in genetic females is prolonged compared with the relatively rapid development of a testis in genetic males (Jost *et al.*, 1973; Haffen, 1977).

Differentiation of a testis

Apart from the precocious growth of its genital ridges, the first visible sign of male development is the elaboration of testes from these indifferent tissues. Expression of the *Sry* gene is associated with the genital (gonadal) rather than the adjacent mesonephric component of the urogenital ridge. As to testis formation, germ cells are drawn towards and into the medulla and the sex cords become a prominent early feature (at about 12.5 days *post coitum* in the mouse and 31 days in sheep). In the mouse, cells that will constitute the sex cords may 'autodifferentiate' from the gonadal blastema (S. Ullmann, personal communication). Germ cells and specialised somatic cells with clear abundant cytoplasm – the sustentacular or presumptive Sertoli cells (Fig. 3.6) – are situated within such testicular cords and intimate contact is ensured by means of long cytoplasmic processes emanating from the pre-Sertoli cells (Magre & Jost, 1983). Differentiation and alignment of Sertoli cells are the first steps in their formation, simultaneous with a rapid proliferation and aggregation (Jost, 1972*b*). The cords become the future seminiferous tubules once a central lumen or canal develops and a basement membrane has formed containing laminin and fibronectin. Such substances may assist polarisation of the Sertoli cells (Vigier *et al.*, 1987). Cord formation confers a characteristic striped appearance on the developing testis (as viewed in whole mounts by transmitted light), distinguishing it from the embryonic or foetal ovary.

The differentiation of Sertoli cells from the supporting cell lineage is thought to result in cells of the other lineages following the male pathway (Jost, 1972*b*; Burgoyne, 1988). The nature of this influence is not known but may depend on one or more key products from the 'pre-Sertoli' cells. One consequence is that the differentiating Sertoli cells organise the germ cells within the future tubules, for the Sertoli cells are initially arranged around

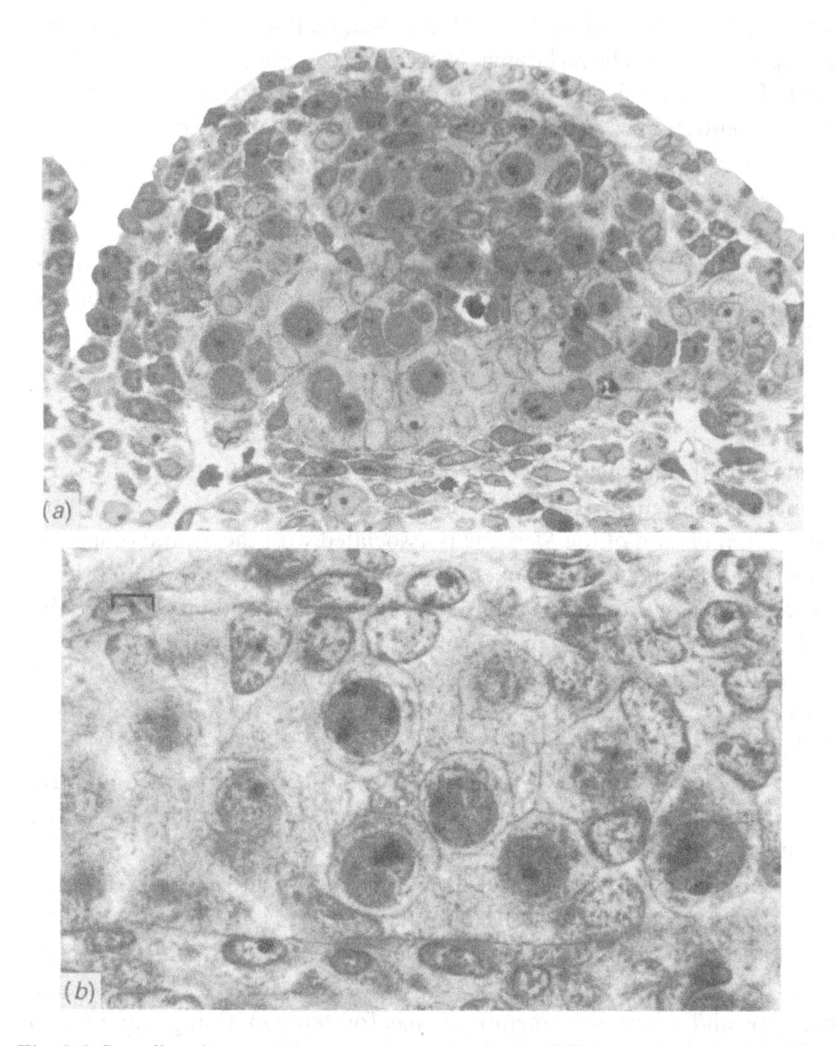

Fig. 3.6. Sertoli and germ cells as seen in section in the differentiating rat testis. The upper photograph (*a*) shows a specimen prepared at 13 days 15 hours of gestation, with differentiating Sertoli cells recognisable in the lower portion (i.e. in the depth of the gonad near the mesonephros) by their large and clear cytoplasm. Photograph (*b*), prepared from a seminiferous tubule of a foetus 14 days 14 hours old, shows that the germ cells are now completely encompassed by Sertoli cell cytoplasm. (Adapted from Jost & Magre, 1993.)

the gonocytes as one of the earliest events (Jost, 1972*b*; Jost & Magre, 1988). Even so, germ cells do not seem essential for cord formation to occur. The Leydig cells of the interstitium appear somewhat later, their morphological differentiation being correlated with the onset of steroidogenesis.

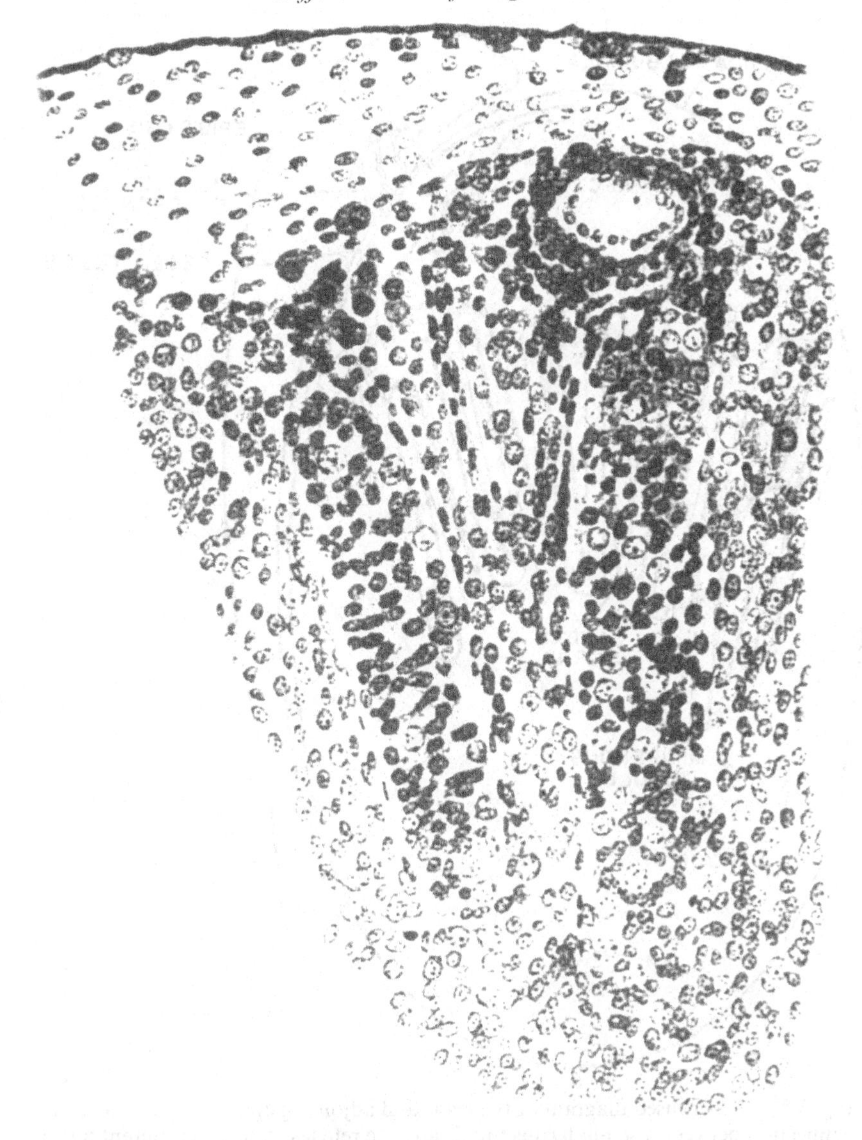

Fig. 3.7. Illustration from the excellent study of Kitahara (1923) to portray some of the relationships between Sertoli and Leydig cells in the early stages of gonadal development.

There is circumstantial evidence that differentiation of Leydig cells stems from an influence of the Sertoli cells (Kitahara, 1923; Fig. 3.7; Taketo-Hosotani *et al.*, 1985; Jost & Magre, 1988; Merchant-Larios & Taketo, 1991). A further vital feature is the connection with ingrowing mesonephric

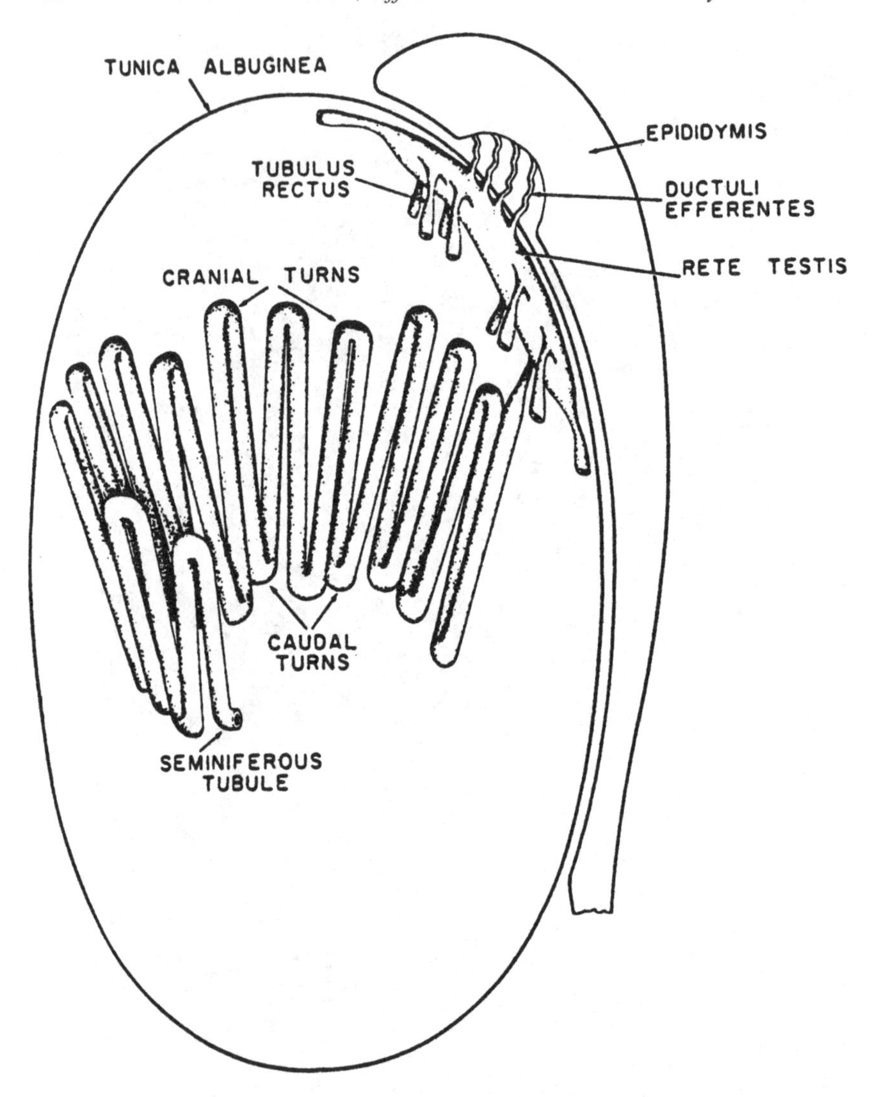

Fig. 3.8. A generalised diagram of the testis and adjoining epididymis to show the connection between a seminiferous tubule and the rete testis, and the efferent ducts connecting the latter to the epididymis. (Taken from Clermont & Huckins, 1961.)

cell cords, the connections being elaborated to form the rete testis which, in adult life, admits the weak sperm suspension collecting from the convoluted seminiferous tubules into the efferent ducts (Fig. 3.8) and so to the adjoining epididymal duct (Clermont & Huckins, 1961).

In the mouse, precursors of the Leydig cells migrate from the mesonephric region into the genital ridge before 12.5 days *post coitum* and differentiation of the Leydig cells is achieved at 13.5 days *post coitum*. This functional aspect has been monitored by activation of the gene for 3β-hydroxysteroid dehydrogenase, which is expressed in the mouse testis by 13.5 days *post coitum* but not by 12.5 days (McLaren, 1991*a*). Normal testis cord formation requires peritubular myoid cells, and such cells also migrate from the mesonephric region. The latter aspect has been clarified in a study involving culture of testes from 11.5-day-old mouse embryos with or without attached mesonephroi and, in the former situation, with or without a permeable filter (Buehr, Gu & McLaren, 1993), an experimental arrangement strongly reminiscent of the work of Byskov & Saxen (1976) and Byskov & Grinsted (1981); however, a transgenic marker was also brought to bear on the problem. The study indicated that cells do indeed migrate from the mesonephric region into the differentiating testis and can contribute to the interstitial cell population and, further, that such a contribution is required for normal cord structure.

Growth of a testis is a function largely of proliferation of cells already within the organ, that is somatic and germ cells, rather than being due to further ingrowths of mesonephric cells. The periphery of the developing gonad becomes occupied by the membranous tunica albuginea, an expansion of collagenous connective tissue (mesenchymatous tissue) under the investing mesothelium. Any germ cells remaining outside the testicular cords are said to degenerate (Zuckerman & Baker, 1977). Once again, perhaps this latter point is open to rather more flexibility than hitherto supposed, especially in instances of anomalous gonadal development (see section below on transdifferentiation).

In addition to its rapid growth, a prominent blood supply is characteristic of the developing testis, assuming a more complex architecture at the periphery than in the foetal ovary. However, although the tunica albuginea and interstitial tissue become extensively vascularised, this is not so for the cellular contents of the seminiferous tubules; hence, the so-called blood–testis barrier (Fig. 3.9), for the germ cells become isolated from direct contact with blood capillaries by a more or less complex basement membrane and by junctional complexes between the Sertoli cells. In the male, the germ cell line remains relatively undifferentiated on the basement membrane as prospermatogonia or spermatogonia after waves of mitotic division until shortly before puberty (Byskov, 1986). Male germ cells do not usually embark upon meiosis until this time, although the preleptotene stage may be attained in some circumstances (Gondos & Byskov, 1981). Thus DNA replication and cell division are not usually resumed until after birth. The first meiotic spermatocytes in the mouse testis are not seen until

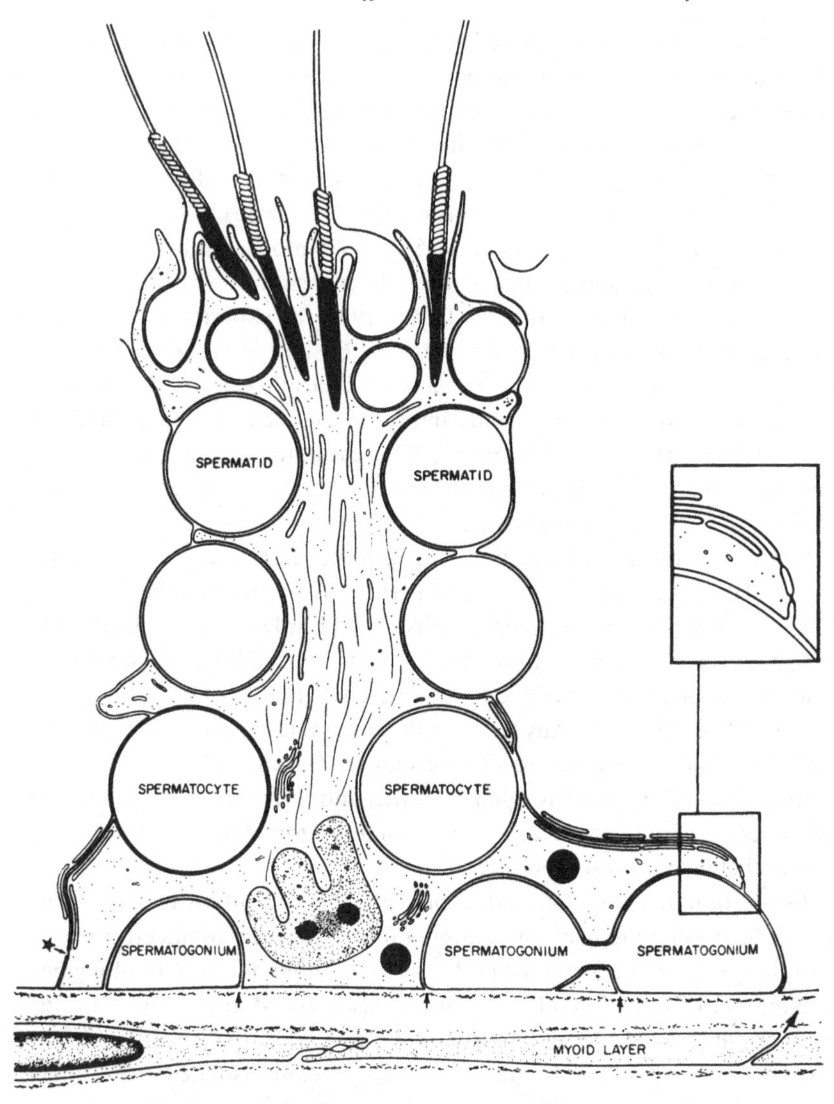

Fig. 3.9. Diagrammatic representation of spermatogenesis, and especially of the localisation of the blood–testis barrier and compartmentalisation of the germinal epithelium by tight junctions between adjacent Sertoli cells. (From Dym & Fawcett, 1970.)

about 7–10 days after birth. The nature of the factor(s) causing such prolonged regulation (inhibition) of meiosis remains uncertain, although it is thought to originate from the surrounding Sertoli cells. Germ cells cannot proceed through differentiation unless in contact with somatic cells.

Viewed in a physiological perspective, a prominent feature of somatic differentiation in the testis is an early ability to synthesise steroid and protein hormones, which have profound influences on the neighbouring duct systems. These remarks focus on both the intra-tubular Sertoli cells and the extensive interstitial tissue of Leydig cells. Leydig cell secretion may play an important rôle in regulating testicular descent to the scrotum, itself a critical step in sexual differentiation. In addition to the steroid and protein hormones of the testis reviewed in many texts (e.g. Knobil & Neill, 1988; Waites & Setchell, 1990), consideration must also now be given to an active synthesis of peptide growth factors.

In most eutherian mammals so far examined, testes can only become fully functional with a normal sperm-producing seminiferous epithelium if they have descended from the abdominal cavity to the scrotal sac (Fig. 3.10), and have thereby been exposed to a temperature lower than that of the abdominal cavity (e.g. 2–5°C cooler). The specialised vasculature of the pampiniform plexus is crucial for temperature regulation of the scrotal gonads. Failure of testicular descent represents the condition of cryptorchidism. If not corrected promptly by surgical means, such failure leads to destruction of the seminiferous epithelium and thereby to sterility and even to testicular cancer. In circumstances in which the cryptorchid condition is unilateral (see Chapter 9), fertility is seldom compromised. Morphological steps in gonadal descent to the scrotum have been described for a number of species (Hunter, 1762, 1786; Wensing & Colenbrander, 1986; Wensing, 1988), and the process delineated into two distinct phases: first, an androgen-independent transabdominal migration to the inguinal region and, second, descent through the inguinal canal to the scrotum (Hutson *et al.*, 1990) – phases previously referred to by Heyns (1987). Special emphasis is given to a key rôle of the gubernaculum in mediating testicular descent (Hunter, 1762). On the basis of meticulous observations in the rat, van der Schoot (1993*a*) has questioned the concept of two phases and noted that there is no evidence for active testis migration from the posterior abdomen towards the inguinal region during foetal life; rather, the testes remain in place at the base of the abdomen in 15- to 16-day-old foetuses.

The specific molecular and endocrine steps that regulate descent still require clarification. An involvement of anti-Müllerian hormone rather than androgens in promoting a first phase of testicular descent has been suggested but remains controversial (Hutson & Donahoe, 1986; and Chapter 4), whereas oestrogen administration can act to inhibit gubernacular development and testis migration and/or descent (Grocock, Charlton & Pike, 1988). Conversely, a low molecular weight fraction of a foetal

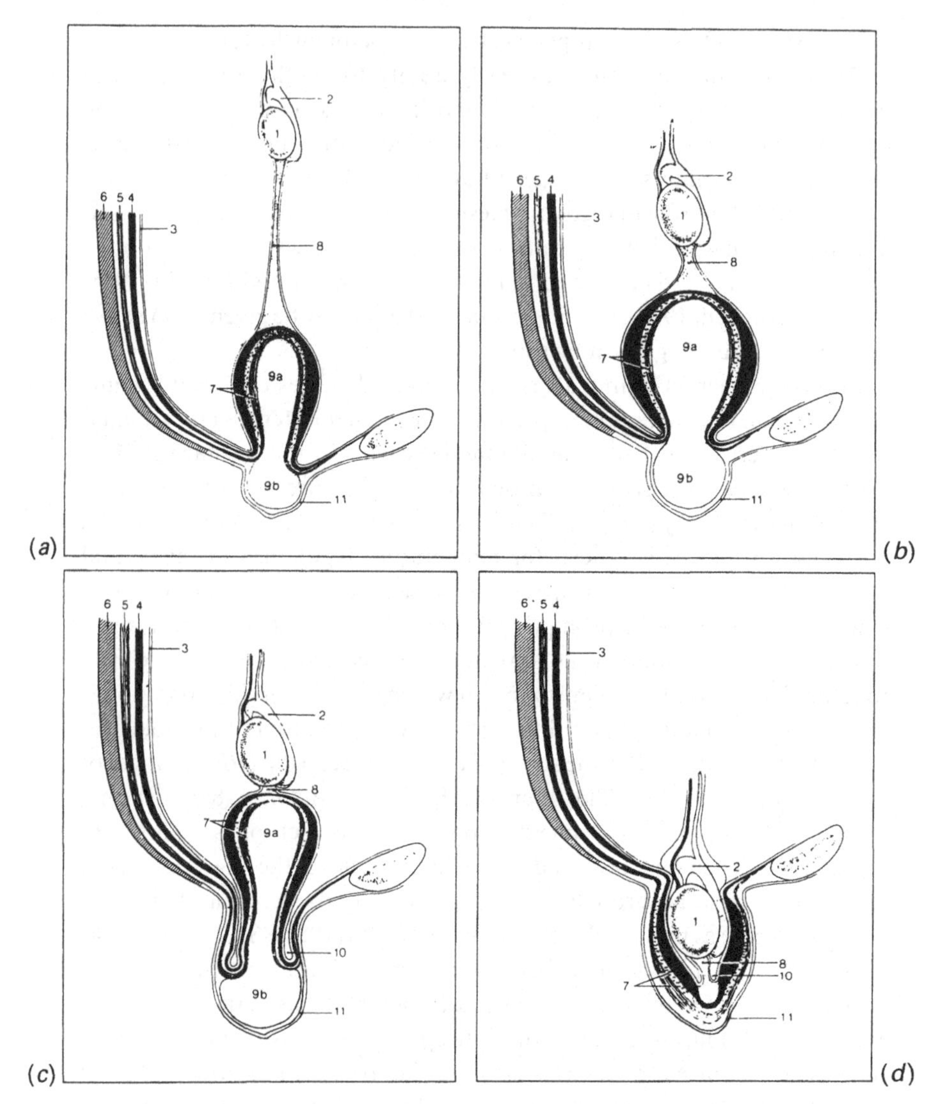

Fig. 3.10. Diagram to illustrate the successive stages of gonadal descent from abdominal to scrotal location in the rat, with a prominent rôle being played by the gubernacular cord. (A) Gubernacular relations and development at 16 days *post coitum* (PC); (B) gubernacular outgrowth at 20 days PC; (C) gubernacular development and vaginal process formation at the day of birth; (D) testicular and cremasteric topography 10–15 days after birth. 1, testis; 2, epididymis; 3, parietal peritoneum; 4, transverse abdominal muscle; 5, internal oblique; 6, external oblique; 7, two layers of the cremaster muscle; 8, gubernacular cord; 9, caudal part of gubernaculum; 9a, mesenchyme of the intra-abdominal segment; 9b, extra-abdominal segment; 10, vaginal process; 11, external spermatic fascia. (From Wensing, 1988.)

testicular extract termed descendin is stimulatory to gubernacular cells during the first phase of testis descent (Wensing, 1988). As far as the rat is concerned, van der Schoot (1992) stated that there was compelling evidence that androgens play no rôle in the prenatal growth or the postnatal inversion and further development of the gubernacular and cremaster muscles, although in rabbits foetal testes certainly control the prenatal growth and differentiation of the gubernacular cones (van der Schoot, 1993*b*). These papilla-like organs invert postnatally and develop into the muscular cremaster sacs, thereby providing space for testicular descent. An appreciation of the rôle of the gubernaculum during testis descent in man was presented by Heyns (1987). In the tammar wallaby, *Macropus eugenii*, *O et al.* (1988) reported that differentiation of the gubernaculum is hormone independent, highlighting the fact that there are sexual dimorphisms in marsupials which precede morphological differentiation of the gonads, suggesting a precocious influence of one or more sex-linked genes.

Differentiation of an ovary

In formation of the female gonad, two overall patterns of differentiation may be distinguished, depending on whether the ovarian germ cells enter *immediate* meiosis without previous steroid synthesis by the neighbouring somatic cells, or *relatively delayed* meiosis with steroid synthesis being demonstrable before meiosis begins. In both instances, Byskov & Hoyer (1988) believe that cells derived from the mesonephros are the principal contributors to the ovarian cell mass, a view not advanced by the earlier works of Witschi (1951) and of Burns (1961) or those of Ullmann (1989) in certain marsupials. The supporting cells differentiate as pre-follicle or pre-granulosa cells under the influence of oocytes.

During formation of an ovary, the germ cells may be enclosed within elongated cell cords proliferated from the coelomic epithelium which are connected to the mesonephros or arranged in a less organised way within the somatic tissues but still connected to the mesonephros during early ovarian growth. Cords of mesonephric cells promote ovarian remodelling by colonising the central part of the presumptive gonad and displacing germ cells towards the periphery (cortex). Germ cells remaining in the medulla usually tend to degenerate (why?) whilst those in the surrounding cortex continue to proliferate and/or to differentiate. Mesenchymatous cells form a thin tunica albuginea. Following establishment of such cortical dominance and the formation of nests of germ cells distinguishable as oogonia, the embryonic ovaries undergo further development and differen-

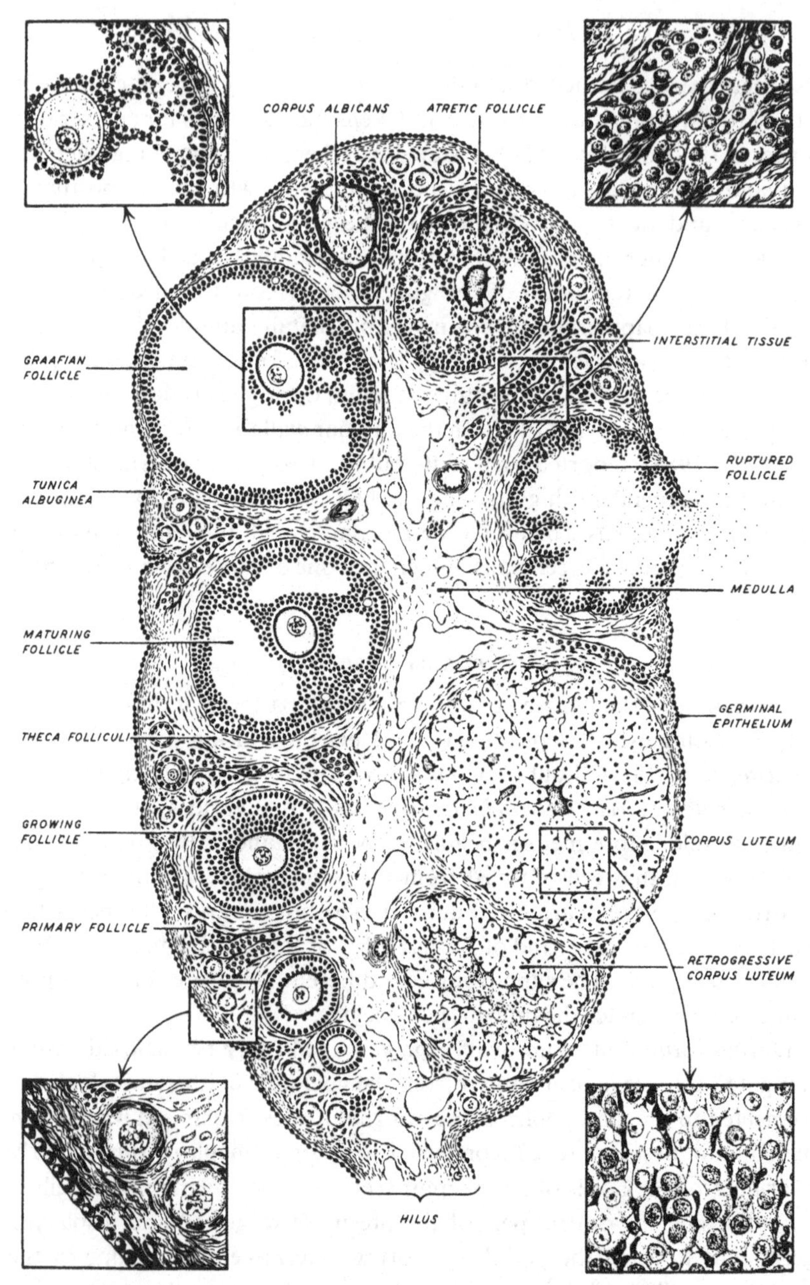

Fig. 3.11. Idealised section through a mammalian ovary after puberty to illustrate the principal structural components. The section is composite in the sense that no distinction has been drawn between the luteal and follicular phases of the oestrous cycle. (From Turner & Bagnara, 1971.)

tiation so that production of potential gametes – the primary oocytes – is essentially complete by the time of birth in most eutherian mammals studied so far. This is due to waves of mitosis causing enormous multiplication in the stock of oogonia, followed by the first phase of meiosis with the nucleus being arrested at the dictyate (diplotene) stage until puberty. Oocytes at diplotene are then surrounded by follicle cells. Extensive atresia of oocytes is a major feature even before the time of birth (reviews by Baker, 1972, 1982; Zuckerman & Baker, 1977). Resumption of meiosis is a preliminary to the events of ovulation after the pre-ovulatory gonadotrophin surge when the basement membrane of the follicle undergoes dissolution and the capillary bed invades. However, an apparent resumption of meiosis may also be a prominent feature of atresia.

Despite the impact of atresia, the foetus at birth will contain a population of oocytes numbered in at least tens or hundreds of thousands rather than in units of simply a hundred. Whether the potential for growth and development is equally distributed amongst the population of oocytes of the neonate is an unresolved question. Certainly, only a minor proportion of the original stock is ever released at ovulation, and a smaller proportion still normally has the opportunity to be fertilised. However, modern procedures of superovulation and autopsy techniques of oocyte extraction by aspiration, dissection or digestion as applied to the ovaries of rodents and domestic farm animals indicate a potential for maturation, fertilisation and early embryonic development in large numbers of artificially released oocytes. Of course, development to the stage of a morula or young blastocyst is not an unequivocal or sufficiently sensitive indicator of the ability to develop into a viable, full-term foetus, and the numbers achieving the latter are much reduced.

Returning to the foetal ovary, in order for functional competence to be demonstrated in due course, there is a dependence upon (1) oocytes entering prophase of the first meiotic division; (2) the formation of follicles, elaborating from the relatively simple primary follicle to the well-organised tertiary or Graafian follicle (Fig. 3.11); and (3) development of an ability in the somatic cell layers of the theca interna and stratum granulosum to synthesise and secrete diverse steroid and peptide hormones. Functional interactions between the theca and granulosa cells and between somatic and germ cells appear of paramount importance for this endocrine rôle. As to the actual origin of follicular cells, the granulosa cells are thought to be derived from rete cells, themselves originating in the mesonephros. Once such rete cells reach the differentiating ovary, they stop migrating and proliferate actively to accumulate as granulosa cells. Oocytes have induced the supporting cell lineage to differentiate as follicle cells rather than as Sertoli cells. The time at which actual follicles commence to form in different species depends on when the oocytes reach the diplotene stage of the first meiotic prophase, and may occur during foetal life or after birth (Byskov & Hoyer, 1988). In contrast to the early organogenesis of the testis, ovarian follicles tend to form postnatally (Jost & Magre, 1988).

Factors regulating the onset of meiosis are still not understood, but a currently fashionable viewpoint is that the dynamic interplay of two somatic cell proteins – meiosis-inducing substance (MIS) and meiosis-preventing substance (MPS) – determines progression into meiotic prophase (Byskov, 1986). Shortcomings in this interpretation have been discussed by Upadhyay & Zamboni (1982) and by Zamboni & Upadhyay (1983). Both X chromosomes must be active before meiosis begins in female germ cells (Byskov & Hoyer, 1988). Whereas germ cells are not required for the formation of a testis (Merchant, 1975), the ovary fails to develop in the absence of germ cells and the persistence and maintenance of ovarian follicles depends strictly on the presence of germ cells (Jost & Magre, 1988). These observations may suggest that ovarian determination is dependent upon gene action within the germ cells (Burgoyne, 1989).

Transdifferentiation of embryonic gonads

The influence of an absence of germ cells in male and female gonadal differentiation is a theme worth expanding on. In the absence or almost complete absence of germ cells in the male, Sertoli cells will still differentiate and testis cords form in the genital ridge. In females, by contrast, ovarian follicles fail to form in the absence of oocytes and a streak gonad emerges. However, if loss of oocytes occurs subsequent to their induction of follicle formation, then transdifferentiation of follicular cells into their testicular counterparts, Sertoli cells, can occur. Such Sertoli cells may, in turn, induce the differentiation of XX Leydig cells. This form of transdifferentiation of follicle cells has been noted in such situations as ageing rat ovaries already depleted of nearly all oocytes (Crumeyrolle-Arias, Schreib & Ascheim, 1976), and also in foetal mouse ovaries grafted under the kidney capsule of male or female recipients. In this situation, the oocytes will degenerate and Sertoli and Leydig cells will then appear (Taketo-Hosotani *et al.*, 1985; Taketo-Hosotani & Sinclair-Thompson, 1987).

As a more extensive and systematic sequel to the study of MacIntyre (1956) in which rat ovaries were grafted next to a testis and several developed seminiferous-like tubules, Taketo-Hosotani *et al.* (1985) trans-

Table 3.5. *Development of testicular structures in ovarian grafts with half mesonephroi or reattached mesonephroi following transplantation into female hosts*

Region or source of the mesonephros attached to ovarian grafts	No. of grafts	Development of		Frequency of ovotestis development (%)
		ovaries	ovotestis	
Whole (control)	31	19	12	39
Cranial half	17	12	5	29
Caudal half	14	9	5	36
Self (intact)	27	15	12	44
Female foetuses	16	8	8	50
Male foetuses	19	10	9	47

Note:
Adapted from Taketo-Hosotani & Sinclair-Thompson (1987).

planted foetal mouse ovaries on the 12th day of gestation beneath the kidney capsules of adult male mice. Their development continued morphologically as ovaries until the 11th day after transplantation, when formation of seminiferous cords commenced in addition to follicular development in ovotestis-like structures. Between the 11th and 14th days after transplantation, the ovarian grafts frequently contained transitional structures consisting of Sertoli cells, pre-granulosa cells, a third cell type with features intermediate between these two, and oocytes enclosed by a common basal lamina. Leydig cells or peritubular myoid cells were not found in the transitional area, whereas these cells were present around seminiferous cords composed only of Sertoli cells. Oocytes were degenerating or absent in the well-developed seminiferous cords. On the basis of these findings, Taketo-Hosotani *et al.* (1985) suggested that pre-granulosa cells can differentiate into Sertoli cells in ovarian grafts, such Sertoli cells then being responsible for organising seminiferous cords, degeneration of oocytes, and differentiation of other testicular somatic cells.

A further study indicated that a proportion of foetal ovaries can develop testicular structures after transplantation (Table 3.5) and irrespective of the sex of the recipient (host) mouse (Taketo-Hosotani & Sinclair-Thompson, 1987). However, the adjacent mesonephros appeared to protect ovarian grafts from masculinising stimuli more efficiently in female than in male hosts, although the interpretation to be placed on this point is not yet clear. Perhaps it is simply that a male host has a greater intrinsic ability to promote testicular development, thereby overcoming an influence of the mesonephros. In a subsequent report, Taketo-Hosotani (1987) noted that direct contact between the graft and the host kidney was essential for the induction of XX testicular differentiation. Whether such an interaction with the host tissue involved primarily an enhanced loss of oocytes or provided a factor required for prompting XX Sertoli cell differentiation – or both – is uncertain. So far, only in this model of the XX ovary transplanted beneath the kidney capsule of adult male mice has ultrastructural evidence for Sertoli cell differentiation from XX cells been presented (Merchant-Larios & Taketo, 1991).

McLaren (1991*a*) has emphasised that the process of transdifferentiation into Sertoli cells occurs only when follicle cells have first been exposed to meiotic oocytes and the oocytes have subsequently been lost. Apparently, transdifferentiation may also occur in the opposite direction, that is Sertoli cells into follicle cells, when Sertoli cells are exposed to the influence of growing oocytes in the tubules of mouse testes or ovotestes (Ward, Baker & McLaren, 1988). Of course, it is worth noting the context in which the word transdifferentiation has been used here. Indeed, it is debatable if this really is a stringent form of transdifferentiation (see Okada, 1986), or whether transformation would not be a more appropriate term, bearing in mind the homologies between Sertoli and follicular granulosa cells and their probable common origin in cells of the mesonephros (Upadhyay & Zamboni, 1982) or those of the gonadal blastema.

Concluding remarks

Despite the ever-increasing amount of detail, the precise mechanisms which determine the organisation of the undifferentiated gonads into ovarian or testicular tissues remain unknown. One is of course struck by the lengthy migration of the primordial germ cells, and it would be valuable to know whether this phenomenon occurs on precisely the same timescale for male and female gonocytes in species other than the mouse. Any modern working hypothesis that seeks to explain gonadal differentiation in mammals must take account of situations within the same animal in which one gonad may be an ovary whilst the other is a testis, or indeed one or both gonads may be ovotestes. Such plasticity in the outcome of gonadal differentiation suggests that the preeminence given to the Y chromosome in conventional explanations may require further consideration, and that the dogma on sex determination in mammals should also offer scope for a primary influence of autosomal genes (see Chapter 6). In both birds and mammals, gonocytes of one sex do not complete gametogenesis in the somatic tissues of the other sex (Haffen, 1977), so it is worth asking how germ cells discern what sex they themselves are and likewise the sex of the somatic cells that surround them. The nature of these interactions should keep molecular geneticists usefully occupied for some little while, and may extend the rather polarised view expressed by van Tienhoven (1983) that 'vertebrates may differ in whether the somatic tissue or the primordial germ cells determine whether the primordial germ cells become oocytes or spermatozoa'.

The fact that ectopically developing germ cells always differentiate as oocytes seems an important clue. Another important clue may be the fact that XX cells in the freemartin ovary can seemingly dedifferentiate and then form Sertoli cells (see Chapter 5), indicating that Sertoli cell phenotype – in this model at least – is not dependent upon genes on the Y chromosome. As suggested above, such transdifferentiation is associated with failure of the germ cells, raising the question as to whether oocytes have a positive rôle to play in maintaining ovarian supporting cells in their differentiated state (Burgoyne, 1988). It would, moreover, be mildly surprising if future explanations do not invoke key rôles for diverse growth factors in the organisation of a mammalian gonad, perhaps being produced primarily by the somatic cells that are in contact with the germinal elements. Finally, further thought might also be given to a predominant rôle of the mesonephros in contributing somatic components of the gonad and in secreting proteins to influence the germinal components. Precisely why the mesonephros has assumed such an important potential is uncertain, but it may be for reasons more subtle than the fact that it is merely a neighbouring embryonic tissue.

References

Alvarez-Buylla, A. & Merchant-Larios, H. (1986). Mouse primordial germ cells use fibronectin as a substrate for migration. *Experimental Cell Research*, **165**, 362–8.

Baker, T. G. (1963). A quantitative and cytological study of germ cells in human ovaries. *Proceedings of the Royal Society of London, Series B*, **158**, 417–33.

Baker, T. G. (1972). Primordial germ cells. In *Reproduction in Mammals*, 1st edn, ed. C. R. Austin & R. V. Short, pp. 1–13. Cambridge: Cambridge University Press.

Baker, T. G. (1982). Oogenesis and ovulation. In *Reproduction in Mammals*, 2nd edn, vol. 1, ed. C. R. Austin & R. V. Short, pp. 17–45. Cambridge: Cambridge University Press.

Balfour, F. M. (1878). On the structure and development of the vertebrate ovary. *Quarterly Journal of Microscopical Science*, **18**, 383–438.

Bennett, D. (1956). Developmental analysis of a mutation with pleiotropic effects in the mouse. *Journal of Morphology*, **98**, 199–233.

Black, J. L. & Erickson, B. H. (1965). Oogenesis and ovarian development in the prenatal pig. *Anatomical Record*, **161**, 45–56.

Blandau, R. J. , White, B. J. & Rumery, R. E. (1963). Observations on the movements of the living primordial germ cells in the mouse. *Fertility and Sterility*, **14**, 482–9.

Borum, K. (1961). Oogenesis in the mouse. A study of the meiotic prophase. *Experimental Cell Research*, **24**, 495–507.

Buehr, M., Gu, S. & McLaren, A. (1993). Mesonephric contribution to testis differentiation in the fetal mouse. *Development*, **117**, 273–81.

Burgoyne, P. S. (1988). Role of mammalian Y chromosome in sex determination. *Philosophical Transactions of the Royal Society of London, Series B*, **322**, 63–72.

Burgoyne, P. S. (1989). Thumbs down for zinc finger? *Nature (London)*, **342**, 860–2.

Burns, R. K. (1961). Role of hormones in the differentiation of sex. In *Sex and Internal Secretions*, 3rd edn, ed. W. C. Young, pp. 76–158. Baltimore: Williams & Wilkins.

Byskov, A. G. (1981). Gonadal sex and germ cell differentiation. In *Mechanisms of Sex Differentiation in Animals and Man*, ed. C. R. Austin & R. G. Edwards, pp. 145–64. London: Academic Press.

Byskov, A. G. (1982). Primordial germ cells and regulation of meiosis. In *Reproduction in Mammals*, 2nd edn, ed. C. R. Austin & R. V. Short, pp. 1–16. Cambridge: Cambridge University Press.

Byskov, A. G. (1986). Differentiation of mammalian embryonic gonad. *Physiological Reviews*, **66**, 71–117.

Byskov, A. G. & Grinsted, J. (1981). Feminizing effect of mesonephros on cultured differentiating mouse gonads and ducts. *Science*, **212**, 817–18.

Byskov, A. G. & Høyer, P. E. (1988). Embryology of mammalian gonads and ducts. In *The Physiology of Reproduction*, ed. E. Knobil, J. Neill *et al.*, ch. 8, pp. 265–302. New York: Raven Press.

Byskov, A. G. & Saxen, L. (1976). Induction of meiosis in fetal mouse testis *in vitro*. *Developmental Biology*, **52**, 193–200.

Chiquoine, A. D. (1954). The identification, origin and migration of the primordial germ cells in the mouse embryo. *Anatomical Record*, **118**, 135–46.

Clark, J. M. & Eddy, E. M. (1975). Fine structural observations on the origin and association of primordial germ cells of the mouse. *Developmental Biology*, **47**, 136–55.

Clermont, Y. & Huckins, C. (1961). Microscopic anatomy of the sex cords and seminiferous tubules in growing and adult male albino rats. *American Journal of Anatomy*, **108**, 79–97.

Crumeyrolle-Arias, M., Schreib, D. & Ascheim, P. (1976). Light and electron microscopy of the ovarian interstitial tissue in the senile rat: normal aspect and response to HCG of 'deficiency cells' and 'epithelial cords'. *Gerontology*, **22**, 185–204.

De Felici, M. & Dolci, S. (1991). Leukemia inhibitory factor sustains the survival of mouse primordial germ cells cultured on TM_4 feeder layers. *Developmental Biology*, **147**, 281–4.

De Felici, M., Dolci, S. & Pesce, M. (1992). Cellular and molecular aspects of mouse primordial germ cell migration and proliferation in culture. *International Journal of Developmental Biology*, **36**, 205–13.

De Felici, M., Dolci, S. & Pesce, M. (1993). Proliferation of mouse primordial germ cells *in vitro*: a key rôle for cAMP. *Developmental Biology*, **157**, 277–80.

Dolci, S., Pesce, M. & De Felici, M. (1993). Combined action of stem cell factor, leukemia inhibitory factor, and cAMP on in vitro proliferation of mouse primordial germ cells. *Molecular Reproduction and Development*, **35**, 134–9.

Dolci, S., Williams, D. E., Ernst, M. K., Resnick, J. L., Brannan, C. I., Lock, L. F., Lyman, S. D., Boswell, S. H. & Donovan, P. J. (1991). Requirement for mast cell growth factor for primordial germ cell survival in culture. *Nature (London)*, **352**, 809–11.

Dym, M. & Fawcett, D. W. (1970). The blood–testis barrier in the rat and the physiological compartmentation of the seminiferous epithelium. *Biology of Reproduction*, **3**, 308–26.

Eddy, E. M., Clark, J. M., Gong, D. & Fenderson, B. A. (1981). Origin and migration of primordial germ cells in mammals. *Gamete Research*, **4**, 333–62.

Everett, N. B. (1945). The present status of the germ-cell problem in vertebrates. *Biological Reviews*, **20**, 45–55.

Falin, L. I. (1969). The development of genital glands and the origin of germ cells in human embryogenesis. *Acta Anatomica*, **72**, 195–232.

Francavilla, S. & Zamboni, L. (1985). Differentiation of mouse ectopic germinal cells in intra- and perigonadal locations. *Journal of Experimental Zoology*, **233**, 101–9.

Fujimoto, T., Yoshinaga, K. & Kono, I. (1985). Distribution of fibronectin on the migratory pathway of primordial germ cells in mice. *Anatomical Record*, **211**, 271–8.

Gardner, R. L., Lyon M. F., Evans, E. P. & Burtenshaw, M. D. (1985). Clonal analysis of X-chromosome inactivation and the origin of the germ line in the mouse embryo. *Journal of Embryology and Experimental Morphology*, **88**, 349–63.

Gardner, R. L. & Rossant, J. (1979). Investigation of the fate of 4.5 day *post-coitum* mouse inner cell mass cells by blastocyst injection. *Journal of Embryology and Experimental Morphology*, **52**, 141–52.

Gillman, J. (1948). The development of the gonads in man, with a consideration of the role of fetal endocrines and the histogenesis of ovarian tumors. *Contributions to Embryology of the Carnegie Institution*, **32**, 81–131.

Ginsburg, M., Snow, M. H. L. & McLaren, A. (1990). Primordial germ cells in the mouse embryo during gastrulation. *Development*, **110**, 521–8.

Glover, D. M. & Hames, B. D., eds (1989). *Genes and Embryos*. Oxford: IRL Press.

Godin, I., Deed, R., Cooke, J. , Zsebo, K., Dexter, M. & Wylie, C. C. (1991). Effects of the *steel* gene product on mouse primordial germ cells in culture. *Nature (London)*, **352**, 807–9.

Godin, I. & Wylie, C. C. (1991). TGFβ_1 inhibits proliferation and has a chemotropic effect on mouse primordial germ cells in culture. *Development*, **113**, 1451–7.

Godin, I, Wylie, C. & Heasman, J. (1990). Genital ridges exert long-range effects on mouse primordial germ cell numbers and direction of migration in culture. *Development*, **108**, 357–63.

Gondos, B. & Byskov, A. G. (1981). Germ cell kinetics in the neonatal rabbit testis. *Cell and Tissue Research*, **215**, 143–51.

Grocock, C. A. Charlton, H. M. & Pike, M. C. (1988). Role of the fetal pituitary in cryptorchidism induced by exogenous maternal oestrogen during pregnancy in mice. *Journal of Reproduction and Fertility*, **83**, 295–300.

Haffen, K. (1977). Sexual differentiation of the ovary. In *The Ovary*, 2nd edn, vol. 1, ch. 3, ed. S. Zuckerman & B. J. Weir, pp. 69–112. New York & London: Academic Press.

Hardisty, M. W. (1967). The numbers of vertebrate primordial germ cells. *Biological Reviews*, **42**, 265–87.

Heyns, C. F. (1987). The gubernaculum during testicular descent in the human fetus. *Journal of Anatomy*, **153**, 93–112.

Hunter, J. (1762). Observations on the state of the testis in the foetus, and on the hernia congenita. In *Medical Commentaries*, part I, by William Hunter, pp. 75–90. London: Hamilton.

Hunter, J. (1786). A description of the situation of the testis in the foetus, with its descent into the scrotum. In *Observations on Certain Parts of the Animal Œconomy*, pp. 1–26. London.

Hutson, J. M. & Donahoe, P. K. (1986). The hormonal control of testicular descent. *Endocrine Reviews*, **7**, 270–83.

Hutson, J. M., Williams, M. P. L., Fallat, M. E. & Attah, A. (1990). Testicular descent: new insights into its hormonal control. *Oxford Reviews of Reproductive Biology*, **12**, 1–56.

Jones, R. E. (ed.) (1978). *The Vertebrate Ovary: Comparative Biology and Evolution*. New York & London: Plenum Press.

Jost, A. (1972*a*). A new look at the mechanisms controlling sex differentiation in mammals. *Johns Hopkins Medical Journal*, **130**, 38–53.

Jost, A. (1972*b*). Données préliminairès sur les stades initiaux de la différenciation du testicule chez le rat. *Archives d'Anatomie Microscopique et de Morphologie Expérimentale*, **61**, 415–38.

Jost, A. & Magre, S. (1984). Testicular development phases and dual hormonal control of sexual organogenesis. In *Sexual Differentiation: Basic and Clinical Aspects*, ed. M. Serio, pp. 1–15. New York: Raven Press.

Jost, A. & Magre, S. (1988). Control mechanisms of testicular differentiation. *Philosophical Transactions of the Royal Society of London, Series B*, **322**, 55–61.

Jost, A. & Magre, S. (1993). Sexual differentiation. In *Reproduction in Mammals and Man*, ed. C. Thibault, M. C. Levasseur & R. H. F. Hunter, pp. 197–212. Paris: Ellipses.

Jost, A., Magre, S. & Agelopoulou, R. (1981). Early stages of testicular differentiation in the rat. *Human Genetics*, **58**, 59–63.

Jost, A., Vigier, B. Prépin, J. & Perchellet, J. P. (1973). Studies on sex differentiation in mammals. *Recent Progress in Hormone Research*, **29**, 1–35.

Kafri, T., Ariel, M., Brandeis, M., Shemer, R., Urven, L., McCarrey, J. , Cedar, H. & Razin, A. (1992). Developmental pattern of gene-specific DNA methylation in the mouse embryo and germ line. *Genes and Development*, **6**, 705–14.

Kitahara, Y. (1923). Über die Entstehung der Zwischenzellen der Keimdrüsen des Menschen und der Säugetiere und über deren physiologische Bedeutung. *Archiv für Entwicklungs-Mechanik der Organismen*, **52**, 550–615.

Knobil, E., Neill, J. *et al.* (1988). *The Physiology of Reproduction*. New York: Raven Press.

McKay, D. G., Hertig, A. T., Adams, E. C. & Danziger, S. (1953). Histochemical observations on the germ cells of human embryos. *Anatomical Record*, **117**, 201–19.

McLaren, A. (1981*a*). *Germ Cells and Soma: A New Look at an Old Problem*. New Haven: Yale University Press.

McLaren, A. (1981*b*). The fate of germ cells in the testis of fetal sex-reversed

mice. *Journal of Reproduction and Fertility*, **61**, 461–7.
McLaren, A. (1983). Studies on mouse germ cells inside and outside the gonad. *Journal of Experimental Zoology*, **228**, 167–71.
McLaren, A. (1984). Chimaeras and sexual differentiation. In *Chimaeras in Developmental Biology*, ed. N. Le Douarin & A. McLaren, pp. 381–99. London: Academic Press.
McLaren, A. (1991*a*). Sex determination in mammals. *Oxford Reviews of Reproductive Biology*, **13**, 1–33.
McLaren, A. (1991*b*). Development of the mammalian gonad: the fate of the supporting cell lineage. *BioEssays*, **13**, 151–6.
McLaren, A. (1992). The quest for immortality. *Nature (London)*, **359**, 482–3.
MacIntyre, M. N. (1956). Effect of the testis on ovarian differentiation in heterosexual embryonic rat gonad transplants. *Anatomical Record*, **124**, 27–45.
Magre, S. & Jost, A. (1983). Early stages of the differentiation of the rat testis: relations between Sertoli and germ cells. In *Current Problems in Germ Cell Differentiation*, ed. A. McLaren & C. C. Wylie, pp. 201–14. Cambridge: Cambridge University Press.
Maitland, P. & Ullmann, S. L. (1993). Gonadal development in the opossum, *Monodelphis domestica*: the rete ovarii does not contribute to the steroidogenic tissues. *Journal of Anatomy*, **183**, 43–56.
Matsui, Y., Toksoz, D., Nishikawa, S., Nishikawa, S-I., Williams, D., Zsebo, K. & Hogan, B. L. M. (1991). Effect of *Steel* factor and leukaemia inhibitory factor on murine primordial germ cells in culture. *Nature (London)*, **353**, 750–2.
Merchant, H. (1975). Rat gonadal and ovarian organogenesis with and without germ cells. An ultrastructural study. *Developmental Biology*, **44**, 1–21.
Merchant-Larios, H. (1979). Origin of the somatic cells in the rat gonad: an autoradiographic approach. *Annales de Biologie Animale, Biochimie et Biophysique*, **19**, 1219–29.
Merchant-Larios, H. & Alvarez-Buylla, A. (1986). The role of extracellular matrix and tissue topographic arrangement in mouse and rat primordial germ cell migration. In *Development and Function of the Reproductive Organs*, ed. A. Eshkol, B. Eckstein, N. Dekel, H. Peters & A. Tsafriri. Serono Symposia Review No. 11. New York: Raven Press.
Merchant-Larios, H. & Taketo, T. (1991). Testicular differentiation in mammals under normal and experimental conditions. *Journal of Electron Microscopy Techniques*, **19**, 158–71.
Mintz, B. (1957). Embryological development of primordial germ cells in the mouse: influence of a new mutation, W^j. *Journal of Embryology and Experimental Morphology*, **5**, 396–403.
Mintz, B. (1959). Continuity of the female germ cell line from embryo to adult. *Archives d'Anatomie Microscopique et de Morphologie Expérimentale* (Supplement), **48**, 155–72.
Mintz, B. (1968). Hermaphroditism, sex chromosomal mosaicism and germ cell selection in allophenic mice. *Journal of Animal Science*, **27**, Supplement 1, 51–60.
Mintz, B. & Russell, E. S. (1957). Gene-induced embryological modifications of primordial germ cells in the mouse. *Journal of Experimental Zoology*, **134**, 207–37.
Mittwoch, U. (1967). Sex differentiation in mammals. *Nature (London)*, **214**, 554–6.

Mittwoch, U. (1970). How does the Y-chromosome affect gonadal differentiation? *Philosophical Transactions of the Royal Society of London, Series B*, **259**, 113–17.
Mittwoch, U. (1985). Males, females and hermaphrodites. *Annals of Human Genetics*, **50**, 103–21.
Mittwoch, U. (1992). Sex determination and sex reversal: genotype, phenotype, dogma and semantics. *Human Genetics*, **89**, 467–79.
Monk, M. (1990). Changes in DNA methylation during mouse embryonic development in relation to X-chromosome activity and imprinting. *Philosophical Transactions of the Royal Society of London, Series B*, **326**, 299–312.
Monk, M., Boubelik, M. & Lehnert, S. (1987). Temporal and regional changes in DNA methylation in the embryonic, extraembryonic and germ cell lineages during mouse embryo development. *Development*, **99**, 371–82.
Moon, Y. S. & Hardy, M. H. (1973). The early differentiation of the testis and interstitial cells in the fetal pig, and its duplication in organ culture. *American Journal of Anatomy*, **138**, 253–68.
O, W-S. & Baker, T. G. (1978). Germinal and somatic cell interrelationships in gonadal sex differentiation. *Annales de Biologie Animale, Biochimie et Biophysique*, **18**, 351–7.
O, W-S., Short, R. V., Renfree, M. B. & Shaw, G. (1988). Primary genetic control of somatic sexual differentiation in a mammal. *Nature (London)*, **331**, 716–7.
Ohno, S. (1979). *Major Sex-Determining Genes*. Berlin: Springer-Verlag.
Okada, T. S. (1986). Transdifferentiation in animal cells: fact or artifact? *Development, Growth and Differentiation*, **28**, 213–21.
Ożdżénski, W. (1967). Observations on the origin of primordial germ cells in the mouse. *Zoologica Poloniae*, **17**, 367–79.
Patten, B. M. (1948). *Embryology of the Pig*. New York, Toronto & London: McGraw-Hill.
Pelliniemi, L. J. (1975). Ultrastructure of the early ovary and testis in pig embryos. *American Journal of Anatomy*, **144**, 89–112.
Pesce, M., Farrace, M. G., Piacentini, M., Dolci, S. & De Felici, M. (1993). Stem cell factor and leukemia inhibitory factor promote primordial germ cell survival by suppressing programmed cell death (apoptosis). *Development*, **118**, 1089–94.
Resnick, J. L., Bixler, L. S., Cheng, L. & Donovan, P. J. (1992). Long-term proliferation of mouse primordial germ cells in culture. *Nature (London)*, **359**, 550–1.
Rogulska, T. (1969). Migration of the chick primordial germ cells from the intracoelomically transplanted germinal crescent into the genital ridge. *Experientia*, **25**, 631–2.
Rogulska, T., Ożdżénski, W. & Komar, A. (1971). Behaviour of mouse primordial germ cells in the chick embryo. *Journal of Embryology and Experimental Morphology*, **25**, 155–64.
Sanford, J. P., Clark, H. J. , Chapman, V. M. & Rossant, J. (1987). Differences in DNA methylation during oogenesis and spermatogenesis and their persistence during early embryogenesis in the mouse. *Genes and Development*, **1**, 1039–46.
Snow, M. H. L. & Monk, M. (1983). Emergence and migration of mouse primordial germ cells. In *Current Problems in Germ Cell Differentiation*, ed. A. McLaren & C. C. Wylie, pp. 115–35. Cambridge: Cambridge University Press.

Taketo-Hosotani, T. (1987). Factors involved in the testicular development from fetal mouse ovaries following transplantation. *Journal of Experimental Zoology*, **241**, 95–100.
Taketo-Hosotani, T., Merchant-Larios, H., Thau, R. B. & Koide, S. S. (1985). Testicular cell differentiation in fetal mouse ovaries following transplantation into adult male mice. *Journal of Experimental Zoology*, **236**, 229–37.
Taketo-Hosotani, T. & Sinclair-Thompson, E. (1987). Influence of the mesonephros on the development of fetal mouse ovaries following transplantation into adult male and female mice. *Developmental Biology*, **124**, 423–30.
Tam, P. P. L. & Snow, M. H. L. (1981). Proliferation and migration of primordial germ cells during compensatory growth in mouse embryos. *Journal of Embryology and Experimental Morphology*, **64**, 133–47.
Tarkowski, A. K. (1969). Are the genetic factors controlling sexual differentiation of somatic and germinal tissues of a mammalian gonad stable or labile? In *Environmental Influences in Genetic Expression*, ed. N. Kretchmer & D. N. Walcher, #2, pp. 49–60. Washington, DC: Fogarty International Center Proceedings.
Turner, C. D. & Bagnara, J. T. (1971). *General Endocrinology*, 5th edn. Philadelphia & London: W. B. Saunders.
Ullmann, S. L. (1989). Ovary development in bandicoots: sexual differentiation to follicle formation. *Journal of Anatomy*, **165**, 45–60.
Upadhyay, S., Luciani, J. M. & Zamboni, L. (1979). The role of the mesonephros in the development of indifferent gonads and ovaries of the mouse. *Annales de Biologie Animale, Biochimie et Biophysique*, **19**, 1179–96.
Upadhyay, S. & Zamboni, L. (1982). Ectopic germ cells: natural model for the study of germ cell sexual differentiation. *Proceedings of the National Academy of Sciences, USA*, **79**, 6584–8.
van der Schoot, P. (1992). Androgens in relation to prenatal development and postnatal inversion of the gubernacula in rats. *Journal of Reproduction and Fertility*, **95**, 145–58.
van der Schoot, P. (1993*a*). Doubts about the 'first phase of testis descent' in the rat as a valid concept. *Anatomy and Embryology*, **187**, 203–8.
van der Schoot, P. (1993*b*). Foetal testes control the prenatal growth and differentiation of the gubernacular cones in rabbits – a tribute to the late Professor Alfred Jost. *Development*, **118**, 1327–34.
van Tienhoven, A. (1983). *Reproductive Physiology of Vertebrates*, 2nd edn. Ithaca & London: Cornell University Press.
Vigier, B., Watrin, F., Magre, S., Tran, D. & Josso, N. (1987). Purified bovine AMH induces a characteristic freemartin effect in fetal rat prospective ovaries exposed to it *in vitro*. *Development*, **100**, 43–55.
Waites, G. M. H. & Setchell, B. P. (1990). Physiology of the mammalian testis. In *Marshall's Physiology of Reproduction*, 4th edn, vol. 2, ed. G. E. Lamming, ch. 1, pp. 1–105. Edinburgh, London & New York: Churchill Livingstone.
Waldeyer, W. (1870). *Eierstock und Ei. Ein Beitrag zur Anatomie und Entwickelungsgeschichte der Sexualorganie*. Leipzig: Engelmann.
Ward, H. B., Baker, T. G. & McLaren, A. (1988). A histological study of the gonads of T16H/X *Sxr* hermaphrodite mice. *Journal of Anatomy*, **158**, 65–76.
Wartenberg, H. (1978). Human testicular development and the role of the mesonephros in the origin of a dual Sertoli cell system. *Andrologia*, **10**, 1–21.

Wartenberg, H. (1981). The influence of the mesonephric blastema on gonadal development and sexual differentiation. In *Development and Function of Reproductive Organs*, ed. A. G. Byskov & H. Peters, pp. 3–12. Amsterdam: Excerpta Medica.
Wartenberg, H. (1989). Differentiation and development of the testis. In *The Testis*, 2nd edn, ed. H. Burger & D. de Kretser, pp. 67–118. New York: Raven Press.
Wensing, C. J. G. (1986). Testicular descent in the rat and a comparison of this process in the rat with that in the pig. *Anatomical Record*, **214**, 154–60.
Wensing, C. J. G. (1988). The embryology of testicular descent. *Hormone Research*, **30**, 144–52.
Wensing, C. J. G. & Colenbrander, B. (1986). Normal and abnormal testicular descent. *Oxford Reviews of Reproductive Biology*, **8**, 130–64.
Wilkins, A. S. (1993). Germ line development. *BioEssays*, **15**, 699–700.
Wilson, J. D. (1978). Sexual differentiation. *Annual Review of Physiology*, **40**, 279–306.
Witschi, E. (1932). Sex deviations, inversions and parabiosis. In *Sex and Internal Secretions*, 1st edn, ed. E. Allen, chap. 5. Baltimore: Williams & Wilkins.
Witschi, E. (1948). Migration of the germ cells of human embryos from the yolk sac to the primitive gonadal folds. *Contributions to Embryology of the Carnegie Institution*, **32**, 67–80.
Witschi, E. (1951). Embryogenesis of the adrenal and the reproductive glands. *Recent Progress in Hormone Research*, **6**, 1–23.
Zamboni, L. & Merchant, H. (1973). The fine morphology of mouse primordial germ cells in extragonadal locations. *American Journal of Anatomy*, **137**, 299–335.
Zamboni, L. & Upadhyay, S. (1983). Germ cell differentiation in mouse adrenal glands. *Journal of Experimental Zoology*, **228**, 173–93.
Zamboni, L., Bézard, J. & Mauléon, P. (1979). The role of the mesonephros in the development of the sheep foetal ovary. *Annales de Biologie Animale, Biochimie et Biophysique*, **19**, 1153–78.
Zuckerman, S. & Baker, T. G. (1977). The development of the ovary and the process of oogenesis. In *Ovary*, 2nd edn, ed. S. Zuckerman & B. J. Weir, ch. 2, pp. 41–67. New York & London: Academic Press.
Zuckerman, S. & Weir, B. J. (eds) (1977). *The Ovary*, 2nd edn. New York & London: Academic Press.

4

Differentiation of the genital duct system

Introduction

The comments that follow are intended to summarise some of the key events taking place during normal differentiation of the reproductive tracts in eutherian mammals rather than presenting an exhaustive description. Comprehensive reviews of the topic or essays containing important detail include those of Jost (1947, 1961, 1970), Burns (1961), Price, Zaaijer & Ortiz (1969), Josso, Picard & Tran (1977), Byskov (1978, 1986), Wilson (1978), Josso (1981), Short (1982), Josso & Picard (1986) and Byskov & Høyer (1988). Material in the major chapter by Glover, D'Occhio & Millar (1990) is also relevant.

The embryological derivation of male or female reproductive tracts is the principal theme of this chapter; further growth of the tracts up to the time of puberty and the nature of underlying endocrine factors are considered only in passing. Whilst elaboration of a distinct duct system characteristic of a male or a female is the usual sequel to gonadal differentiation, portions of both male and female duct systems may be preserved and indeed become prominent in conditions of intersexuality. In other words, perturbations in gonadal development – such as formation of an ovotestis – may find expression in the morphology of the genital tract. Only exceptionally, however, is partial or extensive duplication of the genital systems in mature animals not correlated with some form of anomalous development of one or both gonads.

Duplication of ducts: genetic inferences

Quite apart from a demonstrable plasticity in the gonadal tissues in the form of their development, the embryonic duct system can also be viewed as bipotential in that it is duplicated during the initial stages of organogenesis. Developmental processes in the very young embryo furnish it with two sets of primitive genital tracts: the Wolffian and Müllerian ducts (Fig. 4.1). These, respectively, male and female duct systems can be distinguished

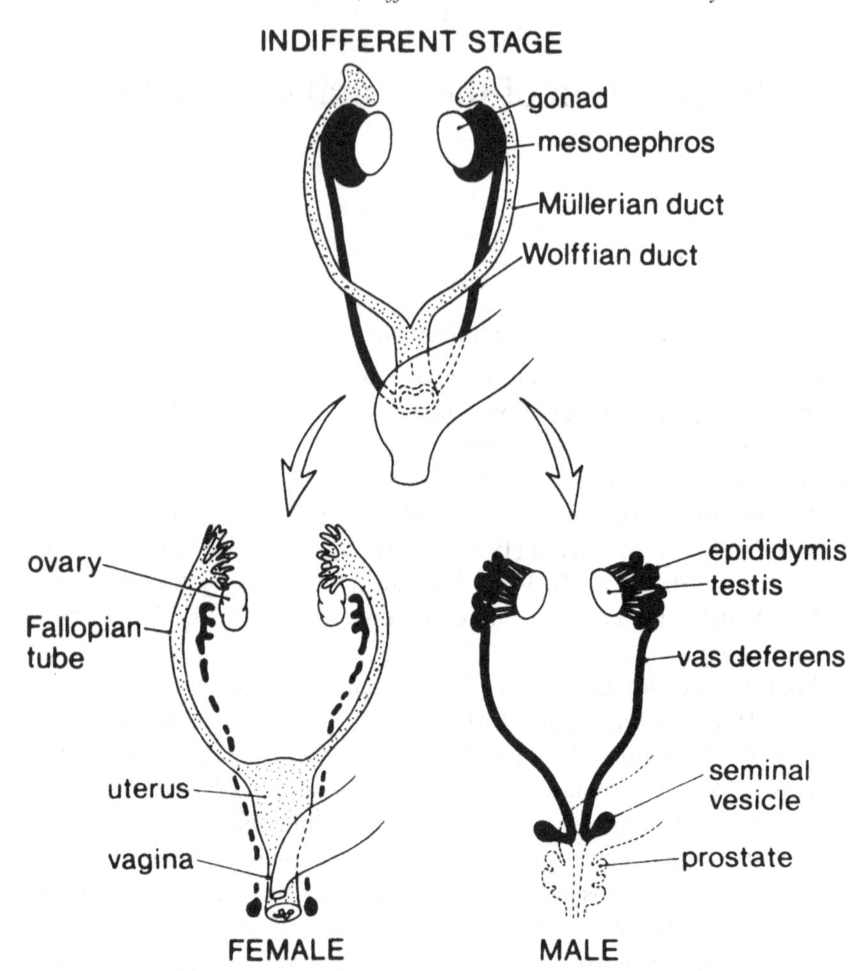

Fig. 4.1. A representation of the indifferent stage of the Wolffian and Müllerian duct systems when both are present in the very young embryo, and then a much simplified illustration of preferential development of one of the duct systems after gonadal sex has been determined. (Derived from Jost, 1961, modified by George & Wilson, 1988.)

alongside each other for a period of time during embryogenesis irrespective of the sex chromosome complement of the individual. Such duplication would imply that there is not yet an awareness or at least an unambiguous expression in these somatic tissues of the genetic sex of the developing embryo, and that there is still the potential to promote formation of either a male or a female duct system. That instructional information should be awaited from the gonads rather than conveyed in the form of a more direct reading of the embryonic genome (i.e. local chromosome information) may serve to emphasise that, even in eutherian mammals with their putative sex chromosomal programming of determination and thereby differentiation, it has been important to retain some element of flexibility.

Sequential formation of duct systems

The fact that the genital system has appropriated embryonic ducts initially performing a quite distinct function was remarked upon by Patten (1948), and considered as a form of developmental opportunism facilitated by the proximity of the growing testes to the degenerating mesonephros. Although both sets of ducts can be distinguished in the very young embryo, this observation is not intended to infer that they are formed simultaneously. The Wolffian ducts appear first as the excretory ducts of the mesonephros; hence the appellation mesonephric duct (Fig. 4.2). Each is arranged lengthwise, receiving transverse tubules associated with primitive capillary networks of the mesonephros. The Müllerian or paramesonephric ducts develop only as the gonads are becoming distinguishable and in proximity to the Wolffian ducts, possibly using the latter as an orientational guide during longitudinal proliferation in a caudal direction. Such caudal growth of the Müllerian duct can be arrested by experimentally interrupting development of the Wolffian duct (Gruenwald, 1941; Didier, 1973), inferring some form of critical dependence. Indeed, some cells of the formative Müllerian system may be derived from the Wolffian ducts (Wilson, 1978) although, in its initial organisation, invagination of the coelomic epithelium is the principal origin of the female system. Thus, mammalian Müllerian ducts do not arise by simple duplication of the Wolffian ducts, which seems to be the case in lower vertebrates. The primitive ducts, which are at first lined with a single-layered epithelium, become more complex due to mesenchymal–epithelium interactions, distinguishable initially as mesenchymal cells arranged in layers around the ducts (Wilson, 1978; Byskov & Høyer, 1988).

In normal circumstances during formation of the embryo, only one of the paired (bilateral) duct systems develops whilst the other undergoes regression; however, this may not be the situation in cases of chromosomal anomalies or hormonal disturbance. Many instances of abnormal development of the genital tract can be traced to incomplete involution of the Wolffian or Müllerian ducts, portions of these systems remaining macroscopically distinguishable or even becoming prominent (see below). In that the dogma, even to this day, is that events within the genital tract are subject

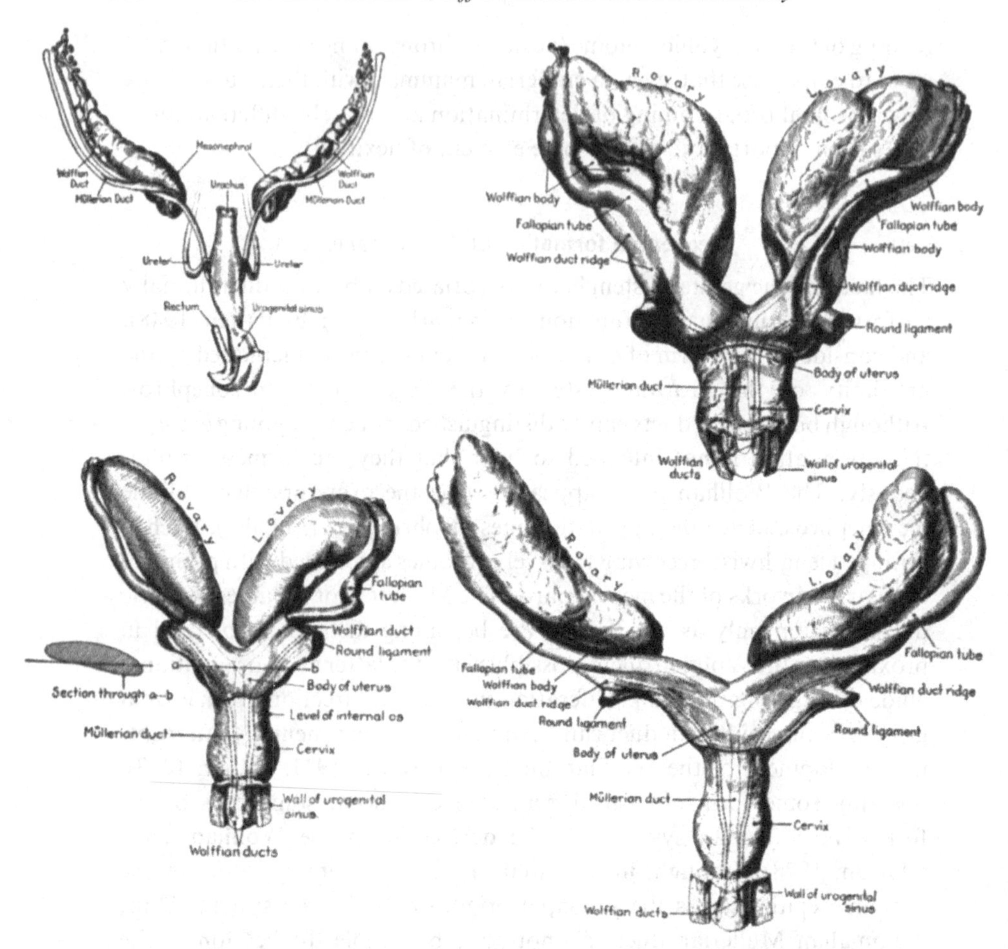

Fig. 4.2. The gonads and genital ducts of the early human embryo. Note the initial presence of both Wolffian and Müllerian duct systems, with the former lying alongside and connected to the mesonephros as its excretory duct. In this illustration, there is preferential development of a female system. (From Hunter, 1930.)

to regulation by secretions of the gonads and, in females, in due course also by those of the extra-embryonic membranes (i.e. by placental hormones), anomalous differentiation and development of the duct system may be an expression of perturbed function in the neighbouring gonad(s) (see Chapters 5, 6 and 7).

Once again, it is therefore worth making the point that degrees of ambivalence in the pattern of development in the genital duct systems may represent a vestige of the more flexible sex-determining and differentiation systems in lower vertebrates in which the sex ratio of the potential offspring can be altered in response to a variety of environmental cues. Nonetheless, as noted in Chapters 2 and 3, maleness is the imposed or induced phenotype in placental mammals whereas femaleness is the permissive state, that is the phenotype that will arise in the absence of maleness. This viewpoint has been prominent at least since the classical experiments of Jost (1947), even though at least two authors consider that this work of Jost has been overinterpreted (Eicher & Washburn, 1983).

Differentiation of a male tract: rôle of anti-Müllerian hormone

The primary controlling factor that dictates which of the genital duct systems will develop is the presence or absence of a Y chromosome, although the effect is mediated through secretion – or absence of secretion – of androgens and another hormone from the embryonic (foetal) testes. Genetic maleness promotes stabilisation and development of the Wolffian (mesonephric) ducts due to the influence of androgens, and almost complete regression of the Müllerian duct system due to a quite distinct inhibitory influence of some other testicular morphogenetic product (Jost, 1947), seemingly protein in nature (Fig. 4.3). The two distinct types of hormone secretion, steroid and protein, are synthesised in different somatic compartments of the foetal testis. Androgens, predominantly in the form of testosterone secreted by the interstitial cells of Leydig, actively stimulate growth and development of the Wolffian duct system, as demonstrated by Greene, Burrill & Ivy (1939), Jost (1947) and reviewed by Jost (1953). A dimeric glycoprotein hormone (140000 daltons, composed of two identical subunits linked by disulphide bonds), referred to as anti-Müllerian hormone (AMH), is produced by Sertoli cells. It acts locally to cause regression of the adjacent paramesonephric ducts in genetic males (Josso *et al.*, 1977; Vigier *et al.*, 1983), although cleavage and dissociation of the molecule appear essential to this rôle (Kuroda *et al.*, 1991). In general terms, such regression involves a narrowing of the duct, a reduction in epithelial height, and formation of densely packed fibroblasts around the ducts (Fig. 4.4). Degeneration usually begins at the cranial end of each duct and proceeds in a caudal direction. On a developmental timescale, Müllerian duct regression in male embryos occurs earlier than the corresponding regression of Wolffian ducts in female embryos.

A more detailed account of regression of the Müllerian ducts in response to AMH would focus on elaborate epithelium–mesenchyme interactions such as (1) dissolution of the basement membrane in lining cells and (2)

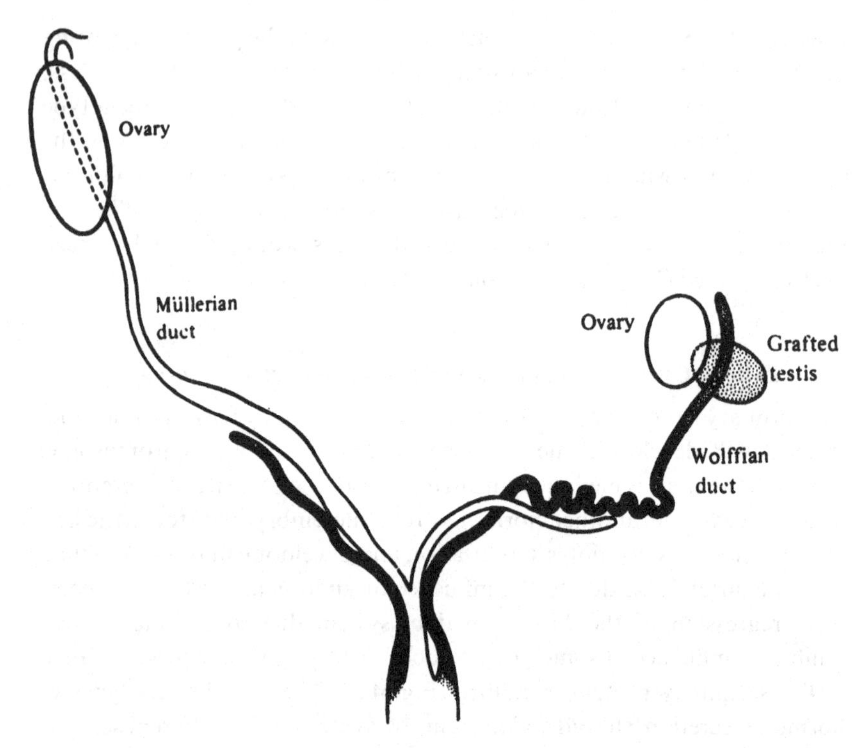

Fig. 4.3. Depiction of Jost's classical experiment on differentiation of the genital duct systems. A portion of testicular tissue was grafted against one ovary of a rabbit foetus, resulting in proliferation of the Wolffian duct and local inhibition of the Müllerian duct. The existence of an anti-Müllerian hormone secreted by the testis was deduced from this elegant experiment, since a crystal of testosterone failed to influence the adjoining Müllerian duct. (From Jost, 1947.)

mesenchymal condensation around the Müllerian duct (Josso & Picard, 1986). In fact, some epithelial cells may actually enter the mesenchymal compartment upon reorientation. Although the exact mechanism of interaction between the Müllerian epithelium and mesenchyme is not fully understood, the mesenchyme is critically involved in AMH-induced regression of these female embryonic ducts in the male. In rats, the periductal mesenchyme condenses as early as 15 days *post coitum* to form a tight fibroblastic ring around the dwindling cluster of epithelial cells (Josso & Picard, 1986), and full regression is achieved within the next 48–72 hours. In man, Müllerian regression has been described at the ultrastructural level by Wartenberg (1985), with a focus on events between weeks 9 and 13 of foetal life. Two stages are distinguished: first, prevention of duct growth by cells of a compact stromal cuff, and second, breakdown of the basal lamina.

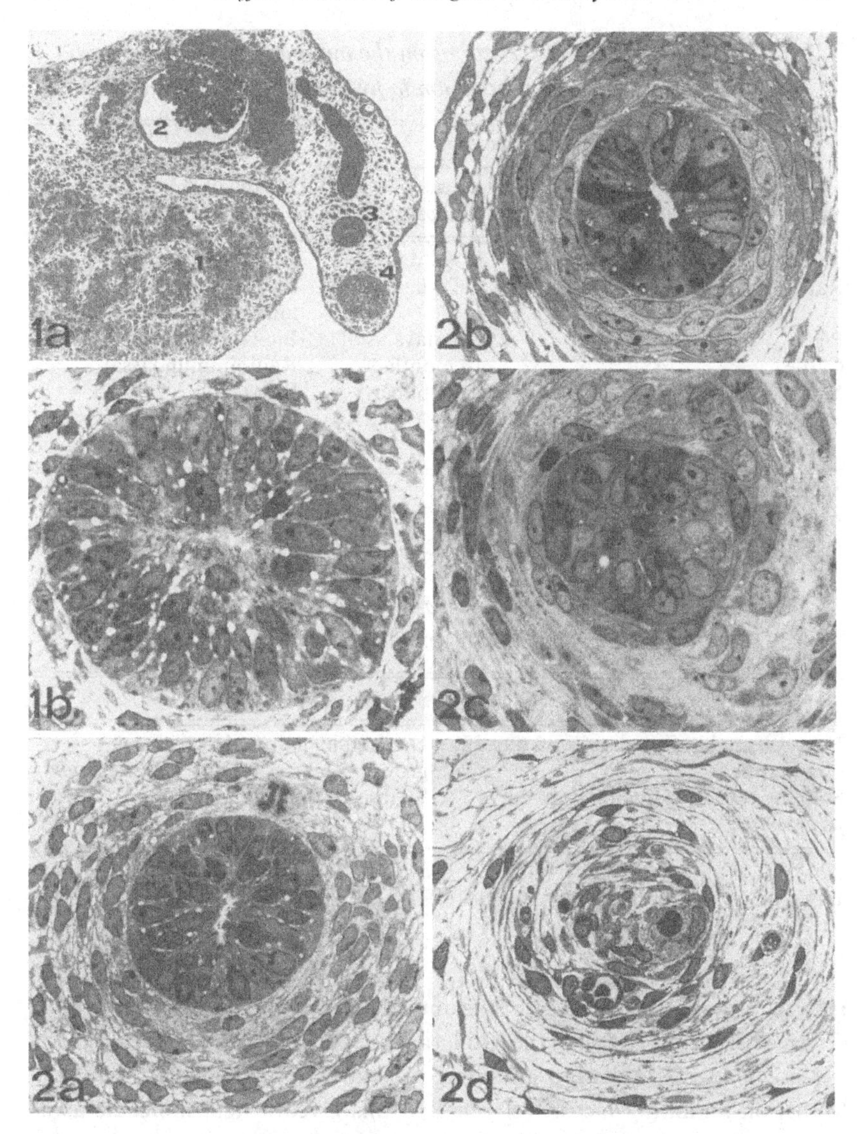

Fig. 4.4. The beautiful illustrations of Wartenberg (1985) to show a cross-section of the testis in a human embryo (*1a*), with the mesonephros, Wolffian duct and Müllerian duct numbered 2, 3 and 4, respectively. The sequence of figures (*2a–2d*) shows stages during regression of the Müllerian duct between days 53 and 91 of human embryonic development.

Table 4.1. *Summarised observations on the ontogeny of testicular anti-Müllerian hormone production in five species of mammal*

	AMH production		
Species	commences (days *post coitum*)	ceases (interval *post partum*)	Means of AMH detection
Mouse	12.5	14–21 days	*In situ* hybridisation
Rat	13.5	4–20 days	Bioassay and immunocytochemistry
Pig	25	60 days	Bioassay
Cattle	43	18 months	Radioimmunoassay
Man	49	2 years–puberty	Bioassay, immunocytochemistry and ELISA[a]

Note:
Adapted from Josso *et al.* (1993*a*).
[a] Enzyme-linked immunoabsorbent assay.

Table 4.2. *Site and timing of anti-Müllerian hormone synthesis in the mouse gonad as detected by immunocytochemical staining*

Male (XY gonad)	Female (XX gonad)
First detected in Sertoli cells on day 12 of gestation	Detectable in granulosa cells of growing follicles on day 7 *post partum*
Staining remained intense until day 4 *post partum*	Staining remained detectable

Note:
Based on observations by Taketo *et al.* (1993).

However, necrosis of cells or programmed cell death is said not to play a decisive rôle in regression of the Müllerian duct in man (Wartenberg, 1985).

Known also as Müllerian-inhibiting substance (Donahoe *et al.*, 1987) or Müllerian-inhibiting factor, reference to this glycoprotein molecule as a hormone has been criticised on the grounds that its concentration in circulating blood is seldom high enough to exert demonstrable long-distance effects in the foetus. Even so, the endocrine product can have systemic influences during organogenesis, as is revealed most convincingly in the case of bovine freemartins (Chapter 5), in which the Müllerian ducts of a genetic female co-twinned to a male in situations of placental vascular anastomosis show abnormal development. In the light of this physiological demonstration, application of the word 'hormone' does not seem unrea-

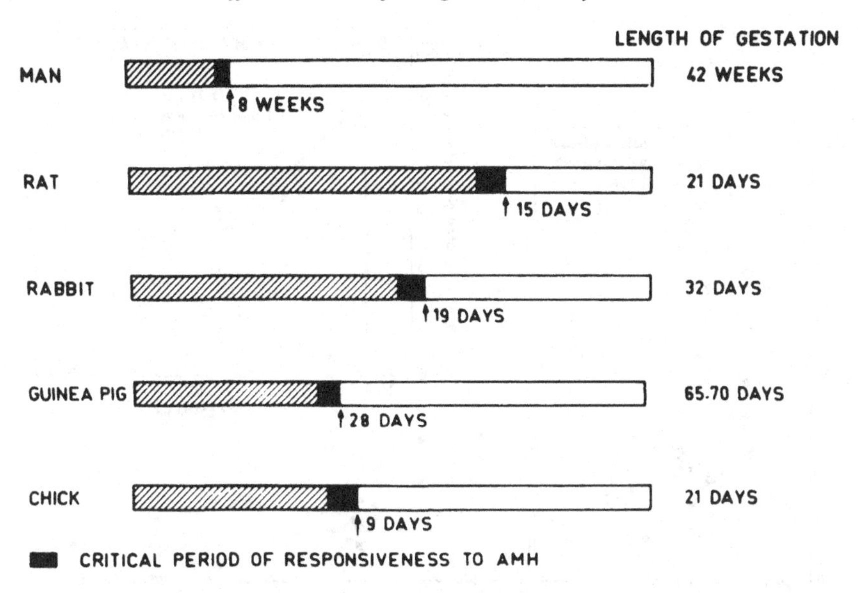

Fig. 4.5. A diagram to emphasise the limited period of sensitivity of the Müllerian ducts in male embryos to the action of anti-Müllerian hormone (AMH) in diverse animal species. The hatched zone represents the duration of the ambisexual stage, the black zone the quite precise period of responsiveness of the ducts to AMH, ending at the arrow. (Taken from Josso *et al.*, 1977.)

sonable and it has been adopted in the present work. AMH belongs to the transforming growth factor-β (TGF-β) family of growth factors and is closely related to the gonadal proteins, inhibin and activin.

AMH is synthesised by foetal Sertoli cells within the proliferating tubules (Blanchard & Josso, 1974), even though the cellular contents of the tubules are still at a relatively early stage of differentiation. Indeed, AMH is the first known molecule to be produced by the Sertoli cells (Tran, Meusy-Dessolle & Josso, 1977; Tran & Josso, 1982), being elaborated from the time of seminiferous tubule organisation through the neonatal period until the commencement of puberty, after which only traces of AMH are detectable in rete testis fluid (Josso *et al.*, 1977; Tran, Meusy-Dessolle & Josso, 1981). Its synthesis begins before the appearance of Leydig cells and has not been demonstrated to require hormonal modulation. These observations are taken to indicate that AMH expression is active in immature Sertoli cells, is probably dependent upon *SRY* (see Chapter 2), and is not totally repressed in the mature testis (Tables 4.1 and 4.2). Nonetheless, the Müllerian ducts of developing male embryos are only sensitive to AMH during a limited early period after gonadal sex differentiation (Fig. 4.5), that is at the end of

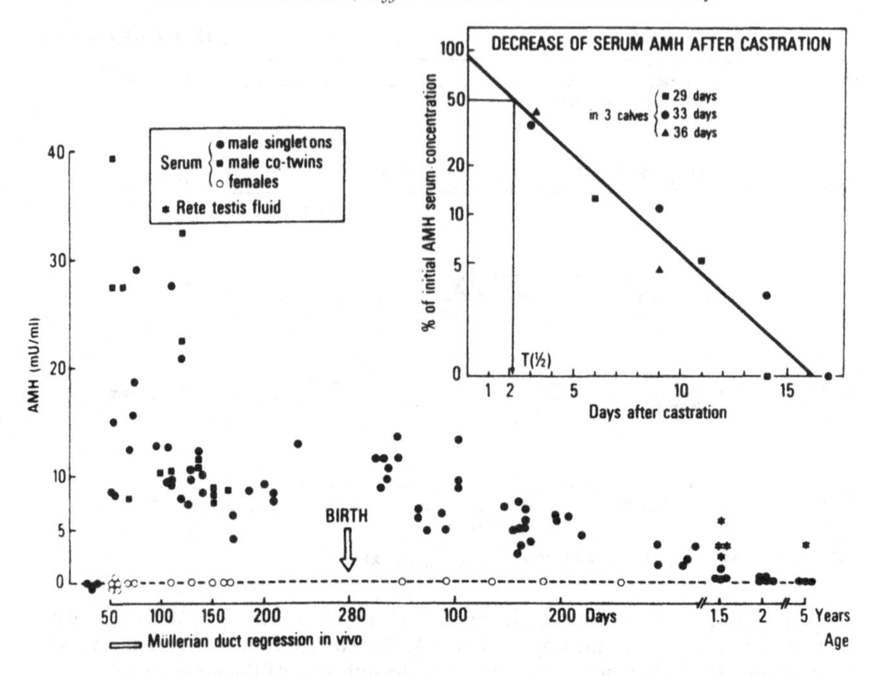

Fig. 4.6. The time-course of anti-Müllerian hormone (AMH) concentrations in calf serum and rete testis fluid, with a description in the inset diagram of the rate of clearance of AMH after castration. (Taken from Josso *et al.*, 1986.)

the ambisexual stage. Exposure before or after this critical period is ineffective, as is the process if the steroid hormone backcloth is modified by oestrogen treatment (Newbold, Suzuki & McLachlan, 1984). However, the influence of oestrogen treatment upon AMH production has yet to be reported (Josso *et al.*, 1991*a*), although it is known that AMH itself can repress cytochrome P_{450} aromatase enzyme biosynthesis (Vigier *et al.*, 1989).

In foetal rats, 24 hours of exposure to AMH are sufficient to induce irreversible Müllerian regression even though the histological lesions will only appear much later (Donahoe *et al.*, 1977*a*). But differing portions of the duct are not simultaneously or equally responsive to the influence of AMH, at least in the rat. Here, sensitivity proceeds in a caudal direction during the interval from 14 to 16 days *post coitum* (Picon, 1969; Josso *et al.*, 1977), although the caudal portion of the duct remains less sensitive to AMH (Josso & Picard, 1986). The basis of this sensitivity is unknown since, at the time of writing, a receptor molecule for AMH has not been characterised (Josso *et al.*, 1993*a*). However, apart from its inferred presence in normal development in males, the putative receptor is presumed to be expressed also in females since untimely exposure of a female foetus to AMH, as in freemartin calves or transgenic mice (Behringer *et al.*, 1990), can provoke regression of the Müllerian ducts, inhibition of aromatase biosynthesis and virilisation of the ovary; this involves killing of the germ cells and inducing formation of seminiferous cord-like structures (Vigier *et al.*, 1987; and Chapter 5). Ovine foetal ovaries exposed to AMH release testosterone instead of oestradiol due to the suppression of aromatase activity (Vigier *et al.*, 1989). The influence of AMH is permanent and irreversible (Taguchi *et al.*, 1984).

A useful hint concerning an AMH receptor comes from a paper by Baarends *et al.* (1994) in which a novel member of the transmembrane serine/threonine kinase receptor family – that includes the TGF-β receptor – was specifically expressed in mesenchymal cells adjacent to the Müllerian duct. The cDNA (C14) encoding the new family member was detected in this location on day 15 of embryonic development in both male and female rats whereas on day 16, C14 mRNA was found only in male embryos in a circular area that included the degenerating Müllerian duct.

One of the most detailed time-course studies on AMH secretion is on bovine material (Fig. 4.6), the 9-month gestation period permitting a useful chronological dissection (Josso *et al.*, 1986). AMH synthesis occurs in the foetal calf at approximately 43 days, at a time when the seminiferous tubules are differentiating. Maximum serum levels and *in vitro* production by testicular tissue are noted during the period of physiological regression of the Müllerian ducts. These two measures of AMH production decrease after 100 days, and remain stable until approximately 1 month after birth. Subsequently, AMH synthesis further decreases and cannot be estimated with accuracy by 18 months although, at that time, AMH is still measurable in rete testis fluid (Vigier *et al.*, 1983; Josso *et al.*, 1986). These same authors have also reported on the time-course of AMH secretion in human tissues, using immunocytochemical methods and a polyclonal antibody raised in rabbits. Testicular tissue was obtained following spontaneous or induced abortion. The results indicated that AMH continues to be produced by human Sertoli cells almost up to the age of puberty, rather than being lost at 2 years of age as reported by Donahoe *et al.* (1977*a*), although serum levels begin to decline as early as 2 years postnatally (Josso *et al.*, 1990). The Müllerian ducts themselves have undergone complete regression by 10 weeks of age in the human male foetus (Jirasek, 1971). In a systematic study of human foetuses using *in situ* hybridisation, AMH transcripts were detected in the testicular tissue of all specimens from 8 weeks onwards, although not in foetal ovaries or in the still undifferentiated tissue of a

7-week-old foetus shown to carry male-determining DNA sequences (Josso *et al.*, 1993*b*).

The extent to which primordial germ cells, diploid cells, may influence full Sertoli cell activity remains to be revealed although, in mice, AMH production in Sertoli cells is not dependent on interaction with germ cells (Münsterberg & Lovell-Badge, 1991). However, in the absence of germ cells, porcine Sertoli cells seem to have limited scope for AMH secretion as judged retrospectively from the extent of Müllerian duct development in one experimental model (Hunter, Cook & Baker, 1985). But AMH is apparently produced by Sertoli cells even in situations in which testis cords fail to be formed (Magre & Jost, 1984). Moreover, Josso *et al.* (1977) report no correlation between the number of surviving germ cells and AMH activity in irradiated preparations. On the other hand, AMH may be implicated in the regulation of maturation of germ cells in both the ovary and the testis. The evidence to date in this regard is tentative and not derived from strictly physiological situations (see Behringer *et al.*, 1990; Kuroda *et al.*, 1991). For example, a bovine AMH preparation can suppress germinal vesicle breakdown in the immature rat ovary, which has been taken to suggest an involvement in the control of meiosis (Takahashi, Koide & Donahoe, 1986). There is also the interesting suggestion of McLaren (1990) that one possible function of AMH in foetal testes could be to eliminate germ cells that had entered meiosis precociously despite an inhibitory effect of the testis cords. And there is the observation of Taketo *et al.* (1991) in a mouse ovotestis model that germ cells were always arrested at the prospermatogonia stage when surrounded by AMH-positive somatic cells, supporting the hypothesis that AMH could be involved in promoting differentiation of germ cells in the male direction. However, the suggestion that AMH plays some essential rôle during testicular differentiation appears difficult to sustain in the light of the observation that the testes differentiate normally in patients lacking AMH due to gene mutations (Knebelmann *et al.*, 1991; see below).

The chronology of AMH secretion stands in marked contrast in the gonads of genetic females on the basis of observations that AMH is present in ovarian follicular fluid (Vigier *et al.*, 1984) and that postnatal granulosa cells (homologous with testicular Sertoli cells, i.e. thought to differentiate from the same precursor cell lineage) also synthesise AMH, but only in the small quantities corresponding to those produced by adult Sertoli cells (Takahashi *et al.*, 1986). AMH secretion was especially prominent in the innermost layer of granulosa cells in ovine antral follicles (Bézard *et al.*, 1987, 1988), that is in relatively mature granulosa cells. But the ovaries do not secrete AMH until shortly before birth, a timing which is of course absolutely essential for normal development of the female duct system, for sensitivity of the Müllerian ducts to an influence of AMH has long since been terminated. In female mice, expression of AMH was first detected at 6 days after birth, and was again restricted to granulosa cells (Münsterberg & Lovell-Badge, 1991). In rats, AMH has been reported in both the developing and mature ovary (Ueno *et al.*, 1989*a*, *b*). So, seen in perspective, there are good correlations in both males and females between the timing of AMH secretion and a potential functional rôle in regulating genital duct development.

However, there remains the question as to why limited AMH production by granulosa cells of the mature Graafian follicle does not lead to some degree of virilisation of the ovary or indeed to damage of primary oocytes. The answer may simply be the absence of appropriate receptor molecules, the window of sensitivity of the gonads being restricted in a manner comparable with that of the Müllerian ducts; hence only the foetal ovary would be susceptible to a conspicuous phase of virilisation, as can be demonstrated *in vitro* (Vigier *et al.*, 1987, 1988). A difference in the species of AMH – that is molecular composition – generated by foetal testes and adult ovaries should also be considered. And an involvement of AMH in regulating waves of follicular development should not be discounted: AMH could be one of the factors released from a dominant Graafian follicle that would act to compromise or suppress those lower in the hierarchy of growth and development, not least by inhibiting aromatase activity.

A further point to be emphasised is that, although involved in key events of sexual differentiation in mammals, the 2.7 kilobase gene containing five exons and coding for AMH is autosomal, having been mapped to the tip of the short arm of human chromosome 19 (Cohen-Haguenauer *et al.*, 1987; Guerrier *et al.*, 1990). The gene coding for TGF-β has been mapped to the tip of the long arm of the same chromosome (Fujii *et al.*, 1986). Knebelmann *et al.* (1991) noted that the human AMH gene contains 2.8 kilobases, arranged in five exons, of which the fifth is most conserved among mammals and shows a marked homology with the TGF-β superfamily. The developmental, tissue and sex specificity of AMH expression implies that gene transcription is controlled by complex regulatory mechanisms, some elements of which are probably located upstream of the gene (Guerrier *et al.*, 1990). As would be anticipated, monoclonal antibodies to AMH will block Müllerian duct regression *in vivo* (Tran *et al.*, 1986) or *in vitro* (Vigier, Picard & Josso, 1982). Moreover, the influence of AMH is not species-specific – the molecule can be used to induce regression of the Müllerian ducts in quite disparate species. Indeed, the *in vitro* AMH bioassay uses Müllerian ducts at the undifferentiated stage from 14.5-day-old foetal rats for various mammalian preparations maintained during 3 days of culture (Picon, 1969; Donahoe *et al.*, 1977*c*; Josso *et al.*, 1977). Rat Müllerian ducts are fully regressed by 17 days of foetal life. Such organ culture assays have been partially superseded by radioimmunoassays involving monoclonal antibodies (Vigier *et al.*, 1984; Josso, 1986; Josso *et al.*, 1986; Necklaws *et al.*, 1986), by enzyme-linked immunoassay techniques (Baker, Metcalfe & Hutson, 1990; Josso *et al.*, 1990), and by a foetal ovary aromatase assay (Vigier *et al.*, 1989; di Clemente *et al.*, 1992).

Some of the most recent findings on AMH focus on the respective activities of different domains within the molecule. For example, Wilson *et al.* (1993) have shown that regression of the Müllerian ducts and inhibition of aromatase biosynthesis in the foetal ovary can be caused by a C-terminal fragment of AMH and that this fragment shares homology with TGF-β. Thus, the C-terminal epitope must contain the receptor binding domain. Both the major physiological activities of the molecule are enhanced by addition of the N-terminal portion. This augmentation of activity reflects a reassociation of the two domains of the AMH molecule (Wilson *et al.*, 1993). A probable explanation here is that the N-terminus influences the conformation of the C-terminus, being required for the correct folding and secretion of the precursor dimer. By contrast, other members of the TGF-β family seem not to require their N-terminal predomain for maintenance of biological activity.

Two other involvements of AMH may have considerable clinical significance. Tumours of Müllerian duct origin can be suppressed with highly purified recombinant human AMH (Donahoe, 1992), although there are reservations as to the extent to which AMH can be regarded as a tumour-suppressing protein (Matzuk *et al.*, 1992). And, late in foetal development, AMH can influence maturation of the lung by suppressing the accumulation of surfactant; this can be demonstrated in the prenatal and early postnatal lung (Catlin *et al.*, 1990).

The rôle of gonadal factors other than AMH in the critical events of differentiation and duct development, especially of other peptide (protein) hormones such as the various growth factors, awaits further research. Similarly, the extent to which such gonadal secretions exert an influence on the genital ducts by local diffusion gradients as distinct from circulation in the progressively elaborating vascular system of the embryo is by no means fully clear. However, of relevance is the finding that AMH is seemingly responsible for the sex reversal of freemartin (female) gonads (Vigier *et al.*,

1987), indicating – as noted above – a circulatory route for the virilising influence from the male foetus (see Chapter 5). Evidence for local diffusion gradients for AMH from the ovotestes of farm animals acting to impede elaboration of the female genital ducts is also presented in Chapter 5. Classical experiments involving culture of embryonic gonads with portions of the genital duct system (e.g. Price *et al.*, 1969) certainly furnish evidence for local diffusible influences from gonadal tissue, even though the spectrum of molecules being synthesised and secreted *in vitro* may not correspond precisely with those being liberated *in vivo*.

Worthy of note here is the finding that AMH injected into a marsupial ovary, that of the tammar wallaby, even in quite massive amounts, fails to bring about masculinisation (R. V. Short, 1992, personal communication). Whilst the timing of this experimental approach may need to be modified, it is possible that the structure of the marsupial AMH molecule differs very slightly from that of eutherians or that there is a rigorous form of genetic control over gonadal determination that is resistant to this particular endocrine intervention.

Under the influence of testosterone secretion from the Leydig cells, and possibly of its metabolite dihydrotestosterone (Blecher & Wilkinson, 1989), the component parts of the male duct system develop from the Wolffian ducts and become distinguishable as the paired and convoluted epididymides, the muscular vasa deferentia and the seminal vesicles. Wilson (1992) stresses that the Wolffian phase of male sexual development does not require dihydrotestosterone formation and, under physiological conditions, is regulated by testosterone itself. By contrast, virilisation of the urogenital sinus and external genitalia during embryogenesis and sexual maturation at puberty do require dihydrotestosterone. The supporting evidence here is that the external genitalia contain 5α-reductase, the enzyme that converts testosterone to dihydrotestosterone. Such dihydrotestosterone binds about tenfold more tightly to the androgen receptor (Fig. 4.7), thereby enabling the external genitalia to be masculinised by very low serum concentrations of testosterone. The action of dihydrotestosterone stimulates the urogenital sinus to form the prostate and prostatic urethra, the penis and scrotum (Fig. 4.8). The paired internal ducts do not contain 5α-reductase at this early stage, but their proximity to the testes may expose them to high local concentrations of testosterone (Griffin & Wilson, 1980; George & Wilson, 1988; Wilson, Griffin & Russell, 1993). Hodgins (1982) suggested that by converting testosterone and progesterone into 5α-reduced metabolites, 5α-reductaste could act to protect the male embryonic urogenital sinus and external genitalia against high concentrations of

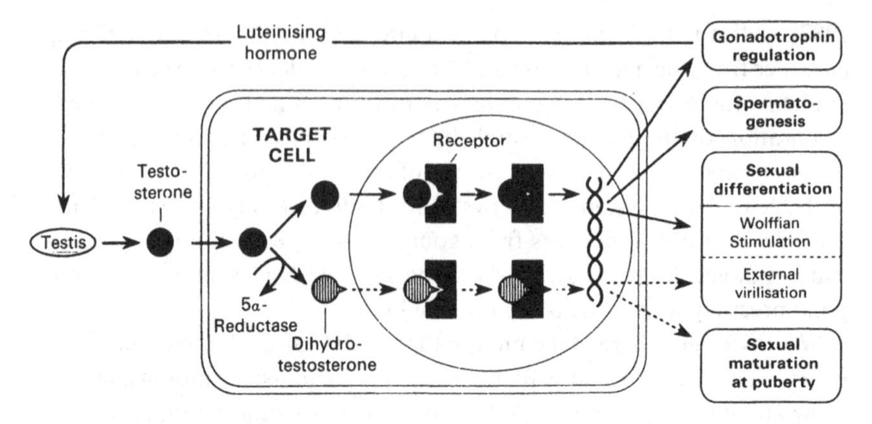

Fig. 4.7. A model of a target cell response to androgens, the major actions of which are listed to the right of the cell. Testosterone enters the cell and binds either to the androgen receptor in the nucleus or is first converted enzymatically to dihydrotestosterone. This latter hormone binds much more avidly to the nuclear receptor. The action of testosterone is indicated by continuous arrows, that of dihydrotestosterone by broken arrows. (From Wilson, 1992.)

progesterone in blood and amniotic fluid. Despite the apparent requirement for dihydrotestosterone in order to complete sexual maturation at puberty, Wilson *et al.* (1993) consider that an action of testosterone over a long period of time might bring about a considerable degree of masculinisation.

Although the timescale of virilisation of the urogenital tract of the tammar wallaby differs markedly from that in eutherian mammals, being much delayed (Table 4.3), testosterone is the androgen found in the neonatal testis and 5α-reductase activity is present in the urogenital sinus and phallus at the time of their virilisation (Renfree *et al.*, 1992). This would suggest comparable endocrine tactics in marsupials and eutherians for promoting full differentiation of the tract.

The precise manner whereby testosterone acts to promote male duct differentiation and development remains uncertain, but recent studies suggest an involvement of prostaglandins, especially PGE_2 in the process. Both anti-PGE_2 antibody and inhibitors of the arachidonic acid–PG pathway block Wolffian duct differentiation *in vitro* in the presence of testosterone, a situation that can be reversed by PGE_2 supplementation (Gupta & Bentlejewski, 1992).

The efferent ducts of the epididymis connecting with the rete testis stem from residual mesonephric tubules in the cranial portion, and provide a true excretory system for spermatozoa and their suspending fluid. Initially,

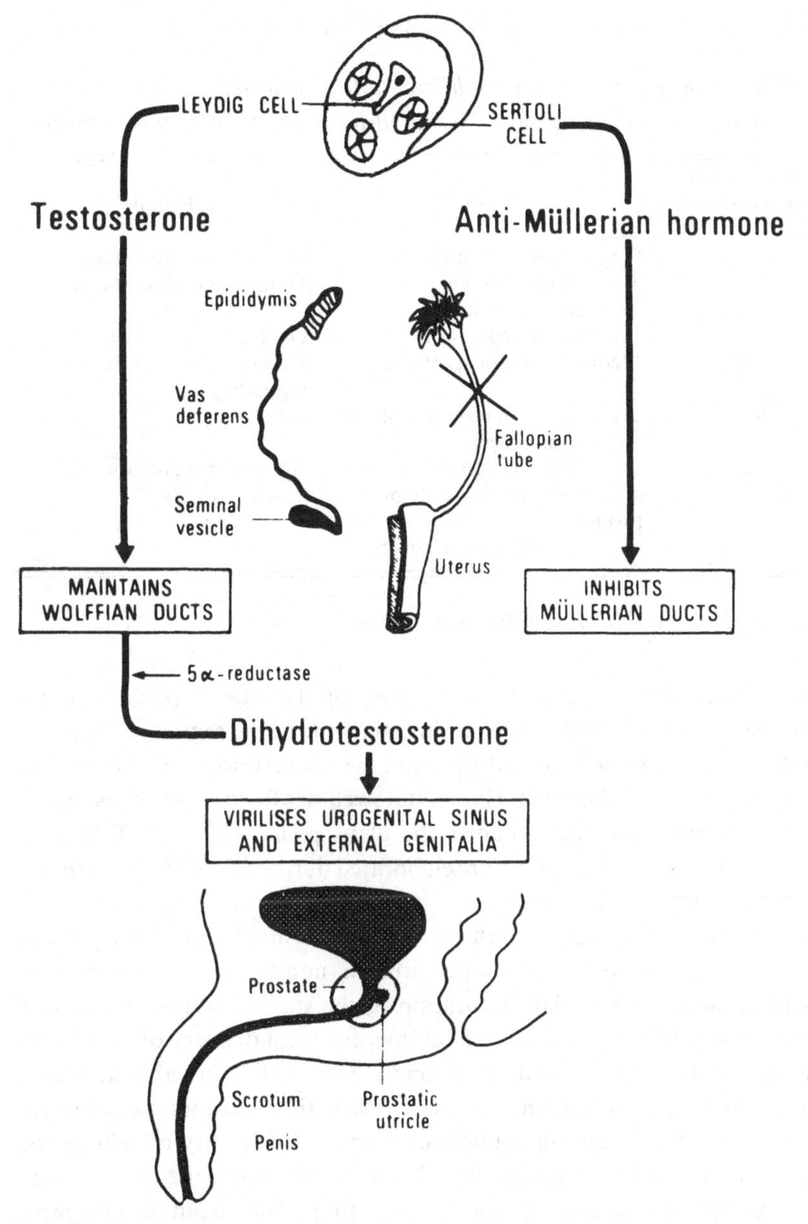

Fig. 4.8. An integrated scheme of male sexual differentiation to highlight the rôle of the two distinct hormones secreted by the foetal testis. Anti-Müllerian hormone suppresses development of the corresponding foetal duct system whereas testosterone promotes growth and differentiation of the Wolffian ducts. Conversion of testosterone to dihydrotestosterone is required for virilisation of the urogenital sinus and external genitalia. (After Josso *et al.*, 1991*a*.)

Table 4.3. *Key stages of sexual differentiation* post partum *in the tammar wallaby,* Macropus eugenii, *observed at the level of the light microscope*

Interval *post partum* (days)	Male	Female
0	Gonads undifferentiated	Gonads undifferentiated
2	Testis cords distinct	Gonads undifferentiated
7	Müllerian ducts have commenced regression	Ovarian cortex distinct
8	Medial fusion of scrotal sacs	Pouch visible; Wolffian ducts regressing
9–10	Transabdominal migration of testes commences	
25	Testes inguinal; Müllerian ducts vestigial; Prostate buds forming	Advanced regression of Wolffian ducts
50–60	Testicular descent complete	

Note:
Adapted from Hutson *et al.* (1988) and Renfree *et al.* (1992)

the seminal vesicles represent a swelling of the caudal portion of the Wolffian ducts. The paired vasa deferentia unite in the region of the pelvic urethra to give a single terminal duct, and the genital folds elongate and fuse to form the shaft of the penis. The genital swellings fuse to form the scrotum and the genital tubercle becomes the glans penis (Fig. 4.9). Whilst all components of the duct system are elaborated during foetal life, growth and rearrangement of the epididymis into an extremely long and highly convoluted duct system is perhaps the most dramatic and conspicuous change. The epididymal duct performs a number of vital functions, focusing on post-testicular maturation of the sperm nucleus and surface membranes under the influence of an incisive local delivery of androgens modulated by specific binding proteins. The epididymis also acts as a storage site for the suspension of spermatozoa that becomes progressively more concentrated during epididymal transit. A majority of cells in the cauda epididymidis will usually have lost the residual spermatid cytoplasm – the kinoplasmic or cytoplasmic droplet – that is shed from the mid-piece of the flagellum. Reviews of the male tract and especially of epididymal function have been presented by Mann (1964), Orgebin-Crist (1969), Glover & Nicander (1971), Glover (1974), Bedford (1975, 1979), Orgebin-Crist, Danzo & Davies (1975), Hammerstedt & Parks (1987), Robaire & Hermo (1988) and Bedford & Hoskins (1990).

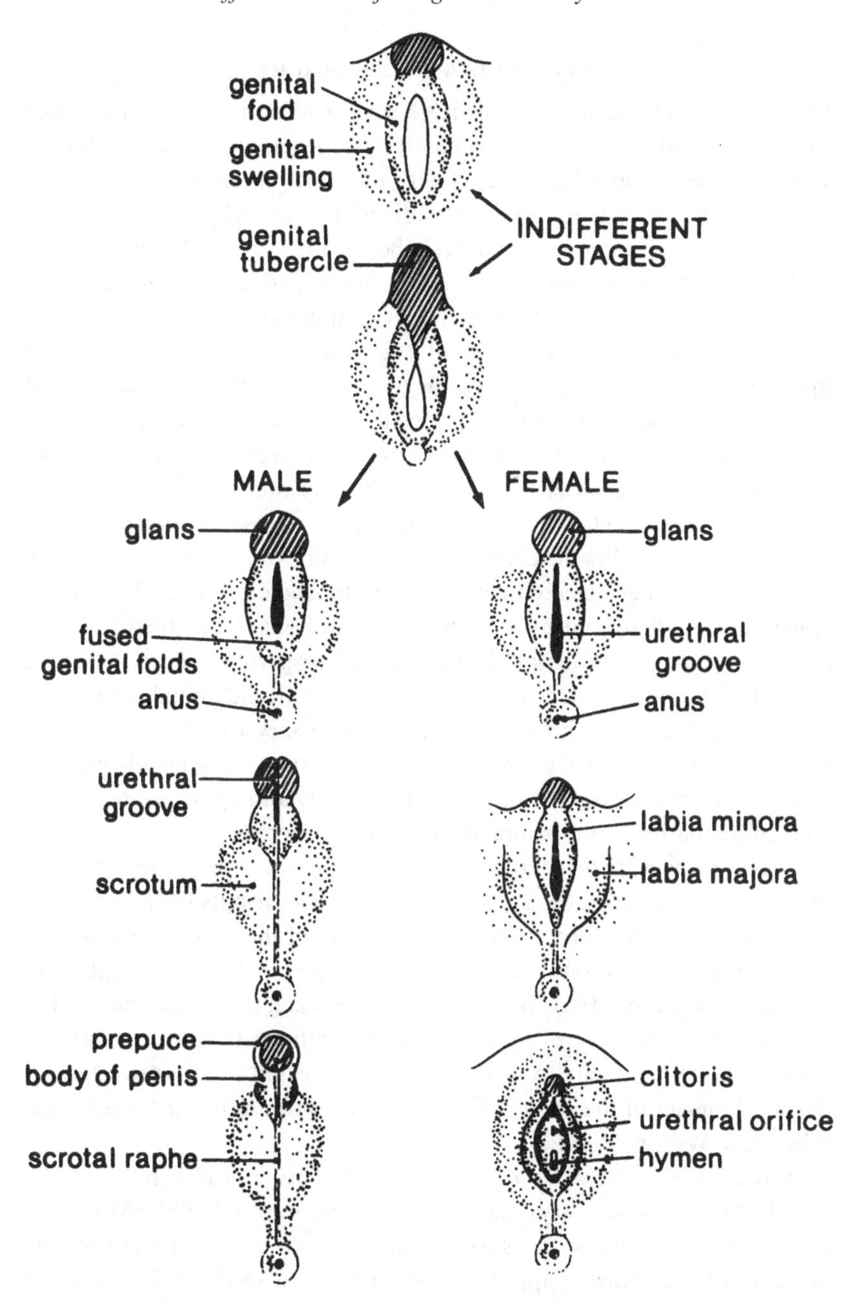

Fig. 4.9. Simplified portrayal of the development of the external genitalia from the indifferent stages until the formation of distinct male and female organs. (From George & Wilson, 1988.)

Differentiation of a female tract

In embryos under the influence of an XX sex chromosome constitution, and with the gradual elaboration of paired ovaries, development of the paramesonephric (Müllerian) ducts occurs with the concomitant degeneration of the Wolffian ducts due to lack of gonadal androgens. These primordia for the female duct system were first described in print by J. Müller (1830) in young embryos of both sexes, whereas the Wolffian ducts were named after C. F. Wolff (1733–1794) who described the mesonephros (Chapter 1). As noted above, Müllerian development is permissive in one sense at least in that it occurs in the absence of specific testicular secretions. At the time of writing, the dogma remains that it is the dearth of Y chromosome activity and the consequential lack of embryonic testicular secretion rather than the positive influence of an X chromosome that permits proliferation of the female duct system. Therefore, it is primarily the absence of androgens and of AMH that underlies development of the female ducts (Jost *et al.*, 1973), but it may no longer be completely correct to suggest – as do Price *et al.* (1969), Price (1970) and many subsequent authors – that retention and early growth of the Müllerian ducts in females, especially of the Fallopian tubes, is fully independent of maternal, placental and foetal hormones. Whilst the absence of testicular factors is necessary for elaboration and growth of the female ducts, the gonads of even the young female embryo could be playing a trophic rôle to ensure a timely and appropriate growth and differentiation of the genital ducts by secretion of peptide and complex protein hormones. Even so, the apparent inability of developing ovaries in several mammals to synthesise oestrogens endorses results from castration experiments which indicate that differentiation of the female genital tract can continue in the absence of any sustained secretion from the foetal ovary (see Jost, 1947, 1953, 1961; Burns, 1961). Elsewhere, it has been noted that in many if not most species, endocrine differentiation of the foetal ovary, as evidenced by its capacity to secrete oestrogen, occurs simultaneously with the development of the ability of the foetal testis to synthesise testosterone (George & Wilson, 1988).

Of relevance is the *in vitro* demonstration of Byskov & Hansen (1986) that the foetal mouse ovary consistently influences growth of the Müllerian duct epithelium and/or its surrounding stromal tissue. However, the timescale of this influence appears critical. Ovaries taken from 13.5-day-old foetuses stimulated the growth of stromal tissue whereas ovaries from 15.5-day-old foetuses inhibited growth of stromal tissue and differentiation of the ampulla and fimbria. The experimental approach did not permit distinction between an influence of cell contact and one of diffusible substances produced by the foetal ovary.

As is frequently noted, sexual differentiation of the gonads and then of their corresponding duct systems implies processes of organogenesis occurring in successive stages. These morphogenetic changes invariably proceed according to an accurately defined schedule of embryonic development which is characteristic for a given species (see Burns, 1961). In the rat, for example, invaginations of the coelomic epithelium, which constitute primordia for the Müllerian ducts, extend only to the caudal tip of the gonad at 14.5 days *post coitum*. During the next 48 hours, the Müllerian ducts proliferate to reach the urogenital sinus (Josso & Picard, 1986).

The component parts of the duct system that arise from persistence and development of the Müllerian ducts are the Fallopian tubes (frequently termed oviducts), the uterus, the cervix and the anterior portion of the vagina; the posterior one-third of the vagina arises from the genital tubercle as do the associated labia minora, labia majora and clitoris. Both Müllerian and Wolffian duct derivatives are thought to contribute to formation of the utero-vaginal plate, the cells of which proliferate to increase the distance between the uterus and urogenital sinus. In due course, the more central cells of this plate undergo dissolution to form the lumen of the vagina. The Fallopian tubes develop from the cephalic (cranial) end of the Müllerian ducts and remain as discrete, paired structures, which is essential if they are to interact closely with the neighbouring ovaries at the time of ovulation. Unlike that of the male, the female duct system is not joined to the gonads. The fimbriated extremities – which will embrace the ovaries – develop relatively late in foetal life, by which time the Fallopian tubes show a characteristic coiling (Price *et al.*, 1969). Varying degrees of fusion are demonstrable in the caudal portion of the ducts to give rise to the uterine tissues and those of the cervix. For example, rabbits have a duplex uterus with two cornua and two cervices, ruminants have a bipartite uterus showing considerable fusion and a large uterine body only partially subdivided by a central septum, and primates have a simplex uterus consisting of a large uterine body. The progressive modification by fusion from a duplex uterus or bicornuate uterus (pig) through the step of a bipartite uterus (cow, sheep) to that of a simplex uterus is associated with a reduction of potential litter size. The musculature of the female tract is derived from the mesenchyme surrounding the Müllerian ducts (Wilson, 1978).

The degree of prenatal differentiation and development achieved in the Müllerian derivatives may be proportional to the duration of gestation.

Thus, in species with a relatively short gestation, such as mouse, rat and hamster, differentiation of the Fallopian tubes continues after birth. Chronological and histological aspects of Müllerian duct differentiation and development have been summarised by Byskov & Hoyer (1988).

Persistent Müllerian duct syndrome

Abnormal persistence of Müllerian duct derivatives (PMDS) is a condition found in human males (Sloan & Walsh, 1976; Josso *et al.*, 1983; Picard *et al.*, 1989; Josso *et al.*, 1991*a*, *b*), and is usually revealed at surgery for inguinal hernia and/or cryptorchidism. It has sometimes been described as a rare form of inherited male pseudo-hermaphroditism, being reported in approximately 150 patients up to 1993 (Josso *et al.*, 1991*a*, 1993*a*; Knebelmann *et al.*, 1991), but it may occur more frequently than previously supposed. The syndrome is characterised by the presence of Fallopian tubes and uterus in otherwise normally virilised males, the Müllerian derivatives being drawn into the inguinal canal by the testis. Because of its association with cryptorchidism, there has been speculation that AMH may play a contributory rôle in promoting descent of the testes (Donahoe *et al.*, 1977*b*), possibly in the initial transabdominal phase of this process by means of an influence on the gubernaculum (Hutson & Donahoe, 1986; Hutson *et al.*, 1990). Direct evidence for AMH involvement in testicular descent has so far not been forthcoming although in (male) mice transgenic for AMH, Behringer *et al.* (1990) did observe failure of testicular descent. They proposed that high levels of AMH in the transgenic mice could lead to a down-regulation of the receptor on the cells that control testicular descent. However, an alternative interpretation of the cryptorchid situation in the persistent Müllerian duct syndrome is one of mechanical obstruction by the retained abdominal uterus and tubes (Josso *et al.*, 1983). In effect, the gonad is trapped by the broad ligament and the vas deferens is tightly applied to the Müllerian derivatives. If the testis does descend, it cannot do so without dragging the Müllerian derivatives into the inguinal canal.

Retention of the Müllerian primordia in genetic males may be found under circumstances of defective virilisation of the external genitalia and urogenital sinus. As discussed by Josso *et al.* (1983), this combined deficiency of androgen and AMH-dependent steps of sex differentiation would point to gonadal dysgenesis as the most probable cause of the syndrome. More rarely, persistence of Müllerian primordia occurs in otherwise normal males, and the condition is sometimes genetically transmitted in a manner suggesting either a dominant sex-limited or a recessive

X-linked disorder (Picard *et al.*, 1989). Since the *AMH* gene is not on the X chromosome (see above), three possibilities appear open to discussion (Fig. 4.10). First, PMDS is due to resistance (insensitivity) of the end organs to the action of AMH, and AMH should be demonstrable by immunocytochemistry in testicular biopsies performed before the age of 5 years (Tran *et al.*, 1987). Second, the condition is due to lack of production of AMH itself, and is expressed in heterozygous males (dominant sex-limited transmission) who have inherited an abnormal chromosome 19, probably from their mother. In this hypothesis, AMH should not be detectable in testicular tissue by immunocytochemistry. The third possibility offered by Picard *et al.* (1989) is that the persistent Müllerian duct syndrome is heterogeneous, due sometimes to lack of production of normal AMH (Guerrier *et al.*, 1989) and sometimes to a receptor defect, bearing in mind that a specific receptor has yet to be demonstrated (Josso *et al.*, 1993*a*).

As a partial response to these possibilities, Knebelmann *et al.* (1991) detected a homozygous stop mutation in the AMH gene in three brothers of a Moroccan family afflicted with PMDS involving a guanine to thymine transversion in the fifth exon, resulting in synthesis of a truncated gene product. But Knebelmann *et al.* (1991) also noted that not all cases of PMDS are linked to a defect of the *AMH* gene itself, for some patients have been demonstrated to express a normal amount of bioactive testicular AMH. Hence, in some instances, the molecular defect responsible for the syndrome in the presence of AMH may be clarified by concentrating research on the putative receptor, although failure of the foetal testis to have secreted AMH at an appropriate time cannot be excluded as a possible explanation. However, there is now sound evidence from a 3-month-old patient with PMDS that modifications of the *AMH* gene involving compound heterozygosity would explain the condition and the failure to detect AMH by enzyme-linked immunoabsorbent assay (Carré-Eusèbe *et al.*, 1992). The mutations in this patient concerned a deletion in the second exon of the maternal allele, disrupting the open reading frame, and a stop mutation in the third exon of the paternal allele. These mutations were shared with a phenotypically normal younger sister and, together with other mutations in the *AMH* gene of this family, would suggest that the gene is highly polymorphic (Carré-Eusèbe, 1992).

En passant, it is worth noting that there is a canine model of PMDS in which the condition is inherited as an autosomal recessive trait with expression limited to homozygous males (Meyers-Wallen *et al.*, 1989). A 1.8 kilobase AMH mRNA transcript was detected in the testes of PMDS-affected males and normal male embryos and neonates, and equal amounts

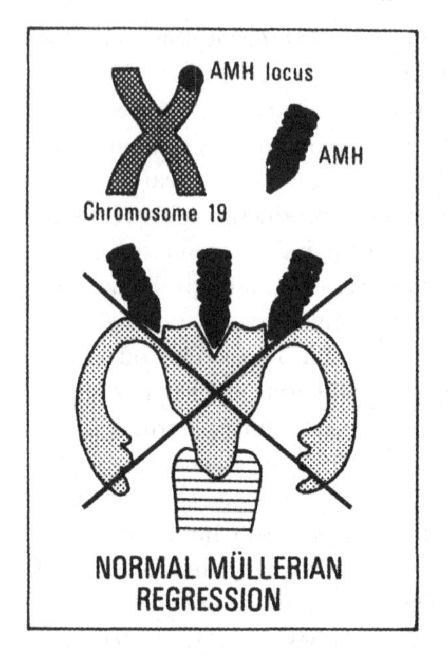

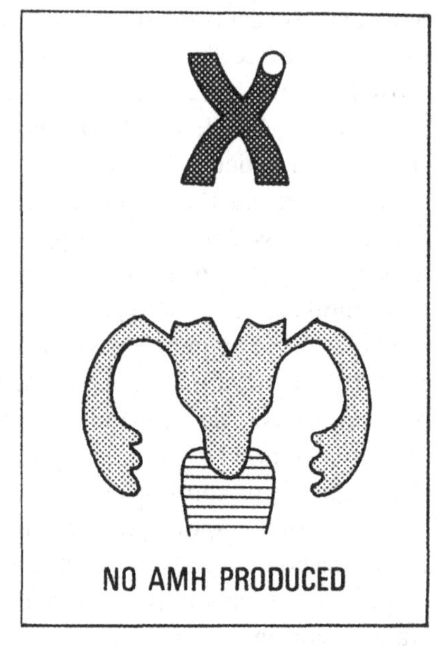

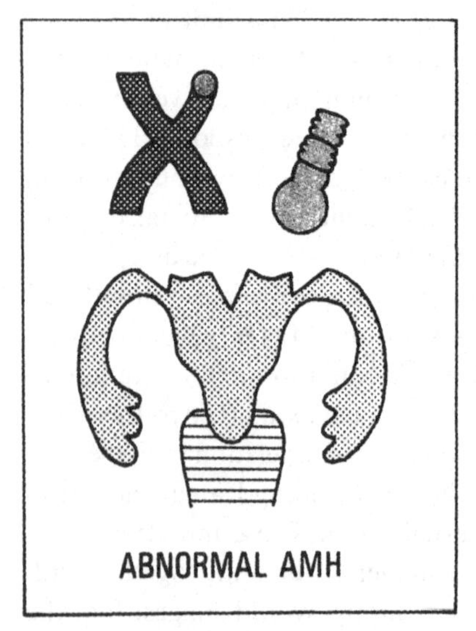

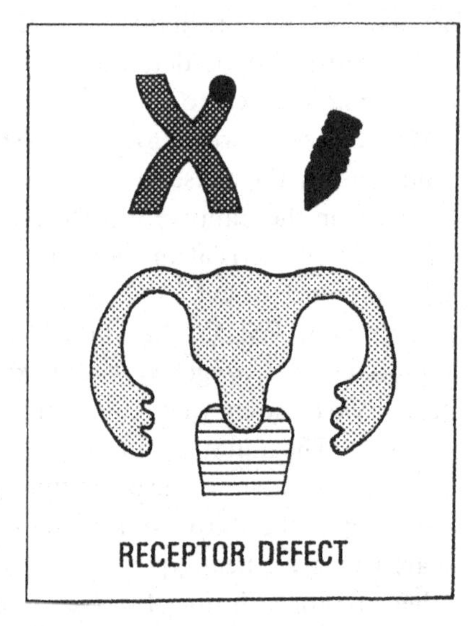

Fig. 4.10. A series of models to explain the possible causes of persistent Müllerian duct syndrome (PMDS) as compared with normal regression of the ducts as shown top left. Top right indicates an absence of anti-Müllerian hormone (AMH) due to a deletion or mutation on the relevant gene. Bottom left suggests synthesis of an anomalous AMH molecule without biological activity whereas bottom right indicates a receptor defect or even an incorrect timing of AMH secretion. (After Josso *et al.* 1991*a*.)

of AMH mRNA transcript were detected in the testes of PMDS-affected embryos and normal male litter mates during the critical period for Müllerian duct regression. Target organ resistance, such as abnormality of the putative AMH receptor, therefore seemed the most appropriate hypothesis to explain the defect (Meyers-Wallen *et al.*, 1993).

Concluding remarks

This chapter has commented on the duplication of the embryonic duct systems – that is their primordia – and the fact that gonadal information is used to instigate preferential development of one set of genital ducts; regression of the other set is the usual but not invariable consequence. A customary interpretation of such endocrine programming of mammalian duct 'selection' by gonadal secretions rather than by a more direct chromosomal influence is that it bestows flexibility. Even so, it remains unclear precisely why such flexibility should have been retained in eutherian mammals and also why expression of chromosomal sex in this regard should be a relatively late event. Of course, the very fact that both male and female duct primordia arise in the first place implies specific gene activity at a stage preceding that of gonadal differentiation.

The key rôle of androgens in male duct development has been appreciated at least since the experiments of Greene *et al.* (1939), but the credit for deducing the existence and secretion of AMH must go to Jost (1947) in Paris. More recent work on this molecule from a Parisian perspective comes from Josso's laboratory, and has contrasted the synthetic abilities of immature Sertoli cells with those of postnatal granulosa cells, highlighting the fact that the Müllerian ducts (of males) are only sensitive to AMH during a very early developmental period. Hence, the postnatal production of AMH by granulosa cells (in females) has no deleterious influence upon the genital tract.

Because the *AMH* gene is activated in male embryos very soon or perhaps almost immediately after *SRY* in the putative sex-determining cascade, the question arises as to whether AMH has major organogenesis effects quite distinct from those of Müllerian suppression. Indeed, there has been the repeated suggestion that it may contribute to normal morphological differentiation of the testis. However, the fact that a mutation leading to the synthesis of inactive AMH does not block the differentiation of functional testicular tissue indicates rather strongly that AMH itself is not specifically required for male gonadal organogenesis (Knebelmann *et al.*, 1991). Clinical instances of boys with persistent Müllerian ducts (see above) and lacking AMH, yet having well-developed cryptorchid testes, also strongly infer no essential rôle for AMH in testis formation.

References

Baarends, W. M. van Helmond, M. J. L., Post, M. *et al.* (1994). A novel member of the transmembrane serine/threonine kinase receptor family is specifically expressed in the gonads and in mesenchymal cells adjacent to the Müllerian duct. *Development*, **120**, 189–97.

Baker, M. L., Metcalfe, S. A. & Hutson, J. M. (1990). Serum levels of Müllerian inhibiting substance in boys from birth to 18 years, as determined by enzyme immunoassay. *Journal of Clinical Endocrinology and Metabolism*, **70**, 11–15.

Bedford, J. M. (1975). Maturation, transport and fate of spermatozoa in the epididymis. In *Handbook of Physiology*, vol. 5, section 7, ed. D. W. Hamilton & R. O. Greep, pp. 303–17. Washington, DC: American Physiological Society.

Bedford, J. M. (1979). Evolution of the sperm maturation and sperm storage functions of the epididymis. In *The Spermatozoon*, ed. D. W. Fawcett & J. M. Bedford, pp. 1–21. Baltimore: Urban & Schwarzenberg.

Bedford, J. M. & Hoskins, D. D. (1990). The mammalian spermatozoon: morphology, biochemistry and physiology. In *Marshall's Physiology of Reproduction*, 4th edn, vol. 2, ed. G. E. Lamming, ch. 5, pp. 379–568. Edinburgh & London: Churchill Livingstone.

Behringer, R. R., Cate, R. L., Froelick, G. J., Palmiter, R. D. & Brinster, R. L. (1990). Abnormal sexual development in transgenic mice chronically expressing Müllerian inhibiting substance. *Nature (London)*, **345**, 167–70.

Bézard, J., Vigier, B., Tran, D., Mauléon, P. & Josso, N. (1987). Immunocytochemical study of anti-Müllerian hormone in sheep ovarian follicles during foetal and post-natal development. *Journal of Reproduction and Fertility*, **80**, 509–16.

Bézard, J., Vigier, B., Tran, D., Mauléon, P. & Josso, N. (1988). Anti-Müllerian hormone in sheep follicles. *Reproduction, Nutrition, Développement*, **28**, 1105–12.

Blanchard, M. G. & Josso, N. (1974). Source of the anti-Müllerian hormone synthesized by the fetal testis: Müllerian-inhibiting activity of fetal bovine Sertoli cells in tissue culture. *Pediatric Research*, **8**, 968–71.

Blecher, S. R. & Wilkinson, L. J. (1989). Non hormone-mediated sex chromosomal effects in development: another look at the Y chromosome–testicular hormone paradigm. In *Evolutionary Mechanisms in Sex Determination*, ed. S. S. Wachtel, ch. 21, pp. 219–29. Boca Raton, Florida: CRC Press.

Burns, R. K. (1961). Role of hormones in the differentiation of sex. In *Sex and Internal Secretions*, 3rd edn, vol. 1, ed. W. C. Young & G. W. Corner, ch. 2, pp. 76–158. Baltimore: Williams & Wilkins.

Byskov, A. G. (1978). The anatomy and ultrastructure of the rete system in fetal mouse ovary. *Biology of Reproduction*, **19**, 720–35.

Byskov, A. G. (1986). Differentiation of mammalian embryonic gonad. *Physiological Reviews*, **66**, 71–117.

Byskov, A. G. & Hansen, J. L. (1986). Ovarian influence on the Müllerian duct differentiation. In *Development and Function of the Reproductive Organs*, ed.

A. Eshkol, B. Eckstein, N. Deckel, H. Peters & A. Tsafriri, pp. 85–91. Serono Symposia Review No. 11. New York: Raven Press.

Byskov, A. G. & Høyer, P. E. (1988). Embryology of mammalian gonads and ducts. In *The Physiology of Reproduction*, ed. E. Knobil, J. Neill *et al.*, ch. 8, pp. 265–302. New York: Raven Press.

Carré-Eusèbe, D., Imbeaud, S., Harbison, M., New, M. I., Josso, N. & Picard, J-Y. (1992). Variants of the anti-Müllerian hormone gene in a compound heterozygote with the persistent Müllerian duct syndrome and his family. *Human Genetics*, **90**, 389–94.

Catlin, E. A., Powell, S. M., Manganaro, T., Hudson, P., Ragin, R., Epstein, J. & Donahoe, P. K. (1990). Sex specific fetal lung development and MIS. *American Review of Respiratory Diseases*, **141**, 466–70.

di Clemente, N., Ghaffari, S., Pepinsky, R. B., Pieau, C., Josso, N., Cate, R. L. & Vigier, B. (1992). A quantitative and interspecific test for biological activity of anti-Müllerian hormone: the fetal ovary aromatase assay. *Development*, **114**, 721–7.

Cohen-Haguenauer, O., Picard, J. Y., Mattéi, M. G., Serero, S., van Cong, N., de Tand, M. F., Guerrier, D., Hors-Cayla, M. C., Josso, N. & Frézal, J. (1987). Mapping of the gene for anti-Müllerian hormone to the short arm of human chromosome 19. *Cytogenetics and Cell Genetics*, **44**, 2–6.

Didier, E. (1973). Recherches sur la morphogenèse du canal de Müller chez les oiseaux. *Wilhelm Roux Archiv für Entwicklungsmechanik der Organismen*, **172**, 271–302.

Donahoe, P. K. (1992). Müllerian inhibiting substance in reproduction and cancer. *Molecular Reproduction and Development*, **32**, 168–72.

Donahoe, P. K., Ito, Y., Morikawa, Y. & Hendren, W. H. (1977*a*). Müllerian inhibiting substance in human testes after birth. *Journal of Pediatric Surgery*, **12**, 323–30.

Donahoe, P. K., Ito, Y., Price, J. M. & Hendren, W. H. (1977*b*). Müllerian inhibiting substance activity in bovine foetal, newborn and prepubertal testes. *Biology of Reproduction*, **16**, 238–43.

Donahoe, P. K., Ito, Y., Marfatia, S. & Hendren, W. H. (1977c). A graded organ culture assay for the detection of Müllerian-inhibiting substance. *Journal of Surgical Research*, **23**, 141–8.

Donahoe, P. K., Cate, R. L., MacLaughlin, D. T., Epstein, J., Fuller, A. F., Takahashi, M., Coughlin, J. P., Ninfa, E. G. & Taylor, L. A. (1987). Müllerian inhibiting substance: gene structure and mechanism of action of a fetal regressor. *Recent Progress in Hormone Research*, **43**, 431–62.

Eicher, E. M. & Washburn, L. L. (1983). Inherited sex reversal in mice: identification of a new primary sex-determining gene. *Journal of Experimental Zoology*, **228**, 297–304.

Fujii, D., Brissenden, J. E., Derynck, R. & Francke, U. (1986). Transforming growth factor-β gene maps to human chromosome 19 long arm and to mouse chromosome 7. *Somatic Cell and Molecular Genetics*, **12**, 281–8.

George, F. W. & Wilson, J. D. (1988). Sex determination and differentiation. In *The Physiology of Reproduction*, ed. E. Knobil, J. Neill *et al.*, ch. 1, pp. 3–26. New York: Raven Press.

Glover, T. D. (1974). Recent progress in the study of male reproductive physiology. In *Reproductive Physiology*, ed. R. O. Greep, MTP International Review of Science, Physiology, Series 1, vol. 8, pp. 221–77. London: Butterworth.

Glover, T. D., D'Occhio, M. J. & Millar, R. P. (1990). Male life cycle and seasonality. In *Marshall's Physiology of Reproduction*, vol. 2, ed. G. E. Lamming, ch. 4, pp. 213–378. Edinburgh, London & New York: Churchill Livingstone.

Glover, T. D. & Nicander, L. (1971). Some aspects of structure and function in the mammalian epididymis. *Journal of Reproduction and Fertility Supplement*, **13**, 39–50.

Greene, R. R., Burrill, M. W. & Ivy, A. C. (1939). Experimental intersexuality. The effect of antenatal androgens on sexual development of female rats. *American Journal of Anatomy*, **65**, 415–69.

Griffin, J. E. & Wilson, J. D. (1980). The syndromes of androgen resistance. *New England Journal of Medicine*, **302**, 198–209.

Gruenwald, P. (1941). The relation of the growing Müllerian duct to the Wolffian duct and its importance for the genesis of malformations. *Anatomical Record*, **81**, 1–19.

Guerrier, D., Boussin, L., Mader, S., Josso, N., Kahn, A. & Picard, J-Y. (1990). Expression of the gene for anti-Müllerian hormone. *Journal of Reproduction and Fertility*, **88**, 695–706.

Guerrier, D., Tran, D., Vanderwinden, J. M., Hideux, S., van Outryve, L., Legeai, L., Bouchard, M., van Vliet, G., de Laet, M. H., Picard, J. Y., Kahn, A. & Josso, N. (1989). The persistent Müllerian duct syndrome: a molecular approach. *Journal of Clinical Endocrinology and Metabolism*, **68**, 46–52.

Gupta, C. & Bentlejewski, C. A. (1992). Role of prostaglandins in the testosterone-dependent Wolffian duct differentiation of the fetal mouse. *Biology of Reproduction*, **47**, 1151–60.

Hammerstedt, R. H. & Parks, J. E. (1987). Changes in sperm surfaces associated with epididymal transit. *Journal of Reproduction and Fertility*, **34** (Supplement 1), 133–49.

Hodgins, M. B. (1982). Binding of androgens in 5α-reductase-deficient human genital skin fibroblasts: inhibition by progesterone and its metabolites. *Journal of Endocrinology*, **94**, 415–27.

Hunter, R. H. (1930). Observations on the development of the human female genital tract. *Contributions to Embryology*, **22**, No. 129, 91–107.

Hunter, R. H. F., Cook, B. & Baker, T. G. (1985). Intersexuality in five pigs, with particular reference to oestrous cycles, the ovotestis, steroid hormone secretion and potential fertility. *Journal of Endocrinology*, **106**, 233–42.

Hutson, J. M. & Donahoe, P. K. (1986). The hormonal control of testicular descent. *Endocrine Reviews*, **7**, 270–83.

Hutson, J. M., Shaw, G., O, W. S., Short, R. V. & Renfree, M. B. (1988). Müllerian inhibiting substance production and testicular migration and descent in the pouch young of a marsupial. *Development*, **104**, 549–56.

Hutson, J. M., Williams, M. P. L., Fallat, M. E. & Attah, A. (1990). Testicular descent: new insights into its hormonal control. *Oxford Reviews of Reproductive Biology*, **12**, 1–56.

Jirasek, J. E. (1971). *Development of the Genital System and Male Pseudohermaphroditism*. Baltimore: Johns Hopkins Press.

Josso, N. (1981). Differentiation of the genital tract: stimulators and inhibitors. In *Mechanisms of Sex Differentiation in Animals and Man*, ed. C. R. Austin & R. G. Edwards, pp. 165–203. London & New York: Academic Press.

Josso, N. (1986). Anti-Müllerian hormone. *Clinical Endocrinology (Oxford)*, **25**, 331–345.

Josso, N., Boussin, L., Knebelmann, B., Fekete, C. N. & Picard, J. Y. (1991*a*). Anti-Müllerian hormone and intersex states. *Trends in Endocrinology and Metabolism*, **2**, 227–33.
Josso, N., Cate, R. L., Picard, J-Y., Vigier, B., di Clemente, N., Wilson, C., Imbeaud, R., Pepinsky, R. B., Guerrier, D., Boussin, L., Legeai, L. & Carré-Eusèbe, D. (1993*a*). Anti-Müllerian hormone: the Jost factor, *Recent Progress in Hormone Research*, **48**, 1–59.
Josso, N., Fekete, C., Cachin, O., Nezelof, C. & Rappaport, R. (1983). Persistence of Müllerian ducts in male pseudohermaphroditism, and its relationship to cryptorchidism. *Clinical Endocrinology*, **19**, 247–58.
Josso, N., Lamarre, I., Picard, J. Y., Berta, P. Davies, N., Morichon, N., Peschanski, M. & Jeny, R. (1993*b*). Anti-Müllerian hormone in early human development. *Early Human Development*, **33**, 91–9.
Josso, N., Legeai, L., Forest, M. G., Chaussain, J. L. & Brauner, R. (1990). An enzyme-linked immunoassay for anti-Müllerian hormone: a new tool for the evaluation of testicular function in infants and children. *Journal of Clinical Endocrinology and Metabolism*, **70**, 23–7.
Josso, N., Picard, J. Y. & Tran, D. (1977). The anti-Müllerian hormone. *Recent Progress in Hormone Research*, **33**, 117–63.
Josso, N. & Picard, J.-Y. (1986). Anti-Müllerian hormone. *Physiological Reviews*, **66**, 1038–90.
Josso, N., Tran, D., Picard, J. Y. & Vigier, B. (1986). Physiology of anti-Müllerian hormone: in search of a new role for an old hormone. In *Development and Function of the Reproductive Organs*, ed. A. Eshkol, B. Eckstein, N. Deckel, H. Peters & A. Tsafriri, Serono Symposia Review No. 11, pp. 73–84. New York: Raven Press.
Josso, N., Vigier, B., Magre, S. & Picard, J.-Y. (1991*b*). Anti-Müllerian hormone and gonadal development. *Seminars in Developmental Biology*, **2**, 285–91.
Jost, A. (1947). Recherches sur la différenciation sexuelle de l'embryon de lapin. III. Rôle des gonades foetales dans la différenciation sexuelle somatique. *Archives d'AnatomieMicroscopique et de Morphologie Expérimentale*, **36**, 271–315.
Jost, A. (1953). Problems of fetal endocrinology: the gonadal and hypophyseal hormones. *Recent Progress in Hormone Research*, **8**, 379–413.
Jost, A. (1961). The rôle of fetal hormones in prenatal development. *Harvey Lecture Series*, **55**, 201–26.
Jost, A. (1970). Hormonal factors in the sex differentiation of the mammalian foetus. *Philosophical Transactions of the Royal Society of London, Series B*, **259**, 119–30.
Jost, A., Vigier, B., Prépin, J. & Perchellet, J. P. (1973). Studies on sex differentiation in mammals. *Recent Progress in Hormone Research*, **29**, 1–35.
Knebelmann, B., Boussin, L., Guerrier, D., Legeai, L., Kahn, A., Josso, N. & Picard, J.-Y. (1991). Anti-Müllerian hormone Bruxelles: a nonsense mutation associated with the persistent Müllerian duct syndrome. *Proceedings of the National Academy of Sciences, USA*, **88**, 3767–71.
Kuroda, T., Lee, M. M., Ragin, R. C., Hirobe, S. & Donahoe, P. K. (1991). Müllerian inhibiting substance production and cleavage is modulated by gonadotropins and steroids. *Endocrinology*, **129**, 2985–92.
McLaren, A. (1990). Of MIS and the mouse. *Nature (London)*, **345**, 111.
Magre, S. & Jost, A. (1984). Dissociation between testicular organogenesis and endocrine cytodifferentiation of Sertoli cells. *Proceedings of the National Academy of Sciences, USA*, **81**, 7831–4.

Mann, T. (1964). *The Biochemistry of Semen and of the Male Reproductive Tract.* London: Methuen.
Matzuk, M. M., Finegold, M. J., Su, J. G. J., Hsueh, A. J. W. & Bradley, A. (1992). α-inhibin is a tumour-suppressor gene with gonadal specificity in mice. *Nature (London)*, **360**, 313–19.
Meyers-Wallen, V. N., Donahoe, P. K., Ueno, S., Manganaro, T. F. & Patterson, D. F. (1989). Müllerian inhibiting substance is present in testes of dogs with Persistent Müllerian Duct Syndrome. *Biology of Reproduction*, **41**, 881–8.
Meyers-Wallen, V. N., Lee, M. M., Manganaro, T. F., Kuroda, T., MacLauchlin, D. & Donahoe, P. K. (1993). Müllerian inhibiting substance is present in embryonic testes of dogs with Persistent Müllerian Duct Syndrome. *Biology of Reproduction*, **48**, 1410–18.
Müller, J. (1830). 'Bildungsgeschichte der Genitalien.' Düsseldorf: Arnz.
Münsterberg, A. & Lovell-Badge, R. (1991). Expression of the mouse anti-Müllerian hormone gene suggests a role in both male and female sexual differentiation. *Development*, **113**, 613–24.
Necklaws, E. D., la Quaglia, M. P., MacLaughlin, D. T., Hudson, P., Mudgett-Hunter, M. & Donahoe, P. K. (1986). Detection of Müllerian inhibiting substance in biological samples by a solid phase sandwich radioimmunoassay. *Endocrinology*, **118**, 791–6.
Newbold, R. R., Suzuki, Y. & McLachlan, J. A. (1984). Müllerian duct maintenance in heterotypic organ culture after in vivo exposure to diethylstilboestrol. *Endocrinology*, **115**, 1863–8.
Orgebin-Crist, M. C. (1969). Studies on the function of the epididymis. *Biology of Reproduction*, **1**, 155–75.
Orgebin-Crist, M. C., Danzo, B. J. & Davies, J. (1975). Endocrine control of development and maintenance of sperm fertilizing ability in the epididymis. In *Handbook of Physiology*, vol. 5, sect. 7, ed. D. W. Hamilton & R. O. Greep, ch. 15, pp. 319–38. Washington, DC: American Physiological Society.
Patten, B. M. (1948). *Embryology of the Pig*, 3rd edn. New York & Toronto: McGraw-Hill.
Picard, J.-Y., Benarous, R., Guerrier, D., Josso, N. & Kahn, A. (1986). Cloning and expression of cDNA for anti-Müllerian hormone. *Proceedings of the National Academy of Sciences, USA*, **83**, 5464–8.
Picard, J.-Y., Guerrier, D., Kahn, A. & Josso, N. (1989). Molecular biology of anti-Müllerian hormone. In *Evolutionary Mechanisms in Sex Determination*, ed. S. S. Wachtel, pp. 209–17. Boca Raton, Florida: CRC Press.
Picon, R. (1969). Action du testicule foetal sur le développement *in vitro* des canaux de Müller chez le rat. *Archives d'Anatomie Microscopique et de Morphologie Expérimentale*, **58**, 1–19.
Price, D. (1970). *In vitro* studies on differentiation of the reproductive tract. *Philosophical Transactions of the Royal Society of London, Series B*, **259**, 133–9.
Price, D., Zaaijer, J. J. P. & Ortiz, E. (1969). Prenatal development of the oviduct *in vivo* and *in vitro*. In *The Mammalian Oviduct*, ed. E. S. E. Hafez & R. J. Blandau, pp. 29–46. Chicago: University of Chicago Press.
Renfree, M. B., Wilson, J. D., Short, R. V., Shaw, G. & George, F. W. (1992). Steroid hormone content of the gonads of the tammar wallaby during sexual differentiation. *Biology of Reproduction*, **47**, 644–7.
Robaire, B. & Hermo, L. (1988). Efferent ducts, epididymis and vas deferens:

structure, functions and their regulation. In *The Physiology of Reproduction*, ed. E. Knobil & J. D. Neill, pp. 999–1080. New York: Raven Press.
Short, R. V. (1982). Sex determination and differentiation. In *Reproduction in Mammals*, 2nd edn, vol. 2, ed. C. R. Austin & R. V. Short, pp. 70–113. Cambridge & London: Cambridge University Press.
Sloan, W. R. & Walsh, P. C. (1976). Familial persistent Müllerian duct syndrome. *Journal of Urology*, **115**, 459.
Taguchi, O., Cunha, G. R., Lawrence, W. D. Robboy, S. J. (1984). Timing and irreversibility of Müllerian duct inhibition in the embryonic reproductive tract of the human male. *Developmental Biology*, **106**, 394–8.
Takahashi, M., Koide, S. S. & Donahoe, P. K. (1986). Müllerian inhibiting substance as an oocyte meiosis inhibitor. *Molecular and Cellular Endocrinology*, **47**, 225–34.
Takahashi, M., Hayashi, M., Manganaro, T. F. & Donahoe, P. K. (1986). The ontogeny of Müllerian inhibiting substance in granulosa cells of the bovine ovarian follicle. *Biology of Reproduction*, **35**, 447–53.
Taketo, T., Saeed, J., Nishioka, Y. & Donahoe, P. K. (1991). Delay of testicular differentiation in the B6.Y^{DOM} ovotestis demonstrated by immunocytochemical staining for Müllerian inhibiting substance. *Developmental Biology*, **146**, 386–95.
Taketo, T., Saeed, J., Manganaro, T., Takahashi, M. & Donahoe, P. K. (1993). Müllerian inhibiting substance production associated with loss of oocytes and testicular differentiation in the transplanted mouse XX gonadal primordium. *Biology of Reproduction*, **49**, 13–23.
Tran, D. & Josso, N. (1982). Localization of anti-Müllerian hormone in the rough endoplasmic reticulum of the developing bovine Sertoli cell using immunocytochemistry with a monoclonal antibody. *Endocrinology*, **111**, 1562–7.
Tran, D., Meusy-Dessolle, N. & Josso, N. (1977). Anti-Müllerian hormone is a functional marker of foetal Sertoli cells. *Nature (London)*, **269**, 411–12.
Tran, D., Meusy-Dessolle, N. & Josso, N. (1981). Waning of anti-Müllerian activity: an early sign of Sertoli cell maturation in the developing pig. *Biology of Reproduction*, **24**, 923–31.
Tran, D., Picard, J. Y., Vigier, B., Berger, R. & Josso, N. (1986). Persistence of Müllerian ducts in male rabbits passively immunized against bovine anti-Müllerian hormone during foetal life. *Developmental Biology*, **116**, 160–7.
Tran, D., Picard, J. Y., Campargue, J. & Josso, N. (1987). Immunocytochemical detection of anti-Müllerian hormone in Sertoli cells of various mammalian species, including man. *Journal of Histochemistry and Cytochemistry*, **35**, 733–43.
Ueno, S., Kuroda, T., MacLaughlin, D. T., Ragin, R. C., Manganaro, T. F. & Donahoe, P, K. (1989*a*). Müllerian inhibiting substance in the adult rat ovary during various stages of the estrous cycle. *Endocrinology*, **125**, 1060–6.
Ueno, S., Takahashi, M., Manganaro, T. F., Ragin, R. C. & Donahoe, P. K. (1989*b*). Cellular localisation of Müllerian inhibiting substance in the developing rat ovary. *Endocrinology*, **124**, 1000–6.
Vigier, B., Picard, J.-Y. & Josso, N (1982). A monoclonal antibody against bovine anti-Müllerian hormone. *Endocrinology*, **110**, 131–7.
Vigier, B., Picard, J. Y., Tran, D., Legeai, L. & Josso, N. (1984). Production of anti-Müllerian hormone: another homology between Sertoli and granulosa cells. *Endocrinology*, **114**, 1315–20.
Vigier, B., Tran, D., du Mesnil du Buisson, F., Heyman, Y. & Josso, N. (1983).

Use of monoclonal antibody techniques to study the ontogeny of bovine anti-Müllerian hormone. *Journal of Reproduction and Fertility*, **69**, 207–14.

Vigier, B., Watrin, F., Magre, S., Tran, D. & Josso, N. (1987). Purified bovine AMH induces a characteristic freemartin effect in fetal rat prospective ovaries exposed to it *in vitro*. *Development*, **100**, 43–55.

Vigier, B., Watrin, F., Magre, S., Tran, D., Garrigou, O., Forest, M. G. & Josso, N. (1988). Anti-Müllerian hormone and freemartinism: inhibition of germ cell development and induction of seminiferous cord-like structures in rat foetal ovaries exposed *in vitro* to purified bovine AMH. *Reproduction, Nutrition, Développpement*, **28**, 113–28.

Vigier, B., Forest, M. G., Eychenne, B., Bézard, J., Garrigou, O., Robel, P. & Josso, N. (1989). Anti-Müllerian hormone produces endocrine sex-reversal of fetal ovaries. *Proceedings of the National Academy of Sciencies, USA*, **86**, 3684–8.

Wartenberg, H. (1985). Morphological studies on the role of the periductal stroma in the regression of the human male Müllerian duct. *Anatomy and Embryology*, **171**, 311–23.

Wilson, C. A., di Clemente, N., Ehrenfels, C., Pepinsky, R. B., Josso, N., Vigier, B. & Cate, R. (1993). Müllerian inhibiting substance requires its N-terminal domain for maintenance of biological activity, a novel finding within the transforming growth factor-β superfamily. *Molecular Endocrinology*, **7**, 247–57.

Wilson, J. D. (1978). Sexual differentiation. *Annual Review of Physiology*, **40**, 279–306.

Wilson, J. D. (1992). Syndromes of androgen resistance. *Biology of Reproduction*, **46**, 168–173.

Wilson, J. D., Griffin, J. E. & Russell, D. W. (1993). Steroid 5α-reductase 2 deficiency. *Endocrine Reviews*, **14**, 577–93.

5

Anomalous sexual development in domestic species

Introduction

Anomalous sexual development in domestic farm animals has been recognised since ancient times, and overtly bizarre conditions appear to have enjoyed special significance in the fertility rites of various societies. Indeed, what might be regarded as selective breeding programmes were even imposed to generate an increased incidence of such animals on islands of the New Hebrides (see Baker, 1925). Because anatomical and behavioural abnormalities have attracted attention for many years, quite detailed reports exist for diverse conditions in cattle (Marcum, 1974), sheep (Bruere & Macnab, 1968), goats (Hamerton *et al.*, 1969) and pigs (Crew, 1924; Baker, 1925; Brambell, 1929; Breeuwsma, 1970). In a majority of these reports, the emphasis was descriptive rather than analytical, well illustrated as long ago as 1779 when John Hunter was drawing attention to the problem of freemartinism in cattle. More recent studies have progressed from morphological and histological observations to karyotyping and chromosome banding studies in a search for possible genetic lesions underlying the aetiology of sexual abnormalities. However, a molecular approach involving, for example, the use of probes for Y-related DNA sequences is still in its relative infancy in farm animals, even though such probes are becoming available in domestic species for the purpose of monitoring the success of treatments to select or predetermine the sex of embryos prior to transplantation (Kirkpatrick & Monson, 1993; Machaty *et al.*, 1993).

The anomalous development referred to in the first sentence of this introduction invariably concerns anatomical aberrations of a female reproductive tract and gonads that have been taken to reflect an intersex condition (reviews by Biggers & McFeely, 1966; Biggers, 1968; Short, 1969; Bishop, 1972). There is especially useful information for farm animals in which apparent infertility or sterility has led to the identification of structural anomalies before slaughter (Deas *et al.*, 1979). Traditionally, this has been by the technique of rectal palpation in cattle to explore the genital structures, and more recently by means of ultrasonic scanning, laparoscopy or even laparotomy. Such identification of anomalies has permitted experimental studies on animals of known gonadal morphology (e.g. Hamerton *et al.*, 1969; Hunter, Cook & Baker, 1985; Chalmers *et al.*, 1989).

Definition of sexual anomalies

An intersex animal shows one or more anatomical features of both sexes (Short, 1969). For present purposes and based primarily upon the author's own studies (Hunter, Baker & Cook, 1982; Hunter *et al.*, 1985; Hunter, Chalmers & Cavazos, 1988), a working definition of intersex animals would regard them grossly as genetic females (XX sex chromosome constitution) with at least one ovotestis or testis-like structure. In more extreme circumstances, both gonads may be testis-like structures, and one or both may assume an inguinal or a scrotal location. There are macroscopically visible disturbances to the neighbouring Müllerian duct system involving one or both Fallopian tubes but seldom extending to the uterus. Development of a proximal portion of the Wolffian duct as an epididymal-like structure is a frequent characteristic, as is a prominent well-developed clitoris. Behavioural traits may include varying degrees of aggressive behaviour, and 'chomping' of the jaws in pigs associated with production of a frothy saliva, a much described activity in mature boars.

The word 'hermaphrodite' stems from Greek mythology in which Hermaphroditus, the son of Hermes and Aphrodite, merged with the nymph Salmacis to form one body with male and female attributes (Chapter 1). The term hermaphrodite has thus become one to describe a creature possessing the reproductive organs of both sexes. Strictly speaking, 'hermaphrodite' should be applied only in circumstances in which ovarian and testicular tissue are present and both eggs and spermatozoa can be produced in the same individual; in other words, a true hermaphrodite. In theory, self-fertilisation is a logical and possible consequence. In the literature dealing with farm animals, 'hermaphrodite' has been somewhat more loosely applied to cases in which both testicular and ovarian tissues are detected in the same animal, even if only one tissue can produce gametes. By contrast, terms such as pseudohermaphrodite have been applied when there are gonads of one sex and parts of the genitalia of the other. A male pseudohermaphrodite is an animal in which the external appearance is female but two testes are present. Conversely, a female pseudohermaphrodite represents an animal with two ovaries but a phenotype perceived as male. Hence, this form of classification is based primarily on the type of gonadal tissue. There are shortcomings in the use of these labels, not the least of which is that they have generally been applicable only retrospectively, that is when the animal has been slaughtered and both gonads subjected to histological examination. Moreover, because a broad spectrum of gonadal anomalies can be unearthed in domestic farm animals, especially in circumstances of inbreeding, the term 'intersex' may be viewed as preferable and is favoured by the author to cover all forms of morphological sexual ambiguity at both gonadal and genital tract levels. Intersexes would therefore embrace a true hermaphrodite. Freemartins, which are certainly not hermaphrodites, are categorised in the section that follows.

Freemartinism and intersexuality in cattle

Although diverse forms of cystic ovarian disease have frequently been described, and likewise various anomalies of the Fallopian tubes and/or uterus (see Laing, 1979; Arthur, Noakes & Pearson, 1983), perhaps the best known derangement syndrome of the reproductive system in cows is that of freemartinism – a genetically female foetus masculinised in the presence of a male co-twin, giving rise to a sterile heifer calf in instances of placental fusion. 'Best known' rather than 'best understood' abnormality is intentional phrasing here, for it is only in recent years that the aetiology of the freemartin condition has received a seemingly adequate explanation. And yet, as noted in the introduction, it has been a topic of scientific interest at least since the publication of Hunter (1779). Whilst this early work receives customary homage in modern writings, it required the enthusiasm and perception of a clinically trained reproductive biologist (R. V. Short) to realise that the classical drawing in Hunter (1779) of 'Mr Wright's free martin' (Fig. 5.1) was probably not of a freemartin at all! Rather, the condition may have been one of testicular feminisation, as first suggested in the work of Nes (1966). A further shortcoming in John Hunter's report is that it remains unknown whether the animal portrayed was co-twinned to a bull calf (Short, 1969). Even so, to maintain some sense of proportion, Hunter was writing at least 10 years before the French Revolution, so there would be no reason to have expected the scientific rigour of the late twentieth century.

Freemartinism has been found most widely in cattle (Keller & Tandler, 1916; Lillie, 1916, 1917; Chapin, 1917), but has also been noted in sheep and

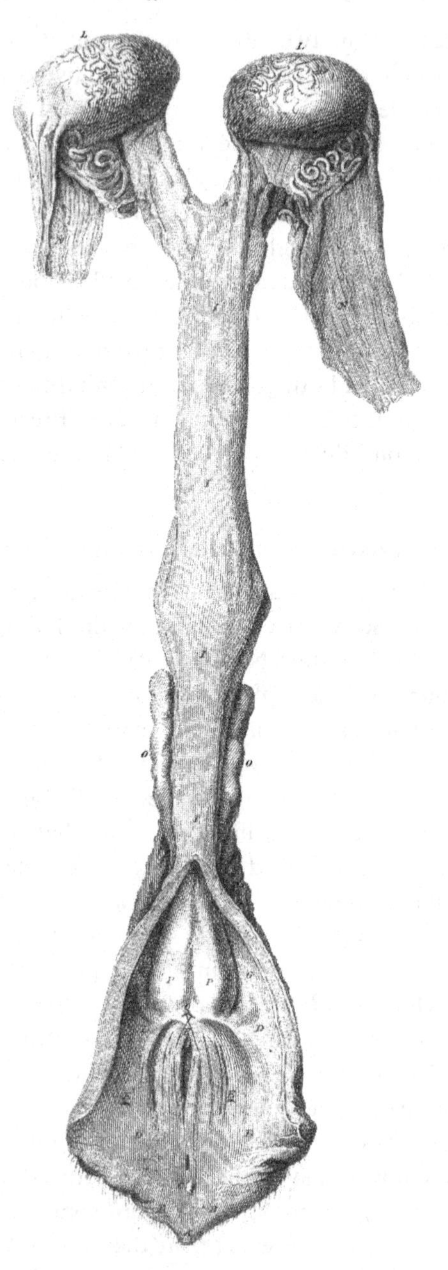

Fig. 5.1. The wonderfully precise drawing by William Bell of the gonads and reproductive tract of 'Mr Wright's free martin' as recorded by John Hunter (1779). In fact, as noted in the present text, this specimen probably represents a situation of testicular feminisation. (Courtesy of Glasgow University Library.)

Table 5.1. *Incidence of freemartinism among heterosexual twins in cattle in a series of 13 separate surveys, each conducted by different authors*

Total examined (no.)	Sample of animals morphologically normal (no.)	Freemartins No.	Freemartins %	Basis for the diagnosis
91	6	85	93.4	Foetal anatomy
24	3	21	87.5	Foetal anatomy
15	3	12	80.0	Foetal anatomy
36	1	35	97.2	Gross anatomy
139	8	131	94.2	Not clear
17	2	15	88.2	Gross anatomy
74	8	66	89.2	Blood typing
37	3	34	91.9	Blood typing
27	2	25	92.6	Homograft tolerance
30	3	27	90.0	XX/XY chimaerism
15	2	13	86.7	XX/XY and red cell chimaerism
8	1	7	87.5	XX/XY chimaerism and foetal anatomy
19	1	18	94.7	XX/XY chimaerism
532	43	489	91.9	

Note:
Adapted from Marcum (1974).

goats and apparently even in pigs (Short, 1969, 1970). There are three essential prerequisites before the freemartin condition can appear in ruminants: (1) that there should be at least two ovulations giving rise to dizygotic twins; (2) that the embryos should be of different sex; and (3) that placental fusion leading to anastomosis of allanto-chorionic blood vessels between the two embryos should have occurred early in development. As a result of the conjoined circulations (Plate 5.1), the genetically female foetus co-twinned to a male usually (>90% in cattle; Table 5.1) shows suppression and disorganisation of ovarian development producing a rudimentary or testis-like gonad depleted of germ cells. Such gonads may reveal varying degrees of dysgenesis of the ovarian cortex associated with primitive tubular structures in the medullary or hilar regions (Chapin, 1917; Willier, 1921; Jost, 1972). If cortical and medullary tissues are both present in the gonad, germ cells survive only in the cortex, where they can develop into primordial follicles; XX germ cells do not survive in the testicular tissue (Willier, 1921). Abnormalities of the reproductive tract are associated with fusion and/or inhibition of development of portions of the Müllerian ducts;

typically, the upper portions of the ducts regress (Jost, Vigier & Prépin, 1972). There is commonly a rudimentary uterus and small vulva terminating in a short vagina sealed off from the rest of the tract. In rare cases, the clitoris may be enlarged but masculinisation of the external genitalia is invariably slight. However, testosterone-dependent growth of the seminal vesicles, epididymis and even prostate may occur (Josso *et al.*, 1992).

Down the years, the most frequently offered explanation for the aetiology of the freemartin has been both an androgenic (masculinising) and an anti-feminising influence of the precociously developing male gonads upon the Müllerian ducts and gonads of the female co-twin. Lillie (1916, 1917, 1922) advanced the theory of a transplacental (vascular) transfer of humoral substances between the conjoined circulations although, of course, steroid hormones had not been characterised at that time. Even if they had been, it should be appreciated that male sex hormones such as androgens cannot induce development of an ovary as a testis and nor can they act to cause regression of the Müllerian ducts. Not surprisingly, therefore, the freemartin condition in cattle was not produced experimentally by this approach (Jost, Chodkiewicz & Mauléon, 1963). So Lillie's hormonal theory for impeded development of the Müllerian ducts and masculinisation of the gonads in freemartins was by no means satisfactory in the context of steroid sex hormones. As will be noted below, however, his intuition turned out to be an accurate guide to physiological events.

Transplacental migration of cells rather than of hormones was also proposed as a possible explanation of the condition (Fechheimer, Herschler & Gilmore, 1963), male cells that enter the freemartin foetus acting to cause sex reversal. Interchange of germ cells has been reported between the freemartin and its co-twin (Ohno *et al.*, 1962; Ohno & Gropp, 1965; Ohno, 1969), migrating germ cells from the male twin crossing the placental anastomoses and homing in on the undifferentiated gonadal ridge of the genetic female. Such twins are erythrocyte chimaeras (Owen, 1945), indicating a transfer of haemopoietic cells between foetuses. Other evidence for exchange of somatic cells is the acceptance of skin grafts between dizygotic twin calves (Anderson *et al.*, 1951). Colonisation of the female with cells from the male foetus might therefore underlie certain instances of disruption and partial sex reversal of the female gonad, although there is no direct evidence indicating that the genotype of germ cells can control the surrounding somatic differentiation. Nonetheless, secretions from the rudimentary testicular cords (the medullary cords) within the disorganised gonad were considered as the principal deleterious influence on the Müllerian ducts rather than a transplacental origin of humoral substances

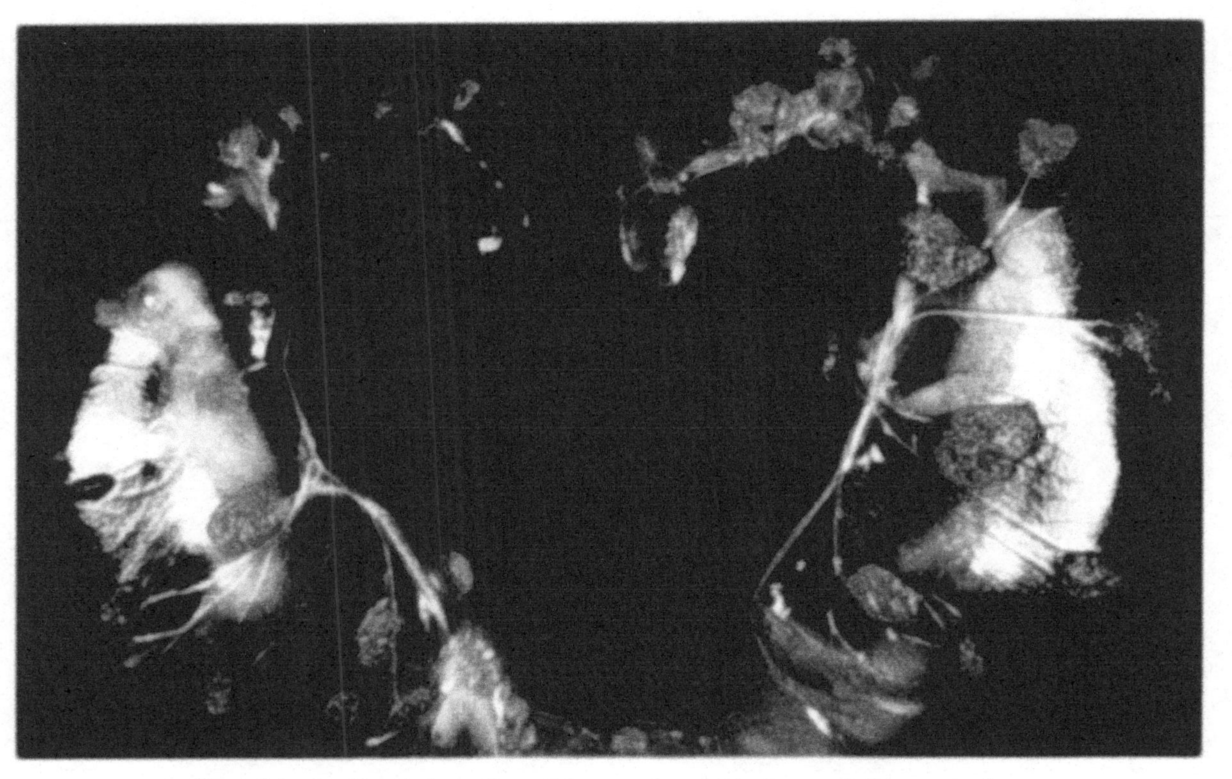

Plate 5.1. An illustration of conditions under which freemartinism may arise. Twin bovine foetuses within fused placental sacs that have conjoined circulations. The prominent structures on the placental surface are the foetal cotyledons which will interdigitate with uterine caruncles to form the organ of exchange – the placentome. (Courtesy of the late Mr L.E.A. Rowson, FRS.)

This image is available for download in colour from www.cambridge.org/9780521182294

(Short, 1969). Although a form of germ cell chimaerism might thus be involved in producing freemartins, other more subtle components such as sensitivity of the XX gonad to H-Y antigen (Ohno *et al.*, 1976) could underlie the aetiology of this condition. Wachtel *et al.* (1980) presented evidence for H-Y antigen circulating in the freemartin and postulated that H-Y antigen of testicular origin could be responsible for the masculinisation of its ovary, without suggesting the time-course of any such involvement.

However, the studies of Vigier *et al.* (1984) on the anti-Müllerian hormone (AMH) have led to a new understanding of the genesis of freemartins, their observations being interpreted on the accepted basis of a male foetus developing precociously when compared with its female twin (Lillie, 1917; Bascom, 1923; Jost, 1972; Mittwoch, 1985), this precocity being found even in very young embryos (Mittwoch, Delhanty & Beck, 1969). In the context of differentiation of gonadal tissues, AMH has been noted to have a virilising and stunting influence on the foetal ovary, so passage of this Sertoli cell protein from the developing male gonads across the conjoined placental circulations into the female could perturb formation of functional ovaries as well as causing abnormal development of the female genital tract. Vigier *et al.* (1984) noted that Müllerian duct regression occurs simultaneously in male and female bovine twins between 50 and 80 days of gestation when these are united by chorionic vascular anastomoses and when serum AMH in freemartins is consistently elevated. During this period, normal testes actively produce AMH whereas freemartin gonads do so only rarely and in limited amounts. Production of AMH by the freemartin gonad is usually much later, at around 100 days, if and when seminiferous tubules develop, and thus too late to play a rôle in Müllerian duct regression. So the conclusion of Vigier *et al.* (1984) was that regression of Müllerian ducts in freemartins is essentially mediated by AMH produced by the testes of the male twin, not least since freemartin gonads *in vitro* produced very low levels of AMH at the appropriate time.

Jost & Magre (1988) summarised this more recent perspective and clarified the timescale for the delayed ovarian masculinisation found in the freemartin condition. They refer to three specific phases:

1 The gonads remain similar to presumptive ovaries for the first 50–60 days of gestation during the period when testes would be differentiating in the male co-twin.
2 Next the gonads become stunted for 6 weeks or more. Indeed, during the period when oogonial proliferation and gonadal growth would occur in normal females, the freemartin gonad is severely inhibited and a tunica

albuginea develops. As a consequence of this inhibition, the freemartin gonad becomes depleted by death of germ cells and very few remain to enter meiosis. In the well-preserved specimens of John Hunter, a much later microscopical study of this material recorded that 'in none of the sexual glands are ova present' (Berry Hart, 1909).

3 Only after this period of inhibition and loss of oocytes do cords of Sertoli-like cells develop adjacent to the intra-ovarian rete. Germ cells may be seen in the newly formed cords but they disappear postnatally (Short *et al.*, 1969), conforming to the prediction that XX germ cells do not survive in an abnormal testicular environment. Leydig cells may also differentiate after the third month of foetal age (Jost *et al.*, 1972, 1973), such development being considered as the active phase of masculinisation.

Testicular organogenesis in presumptive ovaries would be due to the humoral exchange with the male twin, as noted above. Of direct relevance to such gonadal reorganisation, Vigier *et al.* (1987) reported that foetal rat ovaries exposed to purified bovine AMH *in vitro* showed the sequence of (1) ovarian inhibition, (2) tunica albuginea formation and (3) subsequent development of Sertoli-cell cords which characterise the freemartin gonad. Such effects were both time and dose dependent. Sex-reversed gonads produced testosterone instead of oestradiol due to inhibition of aromatase (Vigier *et al.*, 1989). These observations strongly support the earlier suggestion of Jost *et al.* (1972, 1973, 1975) that anti-Müllerian substance (referred to as AMH throughout this monograph: see Chapter 4), from the male twin is the mediator of the freemartin effect in cattle. Moreover, because XX ovarian cells can differentiate into Sertoli or Sertoli-like cells and produce biologically active AMH in culture (Charpentier & Magre, 1990), the pattern of gene activity that defines the Sertoli cell phenotype does not - in one sense - invariably involve genes located uniquely on the Y chromosome. Of course in this instance the gene product from the Y chromosome is presumed to have crossed the placenta. Cells in the ovary apparently retain an ability to transdifferentiate into Sertoli cells (Burgoyne, 1988) but, in the freemartin gonad, this must be viewed as a secondary rather than a primary form of sex reversal. Nonetheless, this would provide a route for the speculation of earlier biologists – the possibility that the freemartin's own gonads also participate in the regression of her Müllerian ducts (Lillie, 1917; Witschi, 1939; Vigier *et al.*, 1981). For there remains the problem of the short range of effective action of AMH as demonstrated in foetal rabbits by Jost's classical 1947 experiment and highlighted in culture by Charpentier & Magre (1990; Table 5.2). Ohno

Table 5.2. *The influence of rat testes at differing stages of prenatal development (13.5 and 17.5 days) upon cultured ovaries and Müllerian ducts taken from 14.5-day-old rat foetuses*

Developmental stage of foetus (days)	Influence upon ovaries			Müllerian inhibiting activity expressed as no. of ducts studied		
	No. of germ cells (mean ± SE)	(No. of cultures)	Structure	None	Partial	Strong
Controls	5050 ± 350	(30)	Ovarian	30	0	0
13.5	4850 ± 450	(8)	Ovarian	0	5	3
17.5[a]	850 ± 150	(21)	Modified	0	0	21
17.5[b]	7650 ± 600	(8)	Ovarian	8	0	0

Notes:
Adapted from the observations of Charpentier & Magre (1990).
Testes from 17.5 days were cultured either in contact [a] or some 5 mm distant [b] from the female genital tracts. All experiments involved 4 days of culture. Female genital systems were cultured alone as controls.

Table 5.3. *Chronology of some key reproductive events during the development of male calves or their freemartin (female) co-twins*

Stage of foetal development (days)	Relevant observations on the gonads and genital duct system
40	Testicular differentiation begins in the male co-twin with the appearance of seminiferous cords (Sertoli cells)
50–58	Principal steps in testicular differentiation become apparent, and male gonads are actively producing AMH. Müllerian regression has been proceeding since day 50 in both twins
50–60	Until approximately day 49, female gonads remain undifferentiated as prospective ovaries. They then come under an inhibitory influence from the male co-twin which: (1) stops growth of the gonads, (2) checks multiplication of the germ cells and (3) provokes disappearance of the tubes
61–80	Gonads of female show regression of the covering epithelium and underlying germ cells
90–100	Ovarian organogenesis and formation of follicles would normally commence in a singleton female. Stunting of freemartin gonads is essentially complete by this stage. Production of AMH by freemartin gonads occurs if and when seminiferous tubules develop – said to be in approximately half the cases. Leydig cells may also be formed

Notes:
The observations come from several sources, but are based extensively on the studies of the late Professor A. Jost and his colleagues (references in text).
NB: AMH has no detectable effect upon foetal ovaries at day 20, indicating a critical (later) phase of sensitivity to this glycoprotein (Vigier *et al.*, 1987).

(1979) suggested resolving this contradiction by assuming that extensive dissemination of AMH from its site of synthesis could be peculiar to the family Bovidae whereas Josso & Picard (1986) considered either a greater Sertoli cell production of AMH in the bovine gonad or a more rapid removal of AMH from the circulation in other species. Even so, in seeking a balance, it might be prudent to consider an influence *in vivo* of secretions from both the male co-twin and the gonads of the freemartin acting together to induce the morphological disturbances (Table 5.3). The contribution from the male co-twin must predominate initially since interruption of placental anastomoses between male and female calves before day 45 of gestation prevents Müllerian regression and ovarian stunting in the female (Vigier, Prépin & Jost, 1976).

Two postscripts can be added to these paragraphs on freemartinism.

First, because there is a transplacental passage of cells in both directions, the male is a blood chimaera. Such males are generally of poorer fertility, having a non-return rate when used in artificial insemination lower than their contemporaries and having poor quality semen (Stafford, 1972). Second, although extensive chorionic vascular anastomoses may occur in dizygotic heterosexual twin pregnancies in marmosets, horses and humans, the ovaries of XX females remain apparently unaffected in the presence of a male co-twin and sexual development is unimpaired (McLaren, 1976; Ohno, 1979, 1989; Benirschke, 1981).

Other forms of intersexuality are rare, at least in Western breeds of cattle. Testicular feminisation has already been mentioned (Nes, 1966) in which there is a failure of the target tissues to respond to testosterone – the so-called syndrome of androgen insensitivity (see Chapters 6 and 7). Gonadal dysgenesis, that is instances in which the gonads are represented by fibrous streaks of tissue, is again exceptional but has been reported in an XY Charolais heifer (Chapman, Bruere & Jaine, 1978).

Anomalous sexual development in sheep and goats

Diverse examples of 'freemartin-like' conditions have been reported in sheep ever since the writings of Hunter (1779), but freemartinism is rare in twin or multiple pregnancies when compared with the situation in cattle (Biggers & McFeely, 1966; Biggers, 1968). The frequency of freemartinism in sheep has been reported as falling between 0.8% and 10% (Stormont, Weir & Lane, 1953; Alexander & Williams, 1964), whilst Dain (1971) found an incidence of 1.2% in twin conceptuses and Long (1980) noted 1.1% in a market sample of 261 ewes. Developmental abnormalities of the reproductive tract are generally similar to those reported in bovine freemartins and, upon cytogenetic studies of peripheral blood samples, have been associated with chimaerism of the sex chromosomes. Gerneke (1965, 1967) reported what he termed chromosomal evidence for a freemartin in sheep on the basis of bone marrow chimaerism and Bruere & McNab (1968) also reported a series of six apparent freemartins on the basis of karyotypes. Tolerance to tissue transplantation in a pair of heterosexual twins was taken by Moore & Rowson (1958) as evidence for freemartinism. Other examples could be cited. The animals are infertile, in fact almost always sterile, but their customary low incidence prevents the problem assuming any practical significance.

If the origin of the freemartin condition in sheep is presumed to lie in conjoined placental circulations of twin foetuses, as is the situation in cattle, then it needs to be established that there have indeed been specific vascular anastomoses between the two placentae rather than simple fusion of the chorionic membranes. Formation of allantoic anastomoses between twin sheep foetuses is very seldom observed in large samples of slaughterhouse specimens (I. Gordon, personal communication). Mellor (1969) reviewed the earlier literature bearing on this point, and concluded from his own observations that intermixing of the bloods of neighbouring foetuses does not occur in 'shared'attachment cotyledons, and nor does any detectable admixture seem to arise from the presence of minor trans-suture vascular anastomosis. In Romanovs, which usually have two or three lambs, Cribiu & Chaffaux (1990) have referred to the 2–3% of infertile lambs in litters of mixed sex, with very occasionally a leucocyte chimaerism of XX/XY although a karyotype of XX in the non-haemopoietic tissues. The gonads and genital tracts of such lambs are masculinised, but there are no germ cells. Even if these were genuine cases of sheep freemartinism, the possibility of placental anastomosis could not be verified, and the incidence in the prolific Romanov breed was only a little over 1%.

The syndrome of testicular feminisation is seemingly quite exceptional in sheep, but an instance was noted by Bruere, McDonald & Marshall (1969). Whilst the animal was phenotypically an overfat but sterile ewe, with a short, blind vagina and no clitoris, there were two large abdominal testes with corresponding portions of the head of the epididymal duct but no other derivatives of the male system. The Müllerian ducts had been suppressed. Other forms of intersexuality are reported in the literature from time to time. A probable case of male pseudohermaphroditism featured in the paper of Dain (1972), in which a genetic female had well-developed testes and the genital apparatus of a male. An XO/XX mosaic sheep was reported by Baylis, Wayte & Owen (1984), the animal exhibiting gonadal dysgenesis. And an XX/XXY mosaic wild sheep that proved to be an intersex was reported by Bunch *et al.* (1991). The clitoris was conspicuously masculinised but otherwise the external genitalia were phenotypically female. Surgical exploration revealed bilateral abdominal ovotestes containing primary follicles but no spermatozoa. There was a female genital tract with twin uterine horns, a cervix and vagina. The aetiology of the condition could only be speculated upon.

Turning to goats (Fig. 5.2), the incidence of freemartinism is thought to be very low, and Keller & Tandler (1916) could find no evidence for placental anastomoses between twin foetuses. On the other hand, intersex conditions in goats are widely recorded and have been the subject of numerous investigations for at least 100 years. In fact, as many intersexes

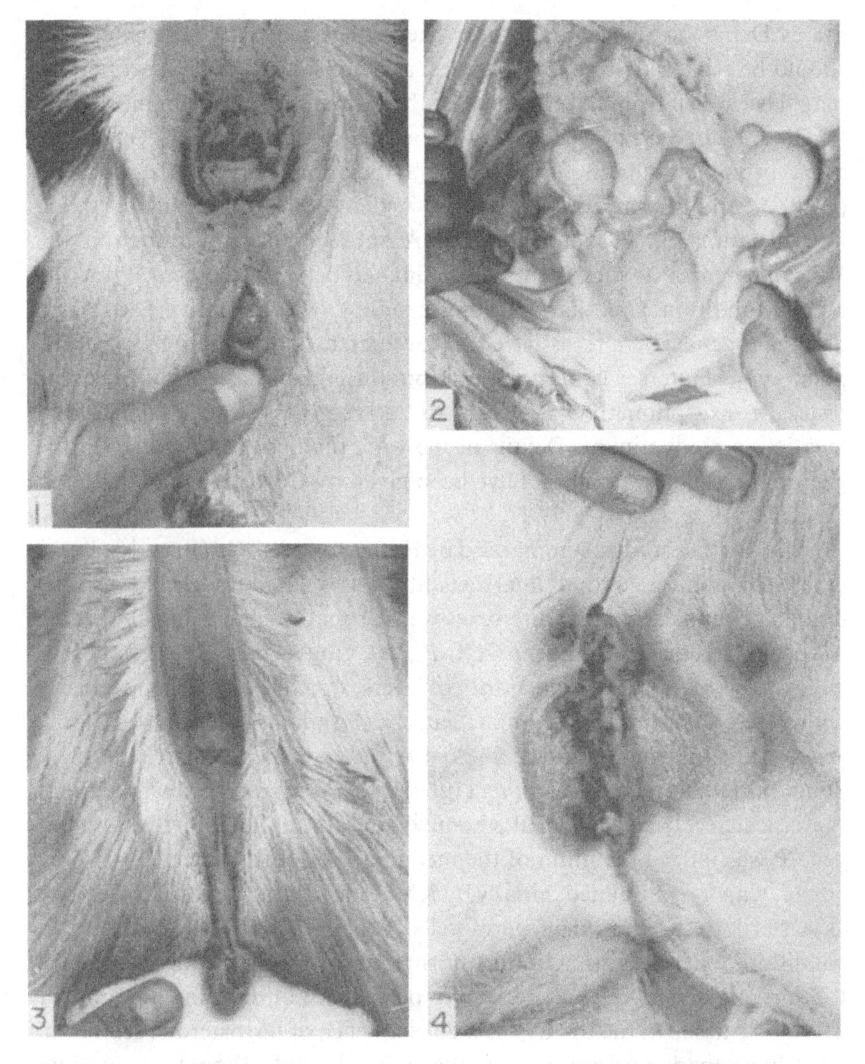

Fig. 5.2. The gonads and genitalia of intersex goats as described by Short *et al.* (1969). Illustration (2) shows testes and epididymides alongside uterine horns, (3) the increased ano-genital distance and penile clitoris, and (4) scrotal testes. (Courtesy of Professor R. V. Short.)

are born co-twin to normal females as to normal males (Laor *et al.*, 1962), an observation that would further argue against freemartinism. Davies (1913) recorded an incidence of 2% intersexuality in kids whilst Eaton & Simmons (1939) noted 6.0% and 11.1% intersexes respectively in the Toggenburg and Saanen breeds within the Beltsville herd of the United

States Department of Agriculture. They suggested that these intersexes should be regarded as genetic females, and considered that the condition may have arisen from development of the primary sex cords in the genetic female and not by growth of the secondary sex cords in the male. One of the more prominent findings was of an association between the hornless (polled) condition and intersexuality, Asdell (1944) noting not a single horned animal in a sample of 200. A subsequent publication on the Beltsville herd also suggested a close linkage between the gene *P* for the polled condition and that for intersexuality or, alternatively, a single dominant gene with a recessive effect for intersexuality (Eaton, 1945). The phenotype of intersex goats is variable but the gonads have usually become testes or exceptionally ovotestes. An abdominal rather than scrotal location is predominant. Details of the wide range of morphology in the genital tract of such animals have been given by Corteel *et al.* (1969) and Hamerton *et al.* (1969).

Many further studies summarised by Hamerton *et al.* (1969) and Soller *et al.* (1969) confirmed that intersexuality was related to the autosomal dominant gene *P*, and that intersexes were probably genetic females which were homozygous for this gene (XX, *P*/*P*). Cytogenetic studies went on to endorse this view, the great majority of intersex goats being genetic females with a 60,XX chromosome complement (Nes, Andersen & Slagsvold, 1963; Basrur & Coubrough, 1964). In line with the suggestion of Eaton & Simmons (1939), Hamerton *et al.* (1969) and Short (1969) concluded that the basic defect in genetic females homozygous for the autosomal dominant gene *P* was indeed retention of the medullary sex cords in the developing gonad, which is populated initially by XX germ cells but goes on to develop as a sterile testis devoid of germ cells. Because there is a spectrum of conditions extending from almost normal male to almost normal female, the variable phenotype may be due to persistence of differing amounts of medullary tissue in the foetus. As a consequence of testosterone secretion, testis-like gonads may descend into the inguinal canals, either to lodge there or to pass to the scrotal location referred to above.

An alternative and more recent interpretation for the aetiology of intersex goats stems from work in mice on the sex reversal gene, *Sxr*. This involves translocation of the testis-determining sequence or gene from the Y chromosome to an X (see especially Chapters 2 and 6). XX *Sxr* animals are thus enabled to develop as phenotypic but sterile males. A comparable gene in goats, if closely linked to the gene *P*, might thereby act to cause testicular development in putative males of XX karyotype. There is also the consideration that intersex goats invariably type positive for H-Y antigen

(Wachtel, Basrur & Koo, 1978). However, molecular analysis of pseudohermaphrodite goats with a 60,XX chromosome complement failed to reveal any Y-derived DNA sequences, including those of *Sry* and *Zfy* (Pailhoux *et al.*, 1994*a*). The importance of autosomal genes in the sex-determining pathway is therefore underlined.

Sex reversal syndrome in horses

Although several forms of sexual anomaly have been reported in the horse, one of the more interesting concerns an XY sex reversal syndrome (Chandley *et al.*, 1975; Hughes & Trommerhausen-Smith, 1977; Kent *et al.*, 1988). Affected animals have a karyotype characteristic of a stallion (64,XY) and yet develop as mares, albeit with a spectrum of phenotypes ranging from almost normal female with ovaries, which may be fertile, to strongly virilised intersex animals secreting high levels of testosterone. In the latter condition, which shows gonadal dysgenesis, abnormal Müllerian development and an enlarged clitoris are common. Mares showing virilisation usually but not always typed H-Y antigen positive whilst those showing a relatively minor degree of sex reversal tended to type H-Y negative (Kent *et al.*, 1988). The range of phenotypic conditions is interpreted as due to variable penetrance of the gene, or, alternatively, to an influence of other modifier genes.

Transmission of the XY sex reversal syndrome apparently involves at least two genes: a primary sex reversing gene and a modifier gene or group of genes. In the study of Kent *et al.* (1988), transmission of an X-linked recessive or autosomal sex-limited dominant, by a carrier female, was suggested in one series of animals. In another series, the conclusion was transmission of an autosomal sex-limited dominant or Y-linked mutation with variable but incomplete penetrance, by a carrier stallion. The carrier stallion harbours the gene but is not directly influenced by it. His sons also transmit the gene. Overall, it was implied that the failure of gonadal differentiation and the consequent perturbation of secondary sex phenotype were due to a failure of testis-determining genes. It is not inferred that the gene for H-Y antigen is itself a testis-determining gene, but the suggestion remains of a close association with testis-determining sequences.

Intersexes have been reported with a normal 64,XX sex chromosome constitution (Bornstein, 1967), but the application of more recent technology might have modified the genetic interpretation. Mosaics (XX/XY) and chimaeras have also been reported on various occasions, whole-body chimaeras presumably being generated by fusion of male and female embryos. Such true hermaphrodites may appear as males suffering from cryptorchidism and poor development of the external genitalia, and having abdominal ovotestes (McIlwraith, Owen & Basrur, 1976; Dunn *et al.*, 1981). On the other hand, normal sexual development and fertility in horses have been found in instances of XX/XY blood chimaeras arising from the presence of heterosexual twins (Podliachouk, Vandeplassche & Bouters, 1974).

Intersexuality in pigs

Whilst development of the freemartin condition in cattle and other ruminants requires anastomosis of the placental circulations between male and female foetuses before it can proceed, a comparable vascular situation is seldom found in pigs and the freemartin condition is only extremely rarely reported (see Hoadley, 1928; Hughes, 1929; Forbes, 1965; Gerneke, 1967; Breeuwsma, 1970). In extensive studies, Breeuwsma (1970) found only one intersex with XX/XY chimaerism that could have been a freemartin. Although the species is polytocous, with frequently as many as 12–16 foetuses or more distributed between the extensive cornua of the uterus, end-to-end fusion of neighbouring placental sacs is a quite exceptional event. In situations in which the proliferating allanto-chorionic sacs overlap in this manner within the uterine lumen early in gestation, the extremities (tips) of these structures become necrotic and shrivel away (Fig. 5.3) so that adjoining sacs do not fuse (Heuser & Streeter, 1929; Hughes, 1927, 1929; Marrable, 1968; Anderson, 1978) and circulations therefore do not become conjoined. Given the anticipated mixture of male and female foetuses within each litter, this would seem essential if the potential fertility of females is not to be compromised. However, overlapping of chorionic membranes later in development is the rule and side-to-side fusion of foetal sacs may occur, although without parabiosis (McFeely & Kressly, 1980).

Despite these remarks, it is a reasonably common occurrence to find anatomical aberrations in presumptive female pigs, not least when judged on the basis of extensive slaughterhouse surveys of reproductive tracts (e.g. Nalbandov, 1976). Aberrations may be distinguishable macroscopically and, for present purposes, concern primarily the gonads rather than congenital malformations of the uterus. As mentioned in the introduction to this chapter, ovotestes or testicular-like structures in presumptive female animals have been taken to represent an intersex condition (Fig. 5.4). They have been the subject of many research publications since the reviews of Crew (1924) and Baker (1925) and the more extensive publications of

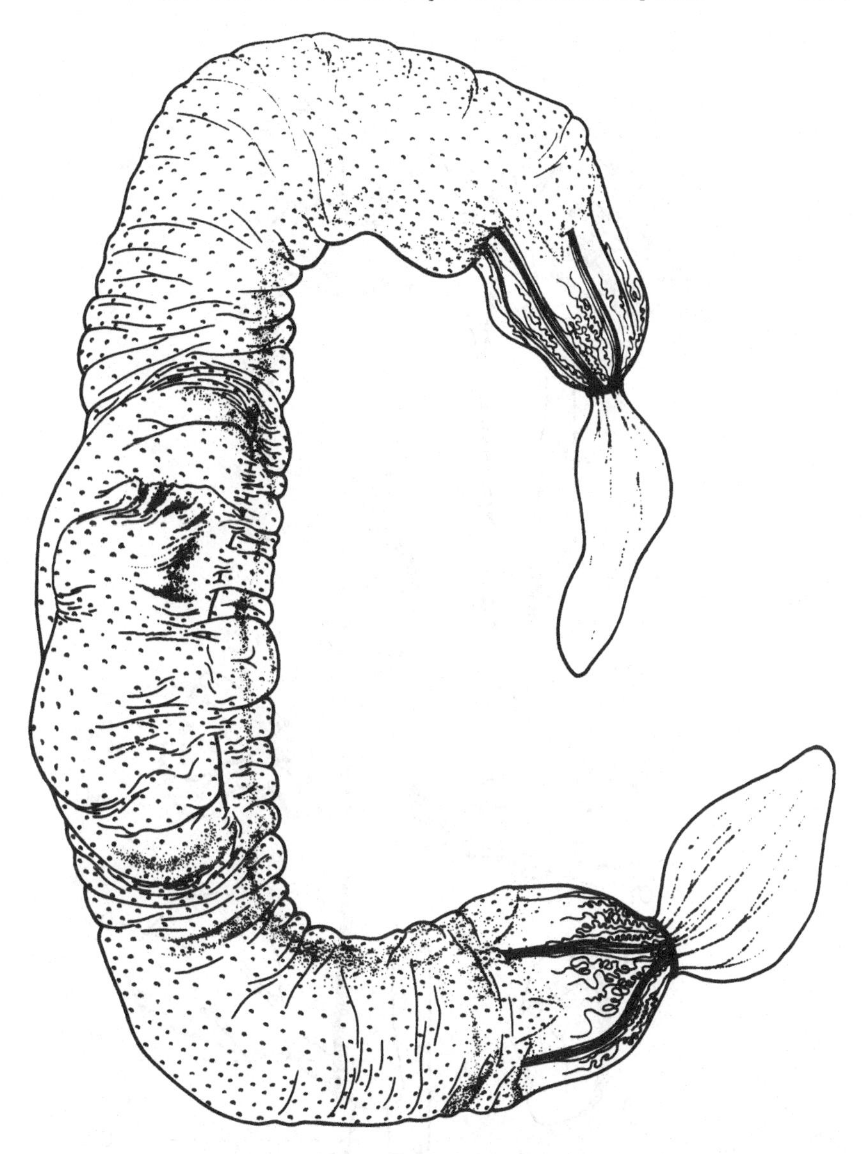

Fig. 5.3. An excellent drawing of the allanto-chorion of a young pig foetus to illustrate the manner in which the extremities tend to shrivel and become necrotic, thereby preventing end-to-end fusion between neighbouring placental sacs. (From Marrable, 1971.)

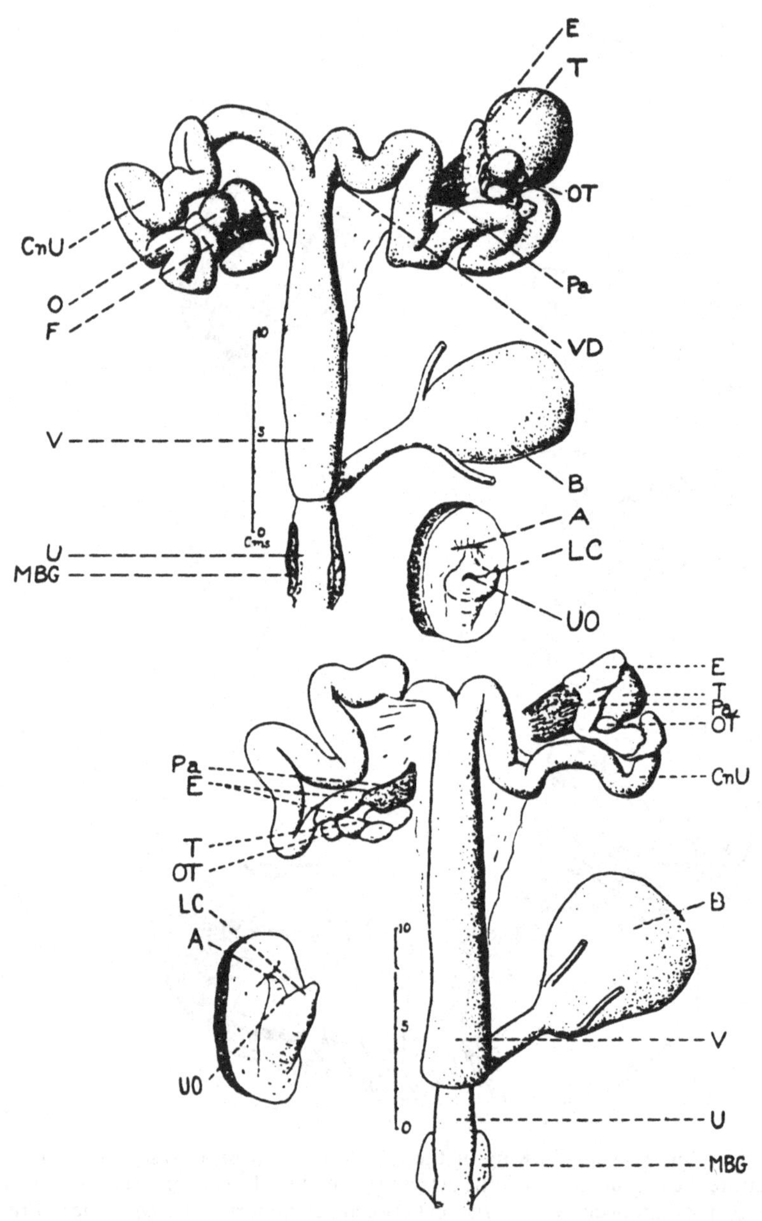

Fig. 5.4. Representation of the gonads and genital tracts in a range of intersex pigs examined on islands of the New Hebrides by Baker (1925, 1926). The characteristic form of the vulva is depicted, and the gonads are present as ovaries, ovotestes or testicular-like tissue. A, anus; B, urinary bladder; CnU, cornu uteri; E, epididymis; F, Fallopian tube; LC, ventral labial commissure; MBG, bulbo-glandularis muscle; O, ovary; OT, ovarian patch on testis; Pa, pampiniform plexus; T, testis; U, urethra; UO, external urethral orifice; V, vagina; VD, vas deferens.

Zuckerman & Groome (1940), Johnston, Zeller & Cantwell (1958), Gerneke (1967) and Breeuwsma (1970). Occasionally, such animals have been termed hermaphrodites (Hammond, 1912; Brambell, 1929), but intersex is the more widely applied word (Scofield, Cooper & Lamming, 1969; Breeuwsma, 1970; Booth & Polge, 1976). The incidence of intersexuality in pigs has been recorded as 0.1–1.4% (Bäckström & Henricson, 1971), but it may achieve 4–5% in a closed-breeding herd employing specific boars. Indeed, Bäckström & Henricson (1971) noted that one Landrace boar generated 12 intersex piglets in a total of 286 offspring, a frequency of 4.2% in 25 litters. And Breeuwsma (1970) recorded a statistically significant difference between boars for the incidence of intersexes generated among their offspring. Approximately 100 intersex animals in the University of Edinburgh pig breeding herd identified at mid-ventral laparotomy during the author's own surgical studies for the 10 years from 1978 to 1987 form the basis of the remarks that follow. Possible explanations for the aetiology of the condition will then be presented. The observations are drawn from Hunter *et al.* (1982, 1985, 1988) and Chalmers *et al.* (1987), Chalmers (1988) and Chalmers *et al.* (1989).

Bearing in mind that intersex animals in the Edinburgh herd were all karyotyped as genetically female (XX sex chromosomes), the prominent tusks and frothy saliva observed in these animals provided immediate clues to the nature of gonadal endocrine activity. The well-developed and frequently much-enlarged clitoris that was always displayed on reflecting the lips of the up-turned vulva similarly inferred androgenic stimulation. Extremely coarse hair and the pungent odour associated with the skin and saliva were both typical of mature males. Scrotal development was conspicuous in about 15% of the intersex animals, and at least one gonad could be palpated therein, invariably on the right-hand side of the scrotal sac (one exception). Animals with scrotal development showed a penile sheath, and penile musculature could be detected by palpation or surgery. During mid-ventral laparotomy to expose the reproductive organs, the texture of the skin – as monitored with the scalpel blade – was consistently tougher than in gilts of comparable age from the same herd. The behaviour of intersex animals has been described in some detail (Hunter *et al.*, 1982, 1985; Chalmers, 1988). A spectrum of sexual activity was noted, ranging from characteristic female traits to the highly aggressive behaviour seen in mature boars.

As to morphology, a wide range of gonadal types was noted in intersex animals (Plates 5.2 and 5.3); these have been illustrated in colour (Hunter *et al.*, 1982, 1985). The most frequent condition was an ovary on the left side and an ovotestis on the right. In such animals, the ovary usually presented evidence of cyclic activity, bearing follicles of preovulatory diameter or mature corpora lutea. Pregnancy was established in a small number of intersex animals, indicating formation of functional corpora lutea (Hunter *et al.*, 1985). Some ovaries showed no evidence of such mature structures, in which case Graafian follicles did not exceed 4–5 mm in diameter. The ovotestis in these animals ranged in proportions from an approximately equal (by volume) distribution of ovarian and testicular tissue to greater than 90% of testicular tissue. Indeed, in some animals, macroscopic evidence of ovarian tissue could not be found in the right gonad. In these situations the gonad was spherical and measured 6–7 cm in diameter.

A more striking condition of intersexuality was the appearance of both gonads as ovotestes. In such instances there was no evidence for cyclic activity in the variable portion of ovarian tissue. Finally, in this gradation, there were instances in which both gonads were testis-like structures, no trace of ovarian tissue being revealed grossly or upon histological examination. Irrespective of whether the gonad was abdominal or scrotal, the testicular tissue was found to be composed of seminiferous tubules and extensive interstitial tissue, but germ cells were never detected in the tubules – only cells that were referred to as Sertoli-like cells. These presumptive Sertoli cells produced a poor staining reaction when compared with tissue from mature boars (Hunter *et al.*, 1982). A functional seminiferous epithelium would not have been expected in an XX gonad. On the other hand, the variable presence of a tunica albuginea was a novel finding.

Turning to the genital tract, a patent bicornuate uterus was observed in all animals. In those showing evidence of cyclic ovarian activity, the dimensions of uterine tissues were comparable with those in normal females but, in the presence of bilateral ovotestes, the uterus was usually of immature appearance, the cervix poorly developed, and its patency frequently in doubt. An evil-smelling yellowish fluid had frequently been retained in the uterus, reminiscent of classical cases of pyometritis. Although the uterus was clearly abdominal in most animals, portions of one uterine horn had passed through the inguinal canal to enter the scrotum in two animals with a scrotal gonad; this bizarre situation had previously been noted by Arthur (1959). Persistence of a fully differentiated bicornuate uterus in the presence of bilateral ovotestes or testicular-like structures would suggest either a sub-optimal production of AMH by the Sertoli-like cells or perhaps an inappropriately timed secretion of this glycoprotein (see Chapter 4). Alternatively, an inadequate and unequal distribution of the putative receptors for AMH between the Fallopian tubes and uterus could be involved in the differential sensitivity. Indeed, regression of the

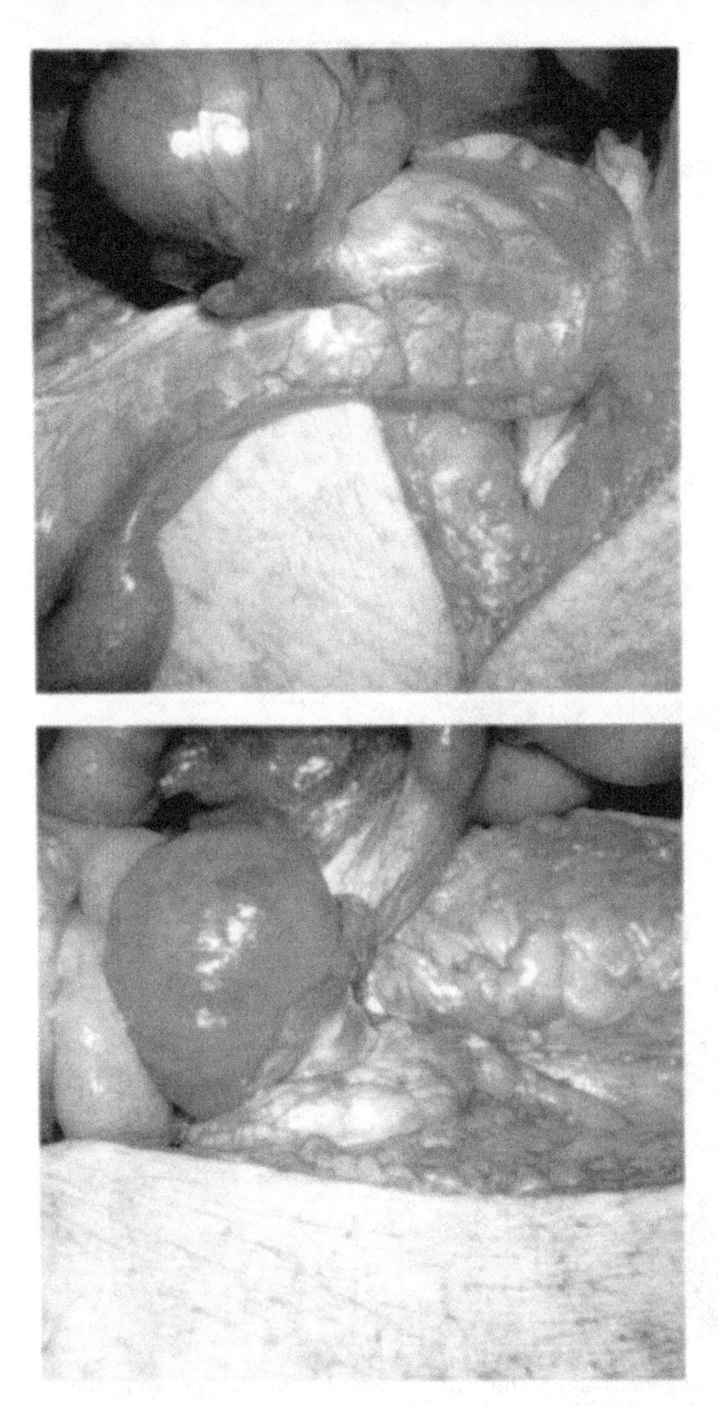

Plate 5.2. Development of the gonad and duct system in two XX intersex pigs from the Edinburgh herd. Both gonads in the illustration have developed as ovotestes, although with only a small portion of ovarian tissue. In the lower picture, the large testicular portion of the gonad is devoid of a tunica albuginea, in contrast to the situation in the upper picture. In both specimens, the Fallopian tube is poorly developed, especially in the ampullary region, whereas the proximal portion of the Wolffian duct has proliferated into a prominent epididymis. (Adapted from Hunter *et al.*, 1982, 1985.)

This image is available for download in colour from
www.cambridge.org/9780521182294

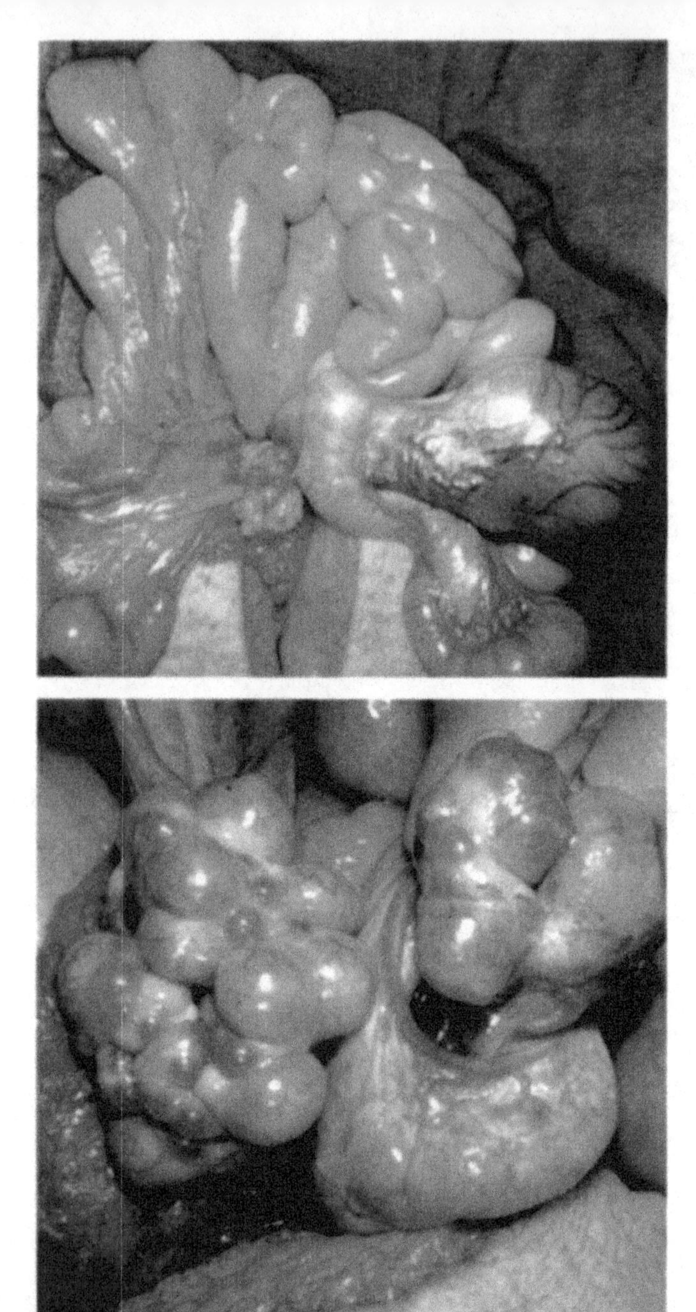

Plate 5.3. The gonads and portions of the reproductive tract in two further XX intersex pigs from the Edinburgh herd. The upper illustration shows an ovary on the left and an enormous testicular-like structure on the right with a neighbouring proliferation of an epididymis. There is suppression of the ampullary portion of the Fallopian tube on this side only and there is normal development of a bicornuate uterus. In the lower specimen, cyclic ovarian activity is demonstrated by the presence of corpora lutea in both gonads, even though the one on the right is an ovotestis that has provoked suppression of the Fallopian tube and caused proliferation of a substantial epididymis. (Taken from Hunter *et al.*, 1985.)

This image is available for download in colour from www.cambridge.org/9780521182294

uterine portion of the Müllerian ducts may not be strictly dependent upon the concentration of AMH (see Josso, Picard & Tran, 1977). In any event, the observations endorse earlier experimental findings in the rat, in which the cranial end of the Müllerian ducts was noted to be more sensitive to AMH (Picon, 1969), a situation also reported in dogs (Meyers-Wallen & Patterson, 1989). If the suggestion of a defective production of AMH is favoured in intersex pigs, then the finding of two animals with scrotal gonads would argue against the view that AMH is critically involved in regulating descent of the testes (see Chapters 3 and 4).

Anomalies were found in the cranial portion of the reproductive tract in all animals. The Fallopian tube adjoining an ovotestis was always underdeveloped, especially the ampullary portion, which was sometimes scarcely visible. However, the isthmus and especially the fimbriated infundibulum never approached the size characteristic of mature animals and the gonad could therefore not be enveloped by the fimbriated extremity. The extent of inhibition of Fallopian tube development was roughly proportional to the amount of testicular tissue. By contrast, the proximal segment of the Wolffian duct had developed into a conspicuous and well-vascularised epididymis adjoining all ovotestes. In animals in which testicular tissue was prominent, the epididymis was distended by a viscous fluid, but histological section confirmed the absence of sperm cells in the lumen of the duct.

Karyotype studies of blood leucocytes were used to establish that intersex animals were genetically female. A modification of the whole blood microtechnique was used, and air-dried chromosome spreads (15–30 per animal) and subsequent banding techniques followed the method of Buckland, Fletcher & Chandley (1976). Whilst there was no ambiguity in the finding of two X chromosomes in all animals, nor was there any evidence of the small metacentric Y chromosome, the standard diploid number of 38 chromosomes was not revealed in all spreads. In one animal examined by Hunter *et al.* (1985), only 35 and 36 chromosomes were counted in two of the spreads. These low counts were associated with a breakage of cells on the microscope slide, and loss of chromosomes from the metaphase plate. The banding techniques failed to produce evidence for a translocated portion of the Y chromosome. Attempts were repeatedly made to culture testicular tissue from the ovotestis of a series of animals, but cells from only one animal revealed metaphase spreads: the testicular tissue had two X chromosomes. Mosaicism was not detected in any of these intersex pigs, despite the statement by Haffen & Wolff (1977) that intersex individuals correspond to mosaics or chromosomal chimaeras, a view excluded by Gerneke (1967).

In animals with two ovotestes and no signs of cyclic activity nor Graafian follicles larger than 1–2 mm diameter, challenges were given with placental gonadotrophins in an attempt to stimulate follicular growth. However, a single subcutaneous injection of 1000 i.u. pregnant mare serum gonadotrophin (PMSG) failed to provoke detectable responses in terms of follicular growth or oestrogenic changes in the genital tract. In a further intersex animal, with a small ovary on the left and a large ovotestis on the right, ovarian tissue in the right gonad failed to respond visibly to a similar dose of PMSG. The vesicular follicles had remained at a diameter of approximately 1 mm when examined 1 week later. Bearing in mind that these Graafian follicles must have undergone enlargement since antral formation, inferring the presence of receptors for endogenous gonadotrophins, it is difficult to explain the apparent lack of response to PMSG. Androgen or protein secretions from the ovotestis may have inhibited expression of the receptors for gonadotrophins in the follicular tissues. If this were so, then the failure of unilateral ovariectomy to prompt growth of follicles in the ovotestis becomes understandable (Hunter *et al.*, 1985).

Intersex animals showing cyclic ovarian activity in at least one of the gonads are presumed to have undergone a preovulatory gonadotrophin surge: in other words, their 'brain sex' could be regarded as female (see Short, 1982). On the other hand, it is unclear whether acyclic animals with ovotestes have the ability to release a surge of gonadotrophic hormones. Accordingly, such animals were given an acute oestradiol challenge, following which secretion of luteinising hormone (LH) into peripheral blood was monitored (Chalmers *et al.*, 1989). Although there was some variability in results, none of five animals showed a positive feedback response of LH comparable to that reported in cyclic females (see Elsaesser & Foxcroft, 1978). Thus, the brain in these animals was thought to have undergone a variable degree of masculinisation, for the hypothalamus did not possess the ability to initiate a full preovulatory LH surge in response to the positive feedback influence of oestradiol.

Because of the condition recorded occasionally in humans of islets of adrenocortical tissue within the gonads forming virilising tumours (Novak & Woodruff, 1967), the question arose as to whether adrenocortical tissue might similarly be found in the gonads of intersex pigs. One means of testing this proposition would be to challenge the animal with a standard clinical dose of adreno-corticotrophic hormone (ACTH; Synacthen, Ciba) and then examine the response in gonadal venous blood. In an intersex pig with abdominal and scrotal gonads (Fig. 5.5), intravenous ACTH increased cortisol and stimulated gonadal steroid secretion by up to tenfold for testosterone, androstenedione, oestradiol and dehydroepiandrostene-

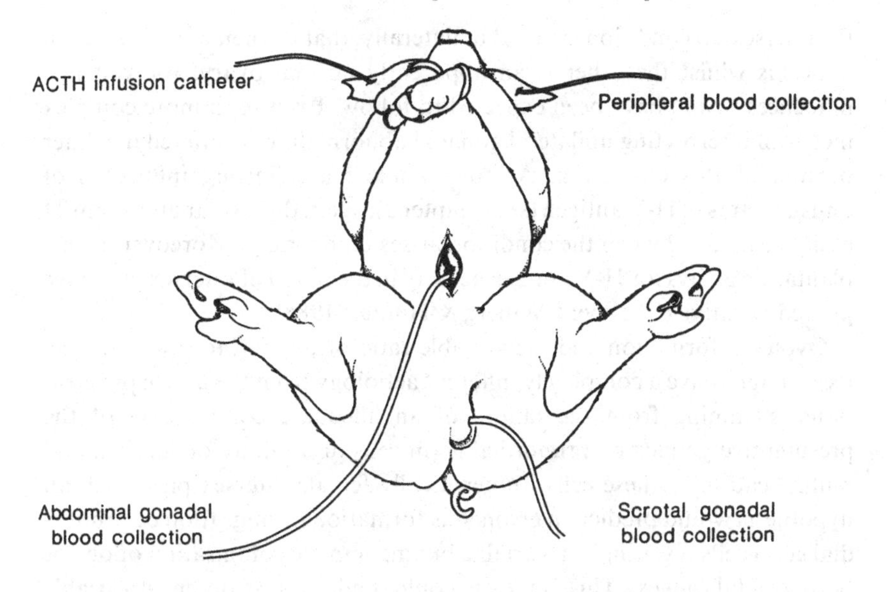

Fig. 5.5. Surgical preparation for examining the response of an Edinburgh intersex pig with abdominal and scrotal gonads to a challenge of ACTH. Some of the results have been reported by Cavazos *et al.* (1987).

dione, whereas the responses in peripheral blood were attenuated (Cavazos *et al.*, 1987). Since the intersex gonad is responding to ACTH stimulation, a conventional interpretation would be that cells within the gonad must possess receptors for ACTH; testicular tissue in normal boars appears to have such receptors (Juniewicz & Johnson, 1984). Thus, endocrine evidence is accumulating for incorporation of adrenocortical or adrenocortical-like cells in the gonads of intersex pigs and possibly in normal boars also. The relevance of this observation is discussed below.

Aetiology of ovotestis formation

Tentative explanations for the condition of intersexuality in pigs have focused on either (1) some degree of vascular anastomosis between fused placentae, with transplacental migration of cells from a neighbouring male embryo or (2) specific genetic errors, such as the presence of XY cells indicating formation of a chimaera, or translocation of a portion of the Y chromosome onto an X chromosome or even onto an autosome (see Breeuwsma, 1970). An influence of one or more recessive genes at a very few autosomal loci has also been invoked (Sittmann, Breeuwsma & te Brake, 1980). However, none of these lines of explanation seems appropriate when the intersexual condition is found unilaterally, that is when one gonad is an ovotestis whilst the other is an apparently normal ovary, for systemic influences would have been expected to follow. Even so, a more complex means of interpreting unilateral gonadal abnormalities is offered in a later portion of this discussion. As to possible masculinising influences of unusual titres of H-Y antigen (see Chapter 2), again this explanation cannot easily be invoked when the condition arises unilaterally. Moreover, transplantation assays for H-Y antigen activity in the blood of intersex pigs have proved negative (Chalmers, Wiberg & Hunter, 1988).

Ovotestis formation and the variable ratio of ovarian to testicular-like tissue might have a completely different aetiology from the above propositions, stemming from the failure of an adequate colonisation of the presumptive gonads by primordial germ cells or a failure of survival and multiplication of these cells. In genetically female intersex pigs, such an hypothesis would predict an erroneous formation or migration of primordial germ cells resulting in a variable but incomplete colonisation of one or both genital ridges. This, in turn could find expression in a variable development of cortical and medullary components of the presumptive gonad(s), either unilaterally or bilaterally. It might be expected, moreover, that with a limited colonisation of the genital ridges by germ cells in an XX animal subsequently found to have developed an ovotestis, the ovarian portion would adjoin the gonadal stalk or pedicle and the testicular portion would be devoid of germ cells. This was almost always the case in the Edinburgh series, except of course in animals in which ovarian tissue was not detected at all.

In XX intersex animals possessing one or two ovotestes, possible inferences to be drawn if the above hypothesis is correct are as follows:

(*a*) The primordial germ cells may indeed exert an influence on differentiation of the gonad, and the extent of such colonisation – at least by XX germ cells in intersex animals – could have acted to determine the relative proportions of cortex and medulla, that is the ratio of ovarian to testicular-like tissue. Even in genetic females, as judged from the Edinburgh studies, there may have been a predisposition for the gonads to become testis-like, except and until colonised by XX germ cells. Expressed in another way, a partial absence of primordial germ cells may lead to a partial formation of testicular tissue. Alternatively, and an interpretation favoured from classical studies in mice, loss of germ cells from the medullary tissue of an ovotestis might simply reflect the inability of these XX cells to survive in a testicular environment beyond the time at which they would normally have entered meiosis (see Chapters 6 and 8).

The opening sentences of the above proposition are not unreasonable when it is recalled that primordial germ cells (diploid cells) must in some way be able to express themselves if they are to migrate from their extra-embryonic site in the yolk sac to the genital ridges of the presumptive gonads. If the primordial germ cells can therefore converse with the systems that influence their migration, they could presumably also interact with the gonadal tissues to influence the nature of cortical and medullary differentiation. This accords with the well-known follicular degeneration if and when the oocyte dies *in situ*. Hence, the mammalian genital ridge may await a germinal source of programming to enable its full and final differentiation (see Hunter *et al.*, 1988; Hunter, 1989).

(*b*) In the absence of any detectable ovarian tissue, suppression of Müllerian duct development and promotion of Wolffian duct derivatives can be quite limited in extent; as noted above, a well-formed bicornuate uterus is invariably observed in intersex pigs. The presence of testicular-like tissue in genetic females would therefore appear insufficient to masculinise most of the duct system. For the latter to occur, an adequate production of AMH by the Sertoli cells may require the influence of germ cells in the seminiferous tubules. Sertoli cells in the ovotestes of intersex pigs have been referred to as 'Sertoli-like cells' due to their poorly staining appearance (Hunter *et al.*, 1982) – hence the argument that a full spectrum of Sertoli cell function may require stimulation from adjacent germ cells, just as normal spermatogenesis requires the intimate support of fully competent Sertoli cells. Such mutual dependence between somatic and germ cells has its female counterpart in the oocyte–granulosa cell relationship.

Outstanding questions arising from these suggestions are the following:

1 Is a principal defect leading to ovotestis formation in pigs simply elaboration of an insufficient number of primordial germ cells?
2 Are there errors in the migration or survival of primordial germ cells leading to an insufficient or unilateral colonisation of the presumptive gonads, perhaps with wayward cells passing to the embryonic adrenal glands or being enveloped by mesonephric tissues?

If the answer to (2) is 'yes', might a lack of chemotactic factors be implicated? Because certain boars are known to produce a relatively high incidence of the intersex condition in their offspring (Breeuwsma, 1970; Bäckström & Henricson, 1971), it should be possible using histological techniques to study the extent of the germ cell population and the pattern of its migration from the yolk sac, and to compare these events with those in embryos generated by boars lacking this hereditary feature. This inferred phenomenon of ectopic germinal elements has already been noted in the mouse (Zamboni & Upadhyay, 1983), even though morphological abnormalities in the reproductive system were not reported.

Because XX cells can form Sertoli cells, it is clear that the pattern of activity that defines the Sertoli cell phenotype does not necessarily involve genes on the Y chromosome. It appears that supporting cells in the ovary retain an ability to transdifferentiate into Sertoli cells (Burgoyne, 1988). The fact that this 'transdifferentiation' is associated with germinal failure raises the question of whether oocytes have a positive rôle to play in maintaining ovarian supporting cells in the differentiated state.

Despite the above arguments that focus on germ cells, an open mind should be maintained as to the involvement of a sex-determining DNA sequence translocated from the Y chromosome onto one of the X chromosomes in the aetiology of the intersex condition. Furthermore, if development of the two gonads were asynchronous (Crew, 1926) and the time of inactivation of the paternal X chromosome were variable in these XX animals, but late enough in development to involve gonadogenesis, then it would be possible for a translocated DNA sequence from the Y chromosome to prescribe a variable degree of testicular formation before losing influence during the inactivation process. This proposition is illustrated in Fig. 5.6. However, this line of reasoning would be compromised by the report that pig embryos first reveal sex chromatin at the stage of about 50 cells (Lyon, 1974) – unless a Y-derived DNA sequence entrains processes that continue after presumptive inactivation.

Finally, a completely novel interpretation for the formation of testicular tissue in XX females can be drawn from the facts that:

1 AMH will act to cause virilisation of ovarian tissue (rats) as noted above (Vigier *et al.*, 1987);
2 granulosa and Sertoli cells have the same embryonic derivation, being mesonephric in origin (Upadhyay & Zamboni, 1982; Byskov & Hoyer, 1988);

and

3 these two lines of gonadal somatic cell secrete closely similar types of protein hormone in the form of inhibin and AMH, respectively. Inhibin is well known to have structural homology with other bioactive glycoproteins (Matzuk *et al.*, 1992; Findlay, 1993). Indeed, AMH may be the first or one of the first proteins secreted by the differentiating Sertoli cell. Accordingly, it appears reasonable to argue that if secretion of inhibin-like molecules in embryonic or foetal XX gonads were to be accompanied

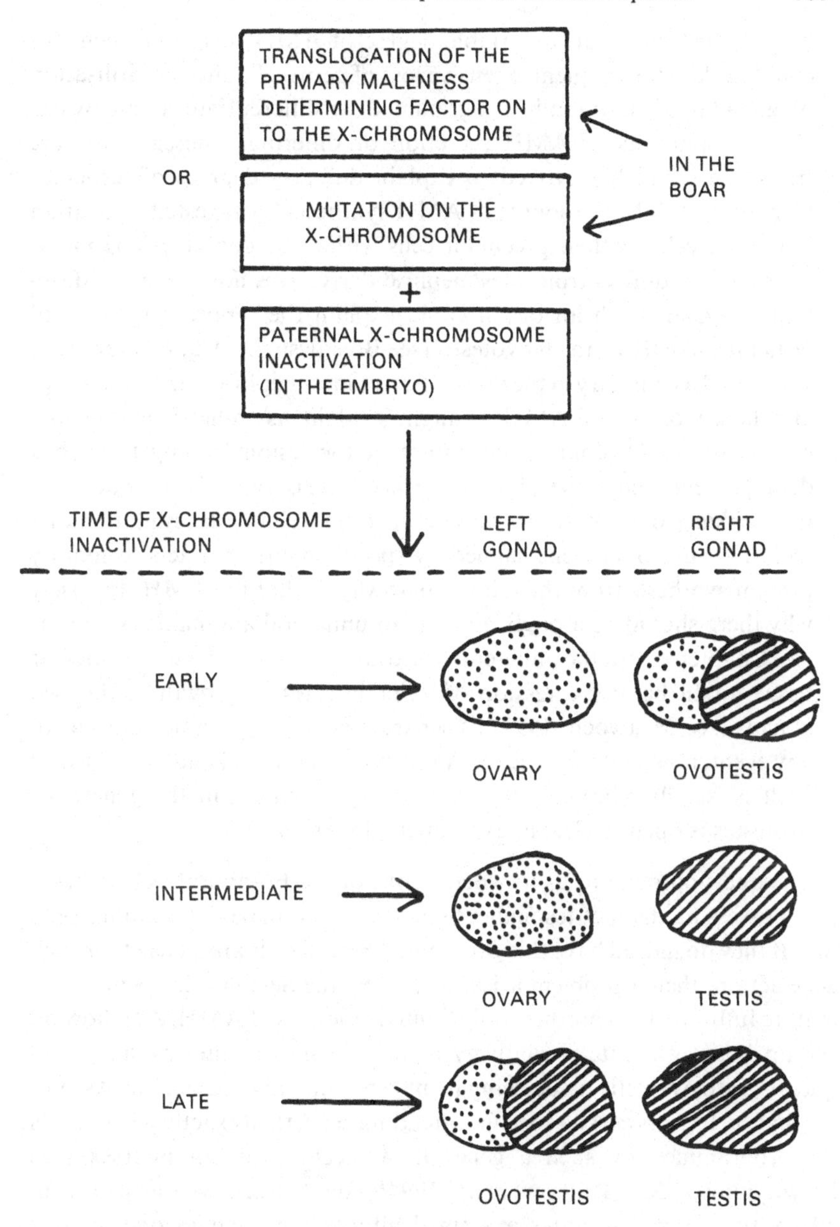

Fig. 5.6. A model to suggest the manner whereby varying gonadal constitutions could arise from translocation of male-determining sequences onto an X chromosome, or by an appropriate mutation on an X chromosome, in the boar followed by inactivation of the paternal X chromosome at different times in the embryo. A presumption in this model is that the right-hand gonad develops at a different rate from the left. (Taken from Hunter *et al.*, 1988.)

by, or substituted for, a variable degree of AMH secretion, then this could explain the frequent extensive loss of germ cells and the virilisation of gonadal tissue initially programmed to differentiate as an ovary. Varying amounts of AMH secretion or differing timescales for the transition could be invoked to explain differing degrees of testicular formation. Whilst the source of AMH synthesis is suggested as ovarian granulosa cells or their precursor cells within the gonad, AMH might alternatively diffuse from mesonephric derivatives adjoining the developing gonad. Such an origin could explain the reported polarity of testicular-like tissue in an ovotestis (see Breeuwsma, 1970; Hunter *et al.*, 1985), and asymmetry in the disposition of mesonephric-derived tissues – and hence origin of AMH – might explain asymmetry in gonadal modifications. Of course, the nature of the lesion leading to such a derangement needs to be addressed. Since there is a genetic component to the incidence of ovotestis formation, it could well involve mutation of a recessive autosomal gene carried by specific boars that acts to modify protein synthesis from the inhibin pathway to that for AMH. Precisely why there should be a predisposition to unilateral anomalies is unclear (see Chapter 9), but the unilateral or variable condition *vis à vis* testicular tissue might be an expression of variable rates of growth in the two gonads (see Mittwoch, 1992) as separate gene programmes unfold. In any event, now that the gene for AMH has been cloned and the molecule itself is readily available, its proposed involvement in the genesis of ovotestes is open to elegant experimental verification.

The question remains as to whether or not the Edinburgh XX intersex pigs possessing distinct testicular tissue carry the male-determining gene *Sry*. If they do indeed have this gene, and if Sertoli cells are a target for such gene action, then a problem arises as to why the Sertoli cells seem not to mature fully and thus produce only limited amounts of AMH. And how do certain boars generate a relatively high incidence of intersexuality, frequently with ovotestis tissue being formed within only one gonad? Assuming that such boars carry a specific, offending gene, then exactly what events are programmed by such a gene? In a recent study of intersex pigs karyotyping 38,XX (Pailhoux *et al.*, 1994*b*), the *Sry* gene was absent from all six true hermaphrodites examined although present in one of four apparent males with genital ambiguities (Table 5.4). In the absence of *Sry*, putative genes acting downstream from *Sry* may trigger testis determination if they undergo an appropriate mutation. Such an explanation would accord with the existence of non-Y chromosome sex-determining genes involved in XX sex reversal in the mole (Jimenez *et al.*, 1993).

Table 5.4. *Molecular findings for XX intersex pigs of quantitatively normal chromosome complement indicating the absence of the testis-determining gene,* Sry, *in the great majority of these animals possessing some testicular tissue*

	Male-determining gene *Sry*	
Summarised nature of animals	present	absent
38,XX male with ambiguities	1	4
38,XX true hermaphrodite	0	6

Modified from Pailhoux *et al.* (1994*b*).

Concluding remarks

Putting their economic significance to one side, and the risk of increasing their incidence by inappropriate breeding policies, including induced twinning in cattle by embryo transplantation, these observations dealing with some of the more common abnormalities in domestic farm animals serve to highlight the vulnerability of sexual differentiation processes to major perturbations. In one sense, at least, the gonad may be reasserting its developmental flexibility as found lower down the evolutionary scale. The observations also suggest that genetic females or, more accurately, those with a perceived XX sex chromosome complement may become especially susceptible to the modifications of intersexuality during pregnancies in which a male foetus is also present. This is not least because of the precocity of embryonic development in males compared with females, a precocity that may be revealed as early as the first cleavage divisions in the young zygote (see Chapter 2).

But congenital abnormalities of the gonads can also arise 'spontaneously' in XX individuals in the absence of a detectable male influence or Y chromosome, and may best be explained by autosomal gene interactions with the programme normally imposed by sex-determining genes. Of course, this statement may require modification if probes for Y-related DNA sequences reveal such male components in the genome of presumptive females. Perhaps it could also be argued that if the normal sequence of gene action were in error, although not involving an identifiable mutation, then this might be a sufficient explanation for wayward formation of gonadal tissue. In any event, as must be clear to the reader by now, anomalous development of the genital tract is invariably a consequence of anomalous development of the gonads, although spontaneous mutation of genes regulating Wolffian or Müllerian development could presumably induce morphological abnormalities in these duct systems. Despite the recent advances described in Chapter 2 concerning *Sry* – the so-called sex determining gene on the Y chromosome – the rôle of autosomal genes in mammalian sex determination still requires further serious consideration, particularly the regulatory influences of these genes during formation of the gonads. It is in this context that the study of abnormalities may prove especially enlightening.

References

Alexander, G. & Williams, D. (1964). Ovine freemartins. *Nature (London)*, **201**, 1296–8.

Anderson, D., Billingham, R. E., Lampson, G. H. & Medawar, P. B. (1951). The use of skin grafts to distinguish between monozygotic and dizygotic twins in cattle. *Heredity*, **5**, 379–97.

Anderson, L. L. (1978). Growth, protein content and distribution of early pig embryos. *Anatomical Record*, **190**, 143–54.

Arthur, G. H. (1959). Some aspects of intersexuality in animals. *Veterinary Record*, **71**, 598–603.

Arthur, G. H., Noakes, D. E. & Pearson, H. (1983). *Veterinary Reproduction and Obstetrics*, 6th edn. London, Philadelphia, Toronto: Baillière Tindall.

Asdell, S. A. (1944). The genetic sex of intersexual goats and a probable linkage with the gene for hornlessness. *Science, N.Y.*, **99**, 124.

Bäckström, L. & Henricson, B. (1971). Intersexuality in the pig. *Acta Veterinaria Scandinavica*, **12**, 257–73.

Baker, J. R. (1925). On sex-intergrade pigs: their anatomy, genetics, and developmental physiology. *Journal of Experimental Biology*, **2**, 247–63.

Baker, J. R. (1926). Asymmetry in hermaphrodite pigs. *Journal of Anatomy*, **60**, 374–81.

Bascom, K. F. (1923). The interstitial cells of the gonads of cattle with especial reference to their embryonic development and significance. *American Journal of Anatomy*, **31**, 223–59.

Basrur, P. K. & Coubrough, R. I. (1964). Anatomical and cytological sex of a Saanen goat. *Cytogenetics*, **3**, 414–26.

Baylis, M. S., Wayte, D. M. & Owen, J. B. (1984). An XO/XX mosaic sheep with associated gonadal dysgenesis. *Research in Veterinary Science*, **36**, 125–6.

Benirschke, K. (1981). Hermaphrodites, freemartins, mosaics and chimaeras in animals. In *Mechanisms of Sex Differentiation in Animals and Man*, ed. C. R. Austin & R. G. Edwards, pp. 421–63. London: Academic Press.

Berry Hart, D. (1909). The structure of the reproductive organs in the free-martin, with a theory of the significance of the abnormality. *Proceedings of the Royal Society of Edinburgh*, **30**, 230–41.

Biggers, J. D. (1968). Aspects of intersexuality in domestic mammals. Proceedings 6th International Congress on Animal Reproduction and Artificial Insemination, Paris, **2**, 841–70.

Biggers, J. D. & McFeely, R. A. (1966). Intersexuality in domestic mammals. *Advances in Reproductive Physiology*, **1**, 29–59.

Bishop, M. W. H. (1972). Genetically determined abnormalities of the

reproductive system. *Journal of Reproduction and Fertility, Supplement*, **15**, 51–78.
Booth, D. W. & Polge, C. (1976). The occurrence of C_{19} steroids in testicular tissue and submaxillary glands of intersex pigs in relation to morphological characteristics. *Journal of Reproduction and Fertility*, **46**, 115–21.
Bornstein, S. (1967). The genetic sex of two intersexual horses and some notes on the karyotype of normal horses. *Acta Veterinaria Scandinavica*, **8**, 291–300.
Brambell, F. W. R. (1929). The histology of an hermaphrodite pig and its developmental significance. *Journal of Anatomy*, **63**, 397–407.
Breeuwsma, A. J. (1970). *Studies on Intersexuality in Pigs*. Doctoral thesis. Rotterdam: Drukkerij Bronder-Offset N.V.
Bruere, A. N. & MacNab, J. (1968). A cytogenetical investigation of six intersex sheep, shown to be freemartins. *Research in Veterinary Science*, **9**, 170–80.
Bruere, A. N., McDonald, M. F. & Marshall, R. B. (1969). Cytogenetical analysis of an ovine male pseudohermaphrodite and the possible role of the Y chromosome in cryptorchidism of sheep. *Cytogenetics*, **8**, 148–57.
Buckland, R. A., Fletcher, J. M. & Chandley, A. C. (1976). Characterisation of the domestic horse (*Equus caballus*) karyotype using G- and C-banding techniques. *Experientia*, **32**, 1146–9.
Bunch, T. D., Callan, R. J., Maciulis, A., Dalton, J. C., Figueroa, M. R., Kunzler, R. & Olson, R. E. (1991). True hermaphroditism in a wild sheep: a clinical report. *Theriogenology*, **36**, 185–90.
Burgoyne, P. S. (1988). Role of mammalian Y chromosome in sex determination. *Philosophical Transactions of the Royal Society of London, Series B*, **322**, 63–72.
Byskov, A. G. & Hoyer, P. E. (1988). Embryology of mammalian gonads and ducts. In *The Physiology of Reproduction*, ed. E. Knobil, J. Neill *et al.*, ch. 8, pp. 265–302. New York: Raven Press.
Cavazos, F., Chalmers, C. A., Cook, B., Kelly, A. S. L. & Hunter, R. H. F. (1987). ACTH affects gonadal steroid synthesis in intersex pigs. *Proceedings of the Society for the Study of Fertility*, No. **46**, 30.
Chalmers, C. A. (1988). The physiology and aetiology of intersexuality in pigs. PhD dissertation, University of Edinburgh.
Chalmers, C. A., Cook, B., Foxcroft, G. R. & Hunter, R. H. F. (1987). The LH response to an oestradiol challenge in intersex pigs. *Proceedings of the Society for the Study of Fertility*, No. **46**, 9.
Chalmers, C., Cook, B., Foxcroft, G. R. & Hunter, R. H. F. (1989). Luteinizing hormone response to an oestradiol challenge in 5 intersex pigs possessing ovotestes. *Journal of Reproduction and Fertility*, **87**, 455–61.
Chalmers, C., Wiberg, U. & Hunter, R. H. F. (1988). H-Y transplantation antigen status of two intersex pigs. *Proceedings of the Society for the Study of Fertility* No. **62**, 36.
Chandley, A. C., Fletcher, J., Rossdale, P. D., Peace, C. K., Ricketts, S. W., McEnery, R. J., Thorne, J. P., Short, R. V. & Allen, W. R. (1975). Chromosome abnormalities as a cause of infertility in mares. *Journal of Reproduction and Fertility, Supplement*, **23**, 377–83.
Chapin, C. L. (1917). A microscopic study of the reproductive system of foetal freemartins. *Journal of Experimental Zoology*, **23**, 453–82.
Chapman, H. M., Bruere, A. N. & Jaine, P. M. (1978). XY gonadal dysgenesis in a Charolais heifer. *Animal Reproduction Science*, **1**, 9–18.
Charpentier, G. & Magre, S. (1990). Masculinizing effect of testes on developing rat ovaries in organ culture. *Development*, **110**, 839–49.

Corteel, J. M., Hulot, F., Courot, M., Attal, J. & Philippon, A. (1969). Examens morphologiques, caryologiques, physiologiques et pathologiques de boucs stériles sans cornes. *Annuaire Génétique de Sélection Animale*, **1**, 341–8.
Crew, F. A. E. (1924). Hermaphroditism in the pig. *Journal of Obstetrics and Gynaecology of the British Commonwealth*, **31**, 369–86.
Crew, F. A. E. (1926). Abnormal sexuality in animals. I. Genotypical. *Quarterly Review of Biology*, **1**, 315–59.
Cribiu, E. P. & Chaffaux, S. (1990). L'intersexualité chez les mammifères domestiques. *Reproduction, Nutrition, Développement, Supplement*, **1**, 51s–61s.
Dain, A. R. (1971). The incidence of freemartinism in sheep. *Journal of Reproduction and Fertility*, **24**, 91–7.
Dain, A. R. (1972). Differences in chromosome lengths between male and female sheep. *Nature, London*, **237**, 455–7.
Davies, C. J. (1913). Caprine freemartins. *Veterinary Journal*, **20**, 62–70.
Deas, D. W., Melrose, D. R., Reed, H. C. B., Vandeplassche, M. & Pidduck, H. (1979). Other non-infectious abnormalities. In *Fertility and Infertility in Domestic Animals*, 3rd edn, ed. A. J. Laing, ch. 7, pp. 137–59. London: Baillière Tindall.
Dunn, H. O., Smiley, D., Duncan, J. R. & McEntee, K. (1981). Two equine true hermaphrodites with 64,XX/64,XY and 63,XO/64,XY chimaerism. *Cornell Veterinarian*, **71**, 123–35.
Eaton, O. N. (1945). The relation between polled and hermaphroditic characters in dairy goats. *Genetics, Princeton*, **30**, 51–61.
Eaton, O. N. & Simmons, V. L. (1939). Hermaphrodism in milk goats. *Journal of Heredity*, **30**, 261–6.
Elsaesser, F. & Foxcroft, G. R. (1978). Maturational changes in the characteristics of oestrogen-induced surges of luteinising hormone in immature domestic gilts. *Journal of Endocrinology*, **78**, 455–6.
Fechheimer, N. S., Herschler, M. S. & Gilmore, L. O. (1963). Sex chromosome mosaicism in unlike sexed cattle twins. In *Genetics Today*, ed. S. J. Geerts, p. 265. Oxford: Pergamon Press.
Findlay, J. K. (1993). An update on the roles of inhibin, activin and follistatin as local regulators of folliculogenesis. *Biology of Reproduction*, **48**, 15–23.
Forbes, T. A. (1965). John Hunter on spontaneous intersexuality. *American Journal of Anatomy*, **116**, 269–300.
Gerneke, W. H. (1965). Chromosomal evidence of the freemartin condition in sheep – *Ovis aries*. *Journal of the South African Veterinary Medical Association*, **36**, 99–104.
Gerneke, W. H. (1967). Cytogenetic investigations on normal and malformed animals, with special reference to intersexes. *Onderstepoort Journal of Veterinary Research*, **34**, 219–300.
Haffen, K. & Wolff, E. (1977). Natural and experimental modification of ovarian development. In *The Ovary*, 2nd edn, vol. 1, ed. S. Zuckerman & B. J. Weir, pp. 430–2. New York: Academic Press.
Hamerton, J. L., Dickson, J. M., Pollard, C. E., Grieves, S. A. & Short, R. V. (1969). Genetic intersexuality in goats. *Journal of Reproduction and Fertility, Supplement*, **7**, 25–51.
Hammond, J. (1912). A case of hermaphroditism in the pig. *Journal of Anatomy and Physiology*, **7**, 307–12.
Heuser, C. H. & Streeter, G. L. (1929). Early stages in the development of pig embryos, from the period of initial cleavage to the time of appearance of limb-buds. *Contributions to Embryology of the Carnegie Institution*, **20**, 1–30.

Hoadley, L. (1928). Twin heterosexual pig embryos (32 mm) found within fused membranes. *Anatomical Record*, **38**, 177–87.
Hughes, J. P. & Trommerhausen-Smith, A. (1977). Infertility in the horse associated with chromosomal abnormalities. *Australian Veterinary Journal*, **53**, 253–7.
Hughes, W. (1927). Sex intergrades in foetal pigs. *Biological Bulletin*, **52**, 121–36.
Hughes, W. (1929). The freemartin condition in swine. *Anatomical Record*, **41**, 213–45.
Hunter, J. (1779). Account of the freemartin. *Philosophical Transactions of the Royal Society of London*, **69**, 279–93.
Hunter, R. H. F. (1989). Une énigme biologique en voie de résolution. *Le Médecin Vétérinaire du Québec*, **19**, 125–35.
Hunter, R. H. F., Baker, T. G. & Cook, B. (1982). Morphology, histology and steroid hormones of the gonads in intersex pigs. *Journal of Reproduction and Fertility*, **64**, 217–22.
Hunter, R. H. F., Chalmers, C. & Cavazos, F. (1988). Intersexuality in domestic pigs: a guide to mechanisms of gonadal differentiation? *Animal Breeding Abstracts*, **56**, 785–91.
Hunter, R. H. F., Cook, B. & Baker, T. G. (1985). Intersexuality in five pigs, with particular reference to oestrous cycles, the ovotestis, steroid hormone secretion and potential fertility. *Journal of Endocrinology*, **106**, 233–42.
Jiménez, R., Burgos, M., Sanchez, A., Sinclair, A. H., Alarcon, F. J., Marin, J. J., Ortega, E. & de la Guardia, R. F. (1993). Fertile females of the mole *Talpa occidentalis* are phenotype intersexes with ovotestes. *Development*, **118**, 1303–11.
Johnston, E. F., Zeller, J. H. & Cantwell, G. (1958). Sex anomalies in swine. *Journal of Heredity*, **49**, 255–61.
Josso, N. & Picard, J-Y. (1986). Anti-Müllerian hormone. *Physiological Reviews*, **66**, 1038–90.
Josso, N., Picard, J. Y. & Tran, D. (1977). The anti-Müllerian hormone. *Recent Progress in Hormone Research*, **33**, 117–63.
Josso, N., Vigier, B., Cate, R. L., Behringer, R., di Clemente, N. & Lyet, L. (1992). Hormonal control of gonadal differentiation. In *Gonadal Development and Function*, ed. S. G. Hillier, pp. 31–39. Serono Symposia Publications No. **94**. New York: Raven Press.
Jost, A. (1947). Recherches sur la différenciation sexuelle de l'embryon de lapin. III. Rôle des gonades foetales dans la différenciation sexuelle somatique. *Archives d'Anatomie Microscopique et de Morphologie Expérimentale*, **36**, 271–315.
Jost, A. (1972). A new look at the mechanisms controlling sex differentiation in mammals. *Johns Hopkins Medical Journal*, **130**, 38–53.
Jost, A., Chodkiewicz, M & Mauléon, P. (1963). Intersexualité du foetus de veau produite par les androgènes. Comparaison entre l'hormone foetale responsable du free-martinisme et l'hormone testiculaire adulte. *Compte Rendu des Séances de l'Académie des Sciences (Paris)*, **256**, 274–6.
Jost, A. & Magre, S. (1988). Control mechanisms of testicular differentiation. *Philosophical Transactions of the Royal Society of London, Series B*, **322**, 55–61.
Jost, A., Perchellet, J. P., Prépin, J. & Vigier, B. (1975). The prenatal development of bovine freemartins. In *Symposium on Intersexuality*, ed. R. Reinboth, pp. 392–406. Berlin: Springer-Verlag.
Jost, A., Vigier, B & Prépin, J. (1972). Freemartins in cattle: the first steps of sexual organogenesis. *Journal of Reproduction and Fertility*, **29**, 349–79.

Jost, A., Vigier, B., Prépin, J. & Perchellet, J. P. (1973). Studies on sex differentiation in mammals. *Recent Progress in Hormone Research*, **29**, 1–35.
Juniewicz, P. E. & Johnson, B. H. (1984). Ability of cortisol and progesterone to mediate the stimulatory effect of ACTH upon testosterone production by the porcine testis. *Biology of Reproduction*, **30**, 134–42.
Keller, K. & Tandler, J. (1916). Über das Verhalten der Eihäute bei der Zwillingsträchtigkeit des Rindes. *Wiener Tierärtzl Wochenschrift*, **3**, 513–27.
Kent, M. G., Shoffner, R. N., Hunter, A., Elliston, K. O., Schroder, W., Tolley, E. & Wachtel, S. S. (1988). XY sex reversal syndrome in the mare: clinical and behavioural studies, H-Y phenotype. *Human Genetics*, **79**, 321–8.
Kirkpatrick, B. W. & Monson, R. L. (1993). Sensitive sex determination assay applicable to bovine embryos derived from IVM and IVF. *Journal of Reproduction and Fertility*, **98**, 335–40.
Laing, A. J. (ed.) (1979). *Fertility and Infertility in Domestic Animals*, 3rd edn. London: Baillière Tindall.
Laor, M., Barnea, R., Angel, H. & Soller, M. (1962). Polledness and hermaphroditism in Saanen goats. *Israel Journal of Agricultural Research*, **12**, 83–8.
Lillie, F. R. (1916). The theory of the free-martin. *Science*, **43**, 611–13.
Lillie, F. R. (1917). The free-martin; a study of the action of sex hormones in the foetal life of cattle. *Journal of Experimental Zoology*, **23**, 371–452.
Lillie, F. R. (1922). The etiology of the free-martin. *Cornell Veterinarian*, **12**, 332–7.
Long, S. E. (1980). Some pathological conditions of the reproductive tract of the ewe. *Veterinary Record*, **106**, 175–6.
Lyon, M. F. (1974). Mechanisms and evolutionary origins of variable X-chromosome activity in mammals. *Proceedings of the Royal Society of London, Series B*, **187**, 243–68.
McFeely, R. A. & Kressly, L. R. (1980). Intersexuality. In *Reproduction in Farm Animals*, 4th edn., ed. E. S. E. Hafez, pp. 494–502. Philadelphia: Lea & Febiger.
McIlwraith, C. S., Owen, R. A. & Basrur, P. K. (1976). An equine cryptorchid with testicular and ovarian tissues. *Equine Veterinary Journal*, **8**, 156–60.
McLaren, A. (1976). *Mammalian Chimaeras*. Cambridge & London: Cambridge University Press.
Machaty, Z., Paldi, A., Csaki, T., Varga, Z., Kiss, I., Barandi, Z. & Vajta, G. (1993). Biopsy and sex determination by PCR of IVF bovine embryos. *Journal of Reproduction and Fertility*, **98**, 467–70.
Marcum, J. B. (1974). The freemartin syndrome. *Animal Breeding Abstracts*, **42**, 227–42.
Marrable, A. W. (1968). The ischaemic extremities of the allanto-chorion of the pig and their relation to the endometrium. *Research in Veterinary Science*, **9**, 578–82.
Marrable, A. W. (1971). *The Embryonic Pig: A Chronological Account*. London: Pitman.
Matzuk, M. M., Finegold, M. J., Su, J. G. J., Hsueh, A. J. W. & Bradley, A. (1992). α-inhibin is a tumour-suppressor gene with gonadal specificity in mice. *Nature (London)*, **360**, 313–19.
Mellor, D. J. (1969). Vascular anastomosis and fusion of foetal membranes in multiple pregnancy in sheep. *Research in Veterinary Science*, **10**, 361–7.
Meyers-Wallen, V. N. & Patterson, D. F. (1989). Sexual differentiation and inherited disorders of sexual development in the dog. *Journal of Reproduction and Fertility, Supplement*, **39**, 57–64.

Mittwoch, U. (1985). Males, females and hermaphrodites. *Annals of Human Genetics*, **50**, 103–21.

Mittwoch, U. (1992). Sex determination and sex reversal: genotype, phenotype, dogma and semantics. *Human Genetics*, **89**, 467–79.

Mittwoch, U., Delhanty, J. D. A. & Beck, F. (1969). Growth of differentiating testes and ovaries. *Nature (London)*, **224**, 1323–5.

Moore, N. W. & Rowson, L. E. A. (1958). Freemartins in sheep. *Nature (London)*, **182**, 1754–5.

Nalbandov, A. V. (1976). *Reproductive Physiology of Mammals and Birds*, 3rd edn. San Francisco: W. H. Freeman.

Nes, N. (1966). Testikulaer feminisering hos storfe. *Nordisk Veterinær Medicin*, **18**, 19–29.

Nes, H. N., Andersen, K. & Slagsvold, P. (1963). Kromosomundersøkelse hos hermafroditte geiter. *Medlemsblad for det Norske Veterinaerförbund*, **7**, 155.

Novak, E. R. & Woodruff, J. D. (1967). *Novak's Gynecologic and Obstetric Pathology*, 6th edn. Philadelphia: W. B. Saunders.

Ohno, S. (1969). The problem of the bovine freemartin. *Journal of Reproduction and Fertility, Supplement*, **7**, 53–61.

Ohno, S. (1979). *Major Sex-Determining Genes*. Berlin: Springer-Verlag.

Ohno, S. (1989). From GATA-GACA repeats to dictyostelium cell adhesion protein to C-CAM, N-CAM and gonad-organizing proteins. In *Evolutionary Mechanisms in Sex Determination*, ed. S. S. Wachtel, ch. 2, pp. 15–22. Boca Raton, Florida: CRC Press.

Ohno, S. Christian, L. C., Wachtel, S. S. & Koo, G. C. (1976). Hormone-like role of H-Y antigen in bovine freemartin gonad. *Nature (London)*, **261**, 597–9.

Ohno, S. & Gropp, A. (1965). Embryological basis for germ cell chimaerism in mammals. *Cytogenetics*, **4**, 251–61.

Ohno, S., Trujillo, J. M., Stenius, C., Christian, L. C. & Teplitz, R. L. (1962). Possible germ cell chimaeras among newborn dizygotic twin calves (*Bos taurus*). *Cytogenetics*, **1**, 258–65.

Owen, R. D. (1945). Immunogenetic consequences of vascular anastomoses between bovine twins. *Science*, **102**, 400–1.

Pailhoux, E., Cribiu, E. P., Chaffaux, S., Darre, R., Fellous, M. & Cotinot, C. (1994*a*). Molecular analysis of 60,XX pseudohermaphrodite polled goats for the presence of *Sry* and *Zfy* genes. *Journal of Reproduction and Fertility*, **100**, 491–6.

Pailhoux, E., Popescu, P. C., Parma, P., Boscher, J., Legault, C., Molteni, L., Fellous, M. & Cotinot, C. (1994*b*). Genetic analysis of 38,XX males with genital ambiguities and true hermaphrodites in pigs. *Animal Genetics*. In press.

Picon, R. (1969). Action du testicule foetal sur le développement *in vitro* des canaux de Müller chez le rat. *Archives d'Anatomie Microscopique et de Morphologie Expérimentale*, **58**, 1–19.

Podliachouk, L., Vandeplassche, M. & Bouters, R. (1974). Gestation gemellaire, chimaerisme et freemartinisme chez le cheval. *Acta Zoologica Pathologica, Antwerp*, **58**, 13–28.

Scofield, A. M., Cooper, K. J. & Lamming, G. E. (1969). The distribution of embryos in intersex pigs. *Journal of Reproduction and Fertility*, **20**, 161–3.

Short, R. V. (1969). An introduction to some of the problems of intersexuality. *Journal of Reproduction and Fertility, Supplement*, **7**, 1–8.

Short, R. V. (1970). The bovine freemartin: a new look at an old problem. *Philosophical Transactions of the Royal Society of London, Series B*, **259**, 141–7.

Short, R. V. (1982). Sex determination and differentiation. In *Reproduction in Mammals*, 2nd edn, vol. 2, ed. C. R. Austin & R. V. Short, pp. 70–113. Cambridge & London: Cambridge University Press.

Short, R. V., Smith, J., Mann, T., Evans, E. P., Hallett, J., Fryer, A. & Hamerton, J. L. (1969). Cytogenetic and endocrine studies of a freemartin heifer and its bull co-twin. *Cytogenetics (Basel)*, **8**, 369–88.

Sittmann, K., Breeuwsma, A. J. & te Brake, J. H. A. (1980). On the inheritance of intersexuality in swine. *Canadian Journal of Genetics and Cytology*, **22**, 507–27.

Soller, M., Padeh, B., Wysoki, M. & Ayalon, N. (1969). Cytogenetics of Saanen goats showing abnormal development of the reproductive tract associated with the dominant gene for polledness. *Cytogenetics*, **8**, 51–67.

Stafford, M. J. (1972). The fertility of bulls born co-twin to heifers. *Veterinary Record*, **90**, 146–8.

Stormont, C., Weir, W. C. & Lane, L. L. (1953). Erythrocyte mosaicism in a pair of sheep twins. *Science*, **118**, 695–6.

Upadhyay, S. & Zamboni, L. (1982). Ectopic germ cells: natural model for the study of germ cell sexual differentiation. *Proceedings of the National Academy of Sciences, USA*, **79**, 6584–8.

Vigier, B., Forest, M. G., Eychenne, B., Bézard, J., Garrigou, O., Robel, P. & Josso, N. (1989). Anti-Müllerian hormone produces endocrine sex-reversal of fetal ovaries. *Proceedings of the National Academy of Sciences, USA*, **86**, 3684–8.

Vigier, B., Picard, J. Y., Bézard, J. & Josso, N. (1981). Anti-Müllerian hormone: a local or long distance morphogenetic factor? *Human Genetics*, **58**, 85–90.

Vigier, B., Prépin, J. & Jost, A. (1976). Chronologie du développement de l'appareil génital du foetus de veau. *Archives d'Anatomie Microscopique et de Morphologie Expérimentale*, **65**, 77–102.

Vigier, B., Tran, D., Legeai, L., Bézard, J. & Josso, N. (1984). Origin of anti-Müllerian hormone in bovine freemartin foetuses. *Journal of Reproduction and Fertility*, **70**, 473–9.

Vigier, B., Watrin, F., Magre, S., Tran, D. & Josso, N. (1987). Purified bovine AMH induces a characteristic freemartin effect in fetal rat prospective ovaries exposed to it *in vitro*. *Development*, **100**, 43–55.

Wachtel, S. S., Basrur, P. & Koo, G. C. (1978). Recessive male-determining genes. *Cell*, **15**, 279–81.

Wachtel, S. S., Hall, J. L., Müller, U. & Chaganti, R. S. K. (1980). Serum-born H-Y antigen in the fetal bovine freemartin. *Cell*, **21**, 917–26.

Willier, B. H. (1921). Structures and homologies of free-martin gonads. *Journal of Experimental Zoology*, **33**, 63–127.

Wislocki, G. B. (1932). Placentation in the marmoset (*Oedipomidas geoffroyi*), with remarks on twinning in monkeys. *Anatomical Record*, **52**, 381–99.

Witschi, E. (1939). Modification of the development of sex in lower vertebrates and in mammals. In *Sex and Internal Secretions*, 2nd edn, ed. E. Allen, C. Danforth & E. A. Doisy, pp. 145–226. Baltimore: Williams & Wilkins.

Zamboni, L. & Upadhyay, S. (1983). Germ cell differentiation in mouse adrenal glands. *Journal of Experimental Zoology*, **228**, 173–93.

Zuckerman, S. & Groome, J. R. (1940). An experimental study of the morphogenesis of intersexuality. *Journal of Anatomy*, **74**, 171–200.

6

Abnormal sexual development in laboratory rodents

Introduction

Because the laboratory mouse has been used so widely in biological research, it will come as little surprise to learn that abnormalities have been observed throughout the reproductive system, not least in inbred strains of mice. These abnormalities extend from an absence of functional gonads to the formation of hermaphrodites in which ovotestes are observed unilaterally or comprising both gonads. The origin of such conditions would appear to reside in spontaneously arising mutant genes. Experimentally generated mice such as chimaeras or transgenics also demonstrate a variety of reproductive abnormalities, and highlight the apparent vulnerability of gonadal and genital tissues to major derangement expressed as modification of structure. Even so, derangement that involves generation of both male and female tissue within the same gonad immediately raises questions concerning sex determination and differentiation. For such a situation to arise, it would seem possible to argue that the downstream gene programmes for testicular and ovarian development are both present in mammalian zygotes, and that sex determination involves active suppression of one of the programmes, a step prompted by a gene located on a sex chromosome. An inappropriate or untimely instigation or regulation of the suppression mechanism could permit generation of an ovotestis.

The objective of this chapter is to comment on some of these abnormal conditions, and to introduce concepts that will be relevant to the ensuing chapter dealing with man. Detailed papers and reviews on sexual anomalies in laboratory rodents include those of Greene, Burrill & Ivy (1939), Hollander, Gowen & Stadler (1956), MacIntyre (1956), Tarkowski (1964), Lyon (1969), Cattanach (1975), Beamer, Whitten & Eicher (1978), Whitten, Beamer & Byskov (1979), Burgoyne & Baker (1981, 1985), Eicher & Washburn (1983, 1986), McLaren (1983), LeBarr & Blecher (1986),

Bradbury (1987), Taketo-Hosotani *et al.* (1989), Lovell-Badge & Robertson (1990) and Mittwoch (1992).

Sex reversal in the mouse

The paragraphs that follow are but a brief introduction to this engaging topic. Detailed descriptions have been given by Lyon, Cattanach & Charlton (1981), Cattanach (1987), Bishop *et al.* (1988, 1989) and an excellent summary by Burgoyne (1988). Some of the essentials are summarised here, although the origin of the condition has been referred to in Chapter 2.

The sex-reversed (*Sxr*) mouse is an inherited condition which, due to a mutation, causes a genetic female to develop as a male. Since these animals are seemingly, but not actually, normal males, LeBarr & Blecher (1986) have termed them pseudomales. Such chromosomally female mice (XX or XO) carry an apparently autosomal dominant mutation that renders them phenotypically and behaviourally male, but with non-germinal testes that are small in adults. Spermatogenesis is absent in XX animals whereas it is present but defective in those of XO constitution. Although the mutation is transmitted by carrier XY males, an autosomal location of *Sxr* could not be demonstrated.

Sex-reversed mice were first described in detail by Cattanach, Pollard & Hawkes (1971), both XX and XO embryos being masculinised by the sex-reversal factor, *Sxr*. What was initially interpreted as an autosomal dominant gene seems not to exert its influence fully in early gestation, for presumptive male-type germ cells may be present at this stage whereas most have disappeared by or shortly after birth although the seminiferous tubule organisation remains throughout life (Cattanach, 1975; McLaren, 1981). A study of the testes of XX *Sxr* mice showed that in the 16-day foetus there was a normal population of XX germ cells – although the testes were smaller (Table 6.1) – but by the time of birth (total gestation length is 19 days), most of the germ cells in these animals were degenerating. Sertoli cells were immature. Occasionally, XX germ cells could be seen up to the 6th day *post partum*, but they had completely disappeared by the 10th day and no germ cells were ever found in the testes of adult XX *Sxr* mice. Thus, there is the seeming inability of XX germ cells to survive in a testicular environment (Burgoyne, 1978). However, the observed presence of the sperm-specific isoenzyme of lactic dehydrogenase in the testes of some post-pubertal *Sxr* mice suggested that cells in late spermatogenic stages were present (Holmes & Jones, 1979). These are almost certainly derived from

Table 6.1. *Observations on mean gonadal volumes in normal and sex-reversed (XX males) mouse embryos at two stages of gestation. Volumes are* $\mu m^3 \times 10^5$

Sex chromosome complement	Nature of gonads	15-day embryos			16-day embryos		
		No. of embryos	Mean gonadal volume	SE[a]	No. of embryos	Mean gonadal volume	SE[a]
XY	Testes	8	933	28.8	21	1333	28.2
XX	Testes	4	846	17.5	9	1152	71.6
XX	Ovaries	5	303	21.0	8	328	17.2

Notes:
Adapted from Mittwoch & Buehr (1973).
[a] SE, standard error of the mean.
The mean testis weights are significantly different between the XY and XX animals ($P < 0.01$).

X*Sxr*O germ cells which have arisen due to a mitotic error prior to the period of XX *Sxr* germ cell loss (Lyon *et al.*, 1981).

In contrast to the XX *Sxr*/+ germ cells which resembled normal germ cells during the foetal period but degenerated a few days after birth, the XO *Sxr*/+ germ cells were able to undergo spermatogenesis and were present in the adult testis (Cattanach *et al.*, 1971). However, the testes were about half the normal size due to the limited germ cell population and the spermatozoa that formed were scarce in number and abnormal in morphology. In fact, XO *Sxr* mice are sterile. Sutcliffe, Darling & Burgoyne (1991) attributed most of the spermatogenic deficiencies in XO Sxr testes to sex chromosome univalence during meiosis, the so-called meiotic pairing site hypothesis of Miklos (1974) – see below – and Burgoyne, Sutcliffe & Mahadevaiah (1992), reviewed by Burgoyne & Mahadevaiah (1993). But Burgoyne *et al.* (1992*b*) also produced evidence to support the hypothesis of Eicher, Phillips & Washburn (1983) that there is a gene or genes on the mouse Y chromosome that is essential for normal sperm development (Table 6.2). McLaren (1981) noted that a few of the XX *Sxr*/+ germ cells enter the prophase of meiosis before birth, as of course would XX germ cells in the normal mouse ovary. Since XO germ cells can survive in the XX *Sxr* testis, it follows that the double chromosome condition leads to the failure of XX germ cells in the testis. Incidentally, the testes of some XX *Sxr*/+ males contain small numbers of growing oocytes within the seminiferous tubules 1–2 weeks after birth (McLaren, 1980).

How did the sex-reversed condition arise? Extensive cytological studies failed to provide satisfactory evidence of a Y–autosome translocation

Table 6.2. *Summarised testis weights and sperm counts as a function of mouse genotype*

Genetic composition of male	Mean testis weight (mg ± SE[a])	Mean sperm per caput epididymidis (× 10^6)
X*Sxr*ᵃ/O	47.3 ± 2.4	0.0
X*Sxr*ᵃ/Yˣ	113.0 ± 8.5	1.6 ± 0.5
X/Y*Sxr*ᵃ	75.4 ± 5.2	1.5 ± 0.4
XY[b]	136.4 ± 8.2	3.9 ± 0.3

Notes:
Data adapted from the work of Burgoyne *et al.* (1992*b*) emphasising X–Y pairing and a Y-chromosomal 'spermiogenesis' gene for fertility.
[a] SE, standard error of the mean.
[b] Outbred males carrying the same strain of Y as the X/Y*Sxr*ᵃ males

(Cattanach *et al.*, 1971). However, Burgoyne proposed that *Sxr* was located in a region of homology between the X and Y chromosomes and was exchanged between them in a single obligatory crossing over during male meiosis (Fig. 6.1). Such a pseudoautosomal inheritance is indistinguishable from a true autosomal inheritance (Burgoyne, 1982). Evidence in support was forthcoming from the use of both DNA probes and more sensitive cytological techniques. *Sxr* is now known to be derived from the normal Y chromosome short arm (Bishop *et al.*, 1988). It appears to have arisen as a duplication of the testis-determining factor of the mouse Y chromosome and the gene for H-Y antigen, repositioned as the result of an inversion to the distal (pseudoautosomal) arm of the Y beyond the pairing and exchange region (Evans, Burtenshaw & Cattanach, 1982; Singh & Jones, 1982). The repositioned *Sxr* region can be visualised cytologically (Evans *et al.*, 1982), and must therefore be several thousand kilobases in length. In meiosis *Sxr*, which includes the zinc-finger protein sequence *Zfy*, is translocated from this chromosome, designated Y *Sxr*, to the paternal X chromosome, now designated X *Sxr*. *Sxr* would therefore reside on the paternal X chromosome in XX *Sxr* pseudomales and on the Y chromosome in XY carriers (Evans *et al.*, 1982). Fertilisation by an X-bearing spermatozoon carrying the gene produces an XX *Sxr* zygote that will develop as a male.

As already noted, *Sxr* induces testicular development and evidently complete phenotypical masculinisation, apparently due to the secretion of gonadal androgens acting on both the internal and external genital tissues. The one anatomical abnormality so far described in these males is a

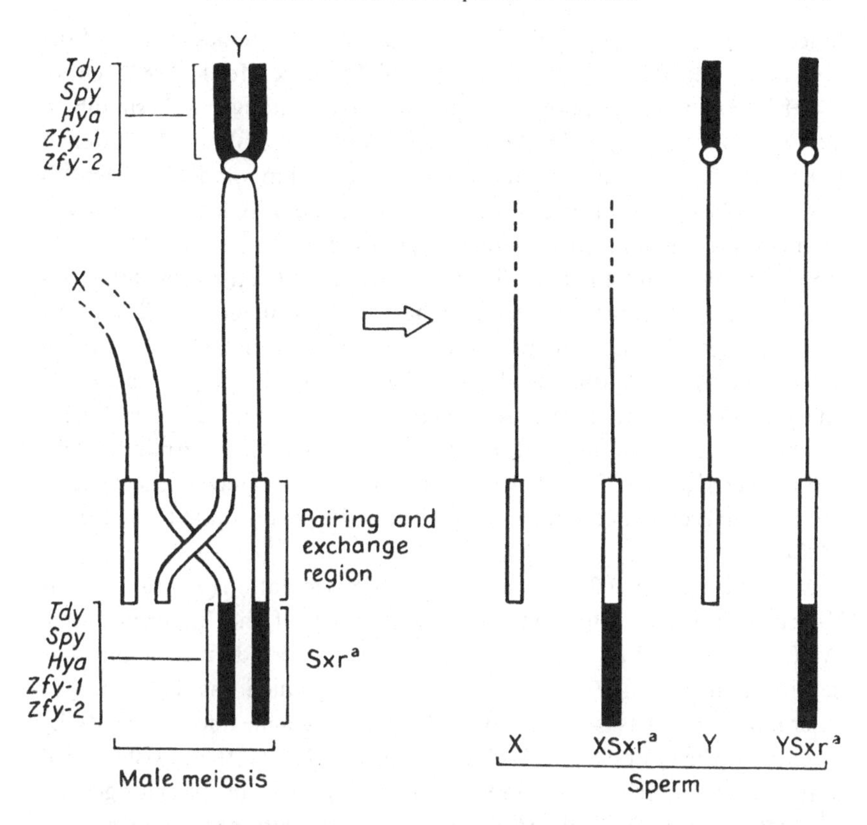

Fig. 6.1. Inherited XX sex reversal in the mouse. The pairing and exchange (pseudoautosomal) region of the normal mouse Y is at the tip of the long arm, whereas the testis-determining gene (*Tdy*) is on the short arm together with *Hya*, *Spy*, *Zfy-1* and *Zfy-2*. XY carriers of the XX sex reversal trait (Sxr^a) have an extra copy of most of the Y short arm located distal to the pairing and exchange region. The obligatory X–Y cross-over during male meiosis results in four types of sperm, including *Tdy*-positive $XSxr^a$ sperm, which generate $XXSxr^a$ sons. Sxr^a is the original form of Sxr. A number of variants have been identified, including the H-Y-negative variant Sxr^b, which arose by deletion of part of Sxr^a. (After Burgoyne, 1992.)

complete lack of the initial segment of the epididymis together with abnormalities in its blood supply (LeBarr & Blecher, 1986). A straightforward molecular explanation for this deficiency has yet to appear in print. The epididymis of X*Sxr*O males, by contrast, includes the initial segment.

Observations concerning the T(X:16)16H translocation, commonly referred to as T16H, seem relevant at this point. This preferentially expressed X–autosome translocation may act to disrupt and partially or fully prevent the masculinising effect of the testis-determining factor.

Indeed, T16H/X*Sxr* embryos may develop as fertile females in spite of the presence of *Sxr* (Cattanach *et al.*, 1982; McLaren & Monk, 1982). Other T16H/X*Sxr* animals develop as sterile males and some even as hermaphrodites (Cattanach *et al.*, 1982; McLaren & Monk, 1982; Ward, McLaren & Baker, 1987). One key to this intriguing spectrum of development is thought to lie in disruption of the random pattern of X inactivation usually found in XX mammals (see Chapter 2). Hence, formation of normal fertile females would be anticipated following preferential inactivation of the *Sxr*-bearing X chromosome and assuming that the *Sxr* sequence itself is thereby inactivated, which may not always be the case (see Cattanach, 1987). As to the formation of hermaphrodites or intersexes, the most reasonable interpretation is that the *Sxr* sequence is inactivated in only some cells, resulting in a mosaic population of male- and female-determining cells (Whitten *et al.*, 1979). Sexual phenotype would be expected to reflect such a proportional cellular composition in the gonadal primordium (Ward *et al.*, 1987).

As noted in Chapter 2, the genes for testis determination (*Tdy*) and for H-Y antigen (*Hya*) are located on the short arm of the Y chromosome (McLaren, 1988; Fig. 6.1). However, in sex-reversed mice a line has developed in which *Tdy* is present but from which the H-Y antigen determining gene has been lost. Such H-Y negative male mice were initially denoted by the symbol *Sxr'* (McLaren *et al.*, 1984), and their existence led to the suggestion that H-Y expression may play a rôle in spermatogenesis rather than being involved directly in testis determination (Burgoyne, Levy & McLaren, 1986). *Sxr* mice that include the sequence required for H-Y antigen expression are now termed Sxr^a whilst the variant that has lost the H-Y transplantation antigen sequence is now termed Sxr^b (Sutcliffe & Burgoyne, 1989). The Sxr^b variant is known to have arisen by deletion of DNA from Sxr^a (Bishop *et al.* 1988). The specific suggestion of Burgoyne *et al.* (1986) was that Sxr^a mice carry a spermatogenesis gene (*Spy*) which is lacking in Sxr^b, the *Spy* gene being expressed cell-autonomously in the germ line and possibly being identical with the gene for H-Y antigen. However, the report of Simpson & Page (1991) shows that Sxr^b differs from Sxr^a as the result of an unequal crossing-over between *Zfy*-1 and *Zfy*-2, leading to a transcribed *Zfy*-2/1 fusion gene and an interstitial deletion (Fig. 6.2).

A different form of sex reversal concerns mice carrying a Y chromosome which leads to the formation of XY phenotypic females which are infertile, seemingly due to a defect inside the XY ovaries. The sex-reversed female mice are generated when the Y chromosome from *Mus poschiavinus* or some strains of *Mus musculus domesticus* (Y^{DOM}) is imposed on a C57BL/

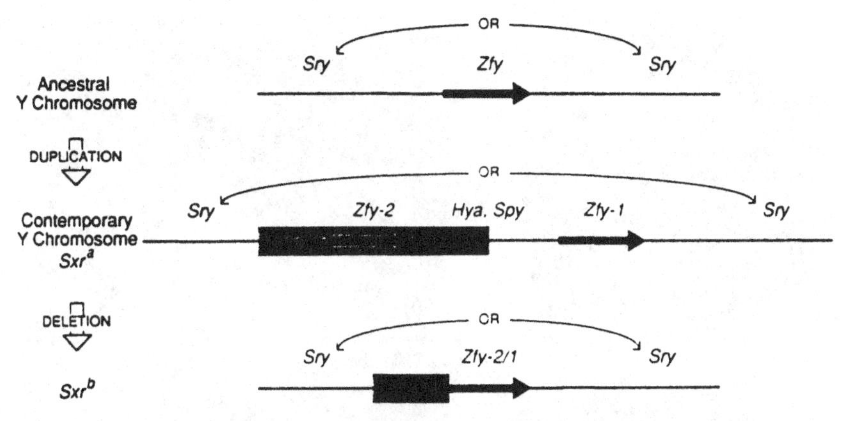

Fig. 6.2. Model depicting evolution and mutation of a block of mouse Y-chromosomal DNA. The single *Zfy* gene on the ancestral Y chromosome was duplicated during rodent evolution, creating *Zfy-1* and *Zfy-2*, both found within *Sxr*^{a}. An interstitial deletion created *Zfy-2/1*, the fusion gene in *Sxr*^{b}. Thick arrows indicate the 5′ to 3′ orientation of the *Zfy* genes. *Hya* and *Spy* may be located anywhere within the deleted region. *Sry* may be located either 5′ or 3′ of the duplicated region. (After Simpson & Page, 1991.)

6J(B6) mouse background. Failure of normal testis development is found in all XY progeny (B6.Y^{DOM}) during foetal life, whilst some develop ovotestes or bilateral ovaries and become phenotypically female (Eicher *et al.*, 1982; Eicher & Washburn, 1983; Nagamine, Taketo & Koo, 1987). These observations led to the presumption that the presence of a Y chromosome in itself is not sufficient to induce differentiation of a testis, and that the *Tdy* gene from *Mus musculus domesticus* (*Tdy*^{DOM}) must interact with another dominant autosomal gene (*Tda-1*) to bring about testicular development; the *Tdy*^{DOM} gene cannot effectively interact with recessive autosomal genes (*tda*-1) from the B6 mouse strain (see Eicher & Washburn, 1983; Taketo-Hosotani *et al.*, 1989).

Because XY females are invariably infertile, even though mating will occur, Eicher *et al.* (1982) suggested that the rapid depletion of oocytes between 4 and 8 weeks of age led to a sterile condition. Taketo-Hosotani *et al.* (1989) were able to demonstrate development of regular oestrous cycles when XX ovaries were grafted into ovariectomised XY females, in contrast to an arrest of oestrous cycles in the reciprocal experiment. They therefore concluded that the principal lesion lies within the XY ovary, and noted that by the time of parturition (19th or 20th day of gestation), oocytes no longer remained within the medullary cords of the XY ovary and thus follicles were not formed in the medullary region, in contrast to the situation in the cortex (Fig. 6.3). This explanation is not universally accepted, and in the

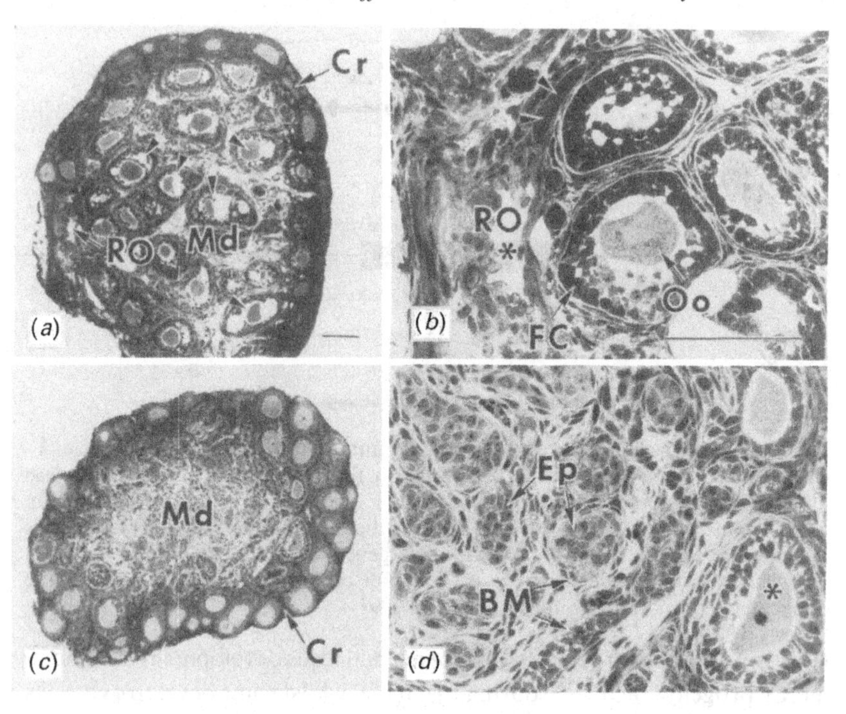

Fig. 6.3. (*a*) An XX ovary at 12 days *post partum*. The section was cut through the centre region including the rete ovary (RO). Follicles have initiated growth in the medullary region (Md). Many are undergoing atresia (arrowheads). The cortex region (Cr) is occupied by primordial follicles. The scale bar represents 0.1 mm. (*b*) Part of the XX ovary shown in (*a*). Follicles in the medullary region, formed with cuboidal follicular cells (FC) and large oocytes (Oo), are undergoing atresia. The rete ovary (RO) is formed by compacted epithelial cells (indicated by arrowheads), which occasionally continue to a lumen (indicated with *). The scale bar represents 0.1 mm. (*c*) An XY ovary at 11 days *post partum*. The medullary region (Md) is completely devoid of follicles, and occupied by remnants of sex cords. Primordial and growing follicles are seen in the cortex region (Cr). Magnification as in (*a*). (*d*) The medullary region of the XY ovary shown in (*c*). Sex cords are composed of epithelial cells (Ep) enclosed in basement membranes (BM). No germ cells are seen inside the sex cord. A follicle near the cortex region is undergoing atresia (indicated with *). Magnification as in (*b*). (Adapted from Taketo-Hosotani *et al.*, 1989. Courtesy of Dr Teruko Taketo.)

report of Mahadevaiah, Lovell-Badge & Burgoyne (1993), emphasis is given in part to the lack of a second X chromosome, in part to the presence of the Y leading to the production of Y oocytes. As a consequence, there would be lethal YY and 'at risk' $X^{P}Y$ conceptuses. In addition, the inefficiency of the Y as a pairing partner for the X during female meiosis leads to increased pachytene atresia.

Other groups have studied comparable or closely similar forms of XY sex reversal in mice in specific strain crosses (e.g. Nagamine *et al.*, 1987; Biddle & Nishioka, 1988; Fig. 6.4). The partial or complete sex reversal was considered due to an inappropriate interaction of different Y chromosomes with autosomal factors. Following on from the observations of Eicher *et al.* (1982) and Eicher & Washburn (1983), ovaries or ovotestes were again found in a proportion of the XY foetuses when the Y chromosome of *Mus musculus poschiavinus* (a local variety of *M. domesticus*) interacted with a putative autosomal gene of C57BL/6J animals – the recessive *tda-1*^{b} (Biddle & Nishioka, 1988). These last authors noted that the *poschiavinus* Y chromosome appears identical to the *domesticus* Y, although with three functionally different classes of *domesticus*-type Y chromosome. These were invoked in a threshold model for the liability and phenotypic expression of gonadal anomalies with different *domesticus*-type Y chromosomes on the C57BL/6J genetic background (see Mittwoch, 1989). At the molecular level, it is now appreciated that the *domesticus*-type Y chromosome lacks a 2.3 kilobase *Taq* I band (Biddle, Eales & Nishioka, 1991).

As to the genesis of XY females, the topic has received an excellent concise review by Burgoyne & Palmer (1991). They note that an obvious route to XY sex reversal would be through a mutation affecting the Y-chromosomal testis-determining factor, and refer to such a mutation that is heritable, Y-linked and fully penetrant (Lovell-Badge & Robertson, 1990). The XY phenotypic females that carry this mutation have lost the primary testis determinant *Sry* due to a deletion, but they retain both copies of *Zfy* and are fertile (see Chapter 2). Capel *et al.* (1993) have recorded three instances of fertile XY females in which the *Sry* locus had remained intact, deletion of Y chromosome sequences occurring outside the minimal testis-determining region. However, as noted by analysis of one of the XY female lines, expression of *Sry* was affected and long-range position effects were therefore invoked to account for such 'silencing'.

Another aetiology for XY females, one that is critically dependent on the timing of gene expression, was offered by Burgoyne & Palmer (1991) on the basis of various previous proposals and their own observations in mice. In essence, their view was that this form of sex reversal could be explained on the basis of a timing mismatch in the programme for gonadal determination. Instead of the *Tdy* programme being imposed relatively early upon the gonadal primordium to order testis formation, its action would in some way be pre-empted by the process of ovary determination. Cattanach (1987) had referred to mutations that delay the timing of *Tdy* action relative to ovary determination rather than by interrupting specific effects

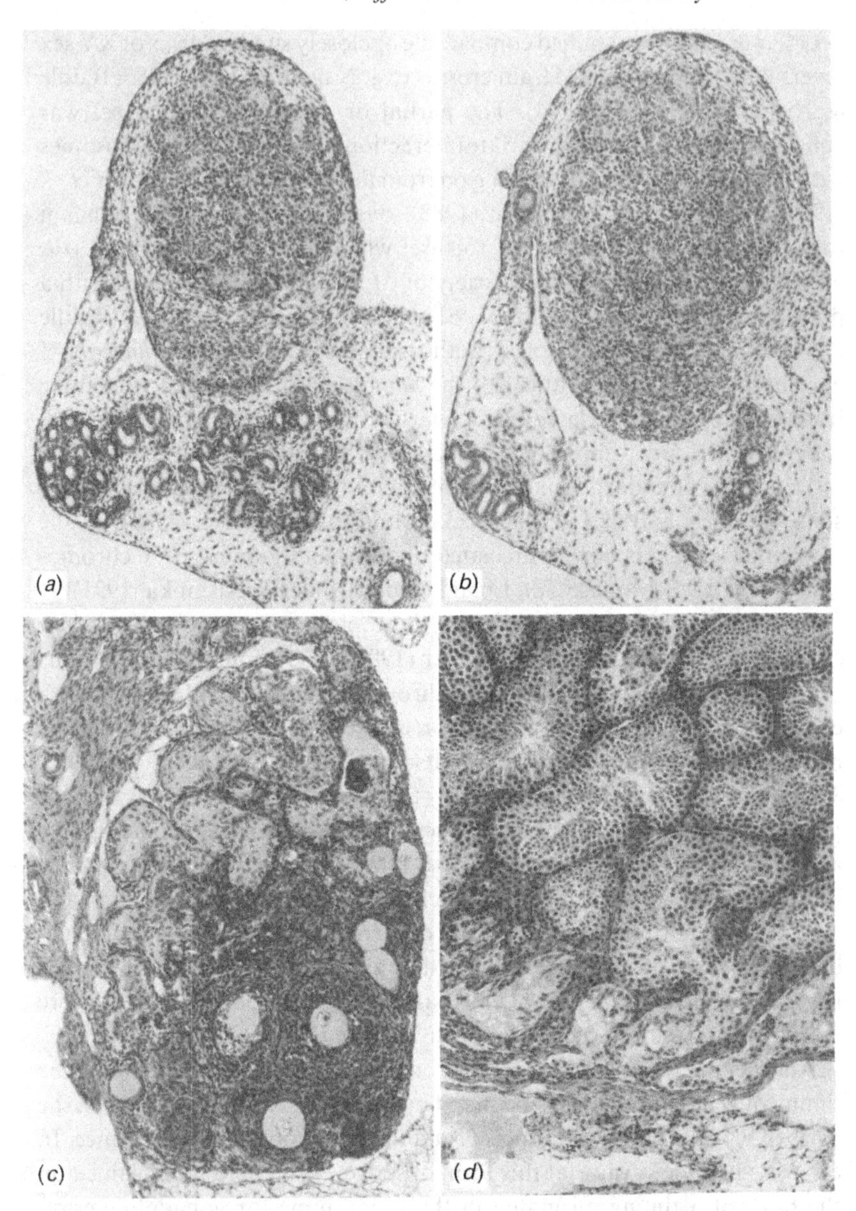

Fig. 6.4. (*a*) and (*b*) B6.Y^{DOM}/Na day 18 mouse ovotestis, cut at slightly different planes, showing clearly the ovarian and testicular regions segregated in (*a*) but contiguous in (*b*). (*c*) Section of a 4-week-old ovotestis from the same strain with primary follicles, disorganised seminiferous tubules, and no clear segregation between the ovarian and testicular regions. (*d*) Section of a testicular gonad, aged 4 weeks 6 days, with an oocyte in the rete testis. (Adapted from Nagamine *et al.*, 1987. Courtesy of Dr Claude Nagamine.)

downstream. Palmer & Burgoyne (1991) were able to show that the *Mus musculus domesticus Tdy* allele acted approximately 14 hours later than the *Mus musculus musculus* allele, and they therefore considered this lapse as a sufficient basis for sex reversal in the appropriate cross. Of course, an outstanding question remains as to precisely what is regulating the time of expression of *Tdy*, thereby compromising the destiny of the gonads. Homozygosity for an autosomal recessive B6 allele is apparently required for the XY sex reversal (Eicher & Washburn, 1983), the *Tda-1* gene so involved being considered by Burgoyne & Palmer (1991) to be implicated in ovary determination as an early-acting variant. Thus, as remarked by McLaren (1991), the difference between the C57BL and other strains could involve the rate of ovarian development rather than effects on the timing of testis determination. Indeed, Eicher & Washburn (1983, 1986) had already proposed a timing mismatch to explain XY females.

Testicular feminisation in mice and rats

The condition of testicular feminisation in mice received description and general analysis from Lyon & Hawkes (1970). The internal reproductive organs of these animals are restricted to testes that remain abdominal (inguinal) and a short, blind-ending vagina. Neither Wolffian nor Müllerian duct derivatives are seen. Although affected animals are of unequivocal XY sex chromosome constitution, externally they resemble phenotypic females. How can this be so? The condition is due to an X-linked mutant gene (*Tfm*) which determines the ability of somatic cells to respond to testosterone; in fact, the target organs fail to respond to androgens. This murine condition is thought to parallel testicular feminisation in rats and humans, in which there is a clear X-linked inheritance. The *Tfm* locus of the X chromosome is specifically responsible for a deficiency in androgen receptors in target tissues, and thus the androgen insensitivity of the sex organ anlagen (Ohno & Lyon, 1970). Only genetic males carrying the *Tfm* locus on their X chromosome are affected, and they are known to have a defective cytosol androgen-binding protein (Attardi & Ohno, 1974; Bullock & Bardin, 1974; Gehring & Tomkins, 1974). During the early stages of embryogenesis, a male duct system and accessory organs are not induced to form and differentiate, although testosterone is secreted by the interstitial cells of abdominal testes. Sertoli cells, dividing spermatogonia and early spermatocytes were always present in mice, which stands in contrast to most instances of this syndrome in man (see Chapter 7). The *Tfm* syndrome is thus a rare example of an extra-gonadal action of the sex chromosomes powerfully influencing the processes of sexual differentiation.

As would be anticipated, the female phenotype of $X^{Tfm}Y$ mice cannot be modified by administration of testosterone to the mother, although female litter mates do become masculinised by this treatment (Goldstein & Wilson, 1972). Even so, and in contrast to the previous understanding, the report of Scott & Blecher (1987) on the *Tfm*/Y mouse indicated that derivatives of the mesonephric duct system may be distinguishable in hemizygous animals. Microscopical epididymides, rete testes and vasa deferentia were found, leading to the question as to whether their elaboration might have been uniquely independent of androgen support or whether there was vestigial ability to respond to androgens. Using the enzyme β-glucuronidase as a marker, there was a suggestion of some degree of androgen sensitivity although a lack of response to exogenous androgen stimulation. Scott & Blecher (1987) speculate that this bizarre survival of microorgans in an otherwise androgen-insensitive male duct system might in some way be related to an evolutionarily conserved rôle in inducing meiosis. This point stems from the studies of Byskov (1978) on the meiosis-inducing action of rete cells in the mesonephric system and, as inferred above, meiosis is initiated in *Tfm*/Y mice. The more general point is that, in normal circumstances, mesenchymal tissue develops androgen receptors before sexual dimorphism becomes apparent (Takeda *et al.*, 1987).

The technique of embryo aggregation enabled chimaeras to be produced containing *Tfm*/Y and normal XY cells, elegantly demonstrating that the cytosol receptor protein is not needed within germ cells for normal spermatogenesis. Chimaeric males could sire offspring from both *Tfm*/Y cells and their normal cells (Lyon & Glenister, 1980). Female mice homozygous for the mutant gene *Tfm*/*Tfm* could thus be generated from fertile chimaeric males, and used to study the rôle of androgen resistance (i.e. the defective receptor) in female reproduction. All seven *Tfm*/*Tfm* females obtained were fertile, clearly indicating that a normal level of androgen-binding protein (and hence androgenic influence) is not essential for reproduction in female mice. Even so, there was a premature cessation of reproduction, with the suggestion of accelerated ageing, perhaps involving altered rates of follicular growth and atresia.

Testicular feminisation has also been described in rats (Bardin *et al.*, 1970) and, as would be anticipated, such animals failed to respond to androgens due, once more, to a deficiency of androgen receptor protein (Bardin *et al.*, 1973). The *Tfm* animal is sometimes referred to as the pseudohermaphrodite rat and has features in common with the *Tfm* mouse, with no external signs of virilisation and no internal genitalia except for small testes.

Quite apart from the intrinsic interest of this *Tfm* syndrome, it demonstrates that for normal maleness, that is the diverse responses to androgens, there is an involvement of a gene on the X chromosome to regulate receptor activity in terms of binding proteins in target tissues. The interrelationships of the X and Y chromosomes are therefore set in a clear developmental context by this example.

Hermaphrodites or intersexes in mice and rats

Reference in the following paragraphs to hermaphrodites is largely to animals that might more appropriately be termed intersexes (see definitions in Chapter 5). Whilst ovarian and testicular tissue may be present in the form of an ovotestis within the same animal, there is no persuasive evidence that both types of tissue are fully functional at puberty. In any event, there is said to be an extremely low incidence of spontaneous hermaphroditism in mice, although so-called true hermaphrodites have been described by Danforth (1927), Fekete (1937) and Hollander, Gowen & Stadler (1956), amongst others. The last authors supposed that chromosomal differences would account for an ovary and testis within the same animal, thereby invoking mosaicism as an explanation for intersexuality. This interpretation was endorsed by Lyon (1969), who demonstrated populations of chromosomally different cells within a sterile, phenotypically male mouse presumed to be an XO/XY mosaic. The reproductive tissues revealed upon dissection were a testis, epididymis, vas deferens and male accessory glands on the left and an ovary, oviduct, uterus and, again, male accessory glands on the right. The presumption was that XY cells can act to determine formation of testicular tissue and XO cells to promote ovarian tissue.

Tarkowski (1961) refers to three cases of intersexuality among mouse chimaerae formed from fused eggs, the judgement being based upon macroscopic examination of the reproductive system. However, a subsequent paper of Tarkowski (1964), again dealing with chimaeric mice derived from fused embryos, refers to instances of true hermaphrodites in which three of 14 newborn intersex animals had gonads containing ovarian and testicular tissue. (It is unclear whether these two publications, some 3 years apart, refer to the same experimental material.) Two animals had an ovary and contralateral ovotestis, although the 'sides' of such gonads were reversed in the two instances; the third animal had two ovotestes. As to the origin of such gonadal derangements, Tarkowski appeared to favour the hypothesis of sex chromosome mosaicism underlying ovotestis formation (i.e. XX and XY cells in the same gonad), but emphasises that the genetic sex of germ cells in such chimaeric hermaphrodites was unknown. As to spontaneously arising hermaphrodites with sex chromosome mosaicism, he envisaged mitotic errors in an XY zygote commencing with non-disjunction of the Y chromosome. In an aside in the same paper, Tarkowski (1964) noted that the genital ducts did not correspond with the sex of the adjacent gonad in a minority of hermaphrodite animals. He concluded that the gonad must have been involved in masculinising activity at the time of sexual differentiation and, further, that testicular tissue had been overlooked or had regressed by the time of investigation.

A much enhanced incidence of hermaphrodites was reported by Beamer, Whitten & Eicher (1978), who noted 3.0% among 1439 inbred BALB/cWt 15-day mouse foetuses compared with only 0.4% among 3310 animals of the same strain found by Whitten *et al.* (1979) at weaning. The novel inference was drawn that a majority of hermaphrodite mice may have 'disappeared' during the late foetal or postnatal development (Whitten *et al.*, 1979). These latter authors seemed to suggest a 'self-correcting' mechanism within the hermaphrodite gonad, with selective tissue proliferation occurring so that the gonad tends to revert to type. But other explanations for the different incidence of hermaphrodites between the two studies would seem possible, especially since no evidence of selective mortality during the period in question was reported. Not least, an influence of different stud males should be considered, since a genetic predisposition to gonadal anomalies can be demonstrated in this respect. In line with the earlier suggestion of Tarkowski (1964), the cause of the so-called hermaphrodite condition was postulated to be sex chromosome mosaicism (Beamer *et al.*, 1978). Indeed, preliminary findings in which the liver of foetal mouse hermaphrodites was karyotyped revealed hermaphrodites to be sex chromosome mosaics (XO/XY or XO/XY/XYY).

When a B6.Y^{DOM} male mouse was crossed with B6 females, half of the XY progeny had paired ovaries with corresponding female genitalia whereas the other half had bilateral testes or had developed as hermaphrodites (Taketo-Hosotani *et al.*, 1989), in line with the report of Eicher & Washburn (1983). Also studying XY sex reversal when a Y chromosome from wild populations of *Mus musculus domesticus* is placed on a C57BL/6J genomic background, Nagamine *et al.* (1987) highlighted hermaphrodites with XY ovaries and contralateral ovotestes or testes. Testicular tissue in ovotestes was invariably concentrated in the mid region of the gonad, with ovarian tissue distributed at the cranial and caudal poles. In such ovotestes, a tunica albuginea was well differentiated only over the region occupied by seminiferous tubules. The ovarian regions were covered by a single layer of

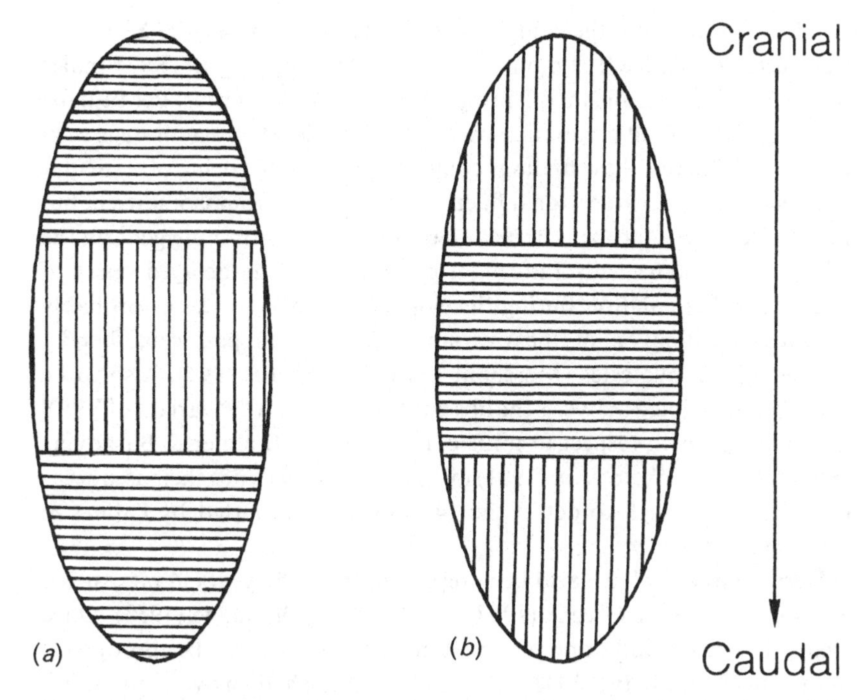

Fig. 6.5. Diagram of gonadal zones in portions of ovotestes in (*a*) mice and (*b*) *Xenopus laevis*, in which the horizontal stripes represent ovarian and vertical stripes testicular tissue. (After Mittwoch, 1992.)

surface epithelium and either lacked or showed poorly differentiated seminiferous tubules, cells in the ovarian regions frequently being arranged as cords. Eicher (1982) had previously noted that foetal ovotestes from *Tda-1* and *Tas XY* sex-reversed hermaphrodites and in hermaphrodites of the BALB/cWt strain were anatomically similar: seminiferous cords tended to be concentrated in the mid region of ovotestes whilst ovarian tissues were concentrated at the cranial and/or caudal poles. Such non-random distribution of tissues received further comment from Mittwoch (1992), drawing in part on well-documented evidence from mosaics and chimaeras (Fig. 6.5).

Initiation of the testis-determination pathway by the *Tdy* allele of *Mus musculus domesticus* had been proposed to occur later in development than when prompted by the *Tdy* of B6 (Eicher & Washburn, 1986). Accordingly, chronological aspects of testicular differentiation in the B6.Y^{DOM} ovotestis were examined by Taketo *et al.* (1991). Using the approach of immunocytochemical staining for anti-Müllerian hormone (AMH), they found that the B6.Y^{DOM} ovotestis initiated differentiation of testicular tissue later in development than did the control B6 testis. However, when the Y^{DOM} was transferred onto an SJL/J mouse background by crossing B6.Y^{DOM} males with SJL/J females, all XY offspring developed normal testes. Because the onset of testicular differentiation was at the same developmental stage as in the B6 male foetus, these findings could suggest that the delay of testicular differentiation is not due to an influence of the Y^{DOM} chromosome itself, but rather due to an improper interaction of the testis-determining gene on the Y^{DOM} chromosome with autosomal genes of B6. The delayed onset of testicular differentiation during development of XY ovotestes probably permits switching on of the ovarian differentiation pathway, thereby producing XY ovotestes (Merchant-Larios & Taketo, 1991). Also to be considered, as noted above, is the observation that the Y^{POS} variant of Y^{DOM} is later acting with respect to testis cord formation (Palmer & Burgoyne, 1991). Perhaps Taketo *et al.* (1991) missed this finding since their assay was not sensitive enough to detect the 14 hour delay reported by Palmer & Burgoyne (1991).

Hermaphrodites have also been reported down the years in rats, as for example by Arey & Greene (1957) and Bradbury & Bunge (1958). Once again, the present author prefers the term intersex, as would be appropriate in Burrill, Greene & Ivy (1941) in which the animal had an ovary on the left and an ovotestis on the right. The ovotestis was sub-perineal in location and the right oviduct was absent. The two rats reported by Greep (1942) consisted, in the first, of a testis on the left and an ovary on the right, whereas in the second there was a scrotal testis on the left and an ovotestis on the right. Bilateral ovotestes were reported in a so-called true hermaphrodite by Arey & Greene (1957), whilst Bradbury & Bunge (1958) documented ovotestes containing oocytes – perhaps the report that prompted the subsequent title of McLaren (1980) on 'oocytes in the testis'. Bradbury & Bunge noted three hermaphrodites among 1500 weanling Sprague–Dawley rats. Two of these had bilateral ovotestes, the third a testis on one side and ovary on the other. Convoluted tubules in the testicular portion contained 'ova' which were thought to have 'migrated' from granulosa areas adjacent to the cortex. In fact, the experimental study of MacIntyre (1956) in mice, in which a testis and ovary were grafted together in the kidney of castrated recipients, led to the development of seminiferous-like tubules in ovaries which, in some cases, contained well-developed oocytes. None of these reports presented evidence of the karyotype, so the genetic sex of the animals was unknown and their provenance was not discussed.

Hypogonadal mice

The hypogonadal (*hpg*) mutant mouse represents a condition that affects both sexes. It arose spontaneously in a colony at the Medical Research Council Radiobiology Unit at Harwell, England. The condition is inherited as an autosomal recessive (Cattanach *et al.*, 1977), and is characterised in the homozygous state by an absence of secondary sexual development.

Although *hpg* animals are born with morphologically intact genital tissues, the testes remain small and the ovaries minute. The intra-abdominal testes contain interstitial tissue of atrophic appearance, and spermatogenesis does not progress beyond diplotene of the first meiotic prophase. The ovaries rarely contain antral follicles and the thread-like uterine tissues show no evidence of oestrogenic stimulation. The vagina may reveal some degree of opening but usually remains imperforate; smears from perforate *hpg* animals are devoid of cellular material. During the interval between weaning and 1 year of age, no further development of the gonads or genital tract in these *hpg* mice was demonstrable. Even so, the apparent morphological normality of the reproductive system at birth would suggest a shortage of functional gonadotrophic hormones or a deficiency in the higher centres as one cause of failure of subsequent development. As an alternative explanation for the *hpg* state, the mutant animal might be completely unresponsive to pituitary hormones at the gonadal level. Further explanations situated in the chain of events between the brain and gonadal endocrine function could be speculated upon.

However, due to a series of elegant experiments, the primary lesion in the hypogonadal mouse is now known to be defective production of the hypothalamic gonadotrophin releasing factor (GnRH) that regulates synthesis and secretion of gonadotrophic hormones in the anterior pituitary gland. The experimental evidence has been summarised systematically by Lyon *et al.* (1981); it runs as follows.

1 The hypothalamic content of GnRH between 30 and 120 days of age – if any – could not be detected by radioimmunoassay in the mutant animal, and the follicle stimulating hormone (FSH) and luteinising hormone (LH) content of the pituitary gland was extremely low, whilst plasma levels of these gonadotrophins were virtually undetectable (Cattanach *et al.*, 1977).
2 The classical technique of electrical stimulation of the median eminence region of the hypothalamus (Harris, 1972) failed to elicit a response in terms of pituitary secretion of LH, suggesting that the *hpg* hypothalamus is not producing GnRH (Iddon, Charlton & Fink, 1980).

3 Pituitary tissue itself can respond to hypothalamic trophic secretion as demonstrated by injecting mutant animals with the synthetic decapeptide form of GnRH, resulting in secretion of LH (Iddon *et al.*, 1980).
4 Using immunocytochemical techniques of appropriate sensitivity based on an antiserum against a GnRH conjugate, immunologically reactive GnRH could not be demonstrated in the median eminence of *hpg* mice, endorsing the previous conclusions.
5 Chronic experiments in which the mutant mice were administered synthetic GnRH either by daily subcutaneous injections or via silastic implants containing a reservoir of the hormone resulted in an elevation of pituitary LH and FSH. Full spermatogenesis has been observed after 60 days of such treatment, whilst female mutants responded to GnRH implants within 5–10 days with vaginal opening and increased uterine weight. These experiments therefore indicate that the *hpg* pituitary gland is able to synthesise and secrete biologically active gonadotrophic hormones in response to a suitable stimulus (Lyon *et al.*, 1981).
6 Finally, transplantation experiments have demonstrated what already seems apparent – that the gonads of both male and female *hpg* mice can show functional responses to appropriate stimulation. Thus, *hpg* testes transplanted under the tunica albuginea of testes of normal males developed to display full spermatogenic activity within 80 days, and *hpg* ovaries transplanted into the periovarian capsule of normal animals (whose own gonads had been removed) developed a fully functional state represented by mature follicles, ovulation and formation of corpora lutea. As a sequel to such transplantation, there has been birth of *hpg* offspring in at least 50% of instances (Bamber *et al.*, 1980).

Hence the overall conclusion is that the aetiology of the *hpg* condition resides in a deficiency of GnRH production, the effects of which are customarily detectable only after birth (Lyon *et al.*, 1981). The pituitary–gonadal axis itself had retained the potential to be physiologically functional. This latter point has been demonstrated most elegantly by means of grafting foetal brain tissue, that is tissue containing GnRH cells from the preoptic area (Krieger *et al.*, 1982). Normal gonadotrophic and gonadal function in *hpg* mice was restored by this approach, and histological examination demonstrated that over 95% of the GnRH axons leaving the grafts innervated the median eminence, with the nerve terminals abutting on the hypothalamo-hypophyseal portal system (Silverman *et al.*, 1985). However, it is now appreciated that *hpg* males restored physiologically following tissue grafting fail to show masculine sexual behaviour in response to receptive females and therefore fail to impregnate them (Livne, Silverman & Gibson, 1992). This is thought to stem from deficient perinatal androgens in the *hpg* mouse which are required to masculinise copulatory behaviour, and this behaviour seemingly cannot be programmed in a defective adult by the transplant approach. Neonatal androgenisation, by contrast, is effective in promoting classical copulatory behaviour (Livne *et al.*, 1992).

As to a more specific analysis of the reason for defective GnRH synthesis, the studies of Mason *et al.* (1986*a*) demonstrated that half of the GnRH gene has suffered deletion in the *hpg* mouse. The deletion or truncation encompasses the distal half of the gene for the common precursor of GnRH and GnRH-associated peptide. It would therefore not be possible for these mutant animals to transcribe and translate the gene in order to produce the appropriate GnRH protein. Reproductive function has been restored in hypogonadal mice by transfer of the gene encoding GnRH with 5 kilobases of 5′ and 3.5 kilobases of 3′ flanking sequence (Mason *et al.*, 1986*b*). Such a transgene showed appropriate tissue-specific expression and the resultant transgenic mice of both sexes reached sexual maturity, mated, and were able to produce healthy offspring. Complete restoration of reproductive function in cases of gene deletion has therefore been demonstrated by this transgenic approach, providing an exciting model system for a more complete analysis of the regulation of GnRH expression and the rôle of GnRH-associated peptide.

Gonadal dysgenesis – streak gonads

Whereas women with an XO sex chromosome complement represent Turner's syndrome and are almost always infertile, with their ovaries represented by streaks of connective tissue (see Chapter 7), a corresponding syndrome in XO mice is seldom – if ever – seen. XO mice have histologically normal ovaries and are invariably fertile (Welshons & Russell, 1959; Cattanach, 1962), but there is a shortened reproductive life and premature onset of sterility due to an accelerated rate of atresia and exhaustion of a reduced pool of oocytes. In such XO mice, which become sterile earlier than their XX sibs, Lyon & Hawker (1973) have concluded that oocyte loss is a consequence of X chromosome monosomy, as is the case in man. And, as summarised by McLaren (1991), oocyte survival is prejudiced if only one X chromosome is present for, in both XO and XY female mice, the number of germ cells in the postnatal ovary is greatly decreased and the reproductive lifespan shortened.

Specific details of the germ cell population and of oocyte depletion in XO mice were presented by Burgoyne & Baker (1981), it being noted that the oocyte population is already approximately halved by the 12th day *post partum*, the earliest stage studied. In a subsequent report, they found that the oocyte deficiency in XO mice was due to an excess atresia of oocytes at the late pachytene stage, 19.5 days *post coitum* (Burgoyne & Baker, 1985), that is in the perinatal period. The explanation underlying this loss may in fact be more subtle than a requirement for two X chromosomes (that is an X dosage deficiency), if the oocyte is to have a normal lifespan (see Short, 1978), and perhaps bears on stringent requirements during meiosis. One suggestion is that there may be a selective destruction of oocytes having unpaired or incompletely paired chromosomes during the stage of pachytene (Miklos, 1974; Burgoyne & Baker, 1984), implying rather complex interactions related to X chromosome activity. Clearly, the X has no formal pairing partner in XO females, but Burgoyne & Baker (1985) have made the intriguing observation that in XO oocytes at 19.5 days *post coitum*, many of the healthy pachytene cells have the X present as a 'hairpin', that is it has paired (non-homologously) with itself (Speed, 1986), and it may be this form of pairing which permits some XO oocytes to survive. Because XO mice are fertile, it has been possible to examine reproduction in XO females. Various studies have indicated that XO females produce fewer than expected XO offspring. A recent explanation for this reduction bears on the origin of the X chromosome. A highly significant prenatal loss of females was found when the single X was of paternal origin, whereas there was no comparable loss of females with a single X chromosome of maternal origin (Hunt, 1991). Accordingly, the reduced viability of females with a paternally derived X could be mediated by the parental origin of the X, that is as an expression of X chromosome imprinting. Alternatively, because the mothers of females with a single paternally derived X have only a single X chromosome, the effect might be mediated by the genotype of the mother, that is a maternal uterine effect (Hunt, 1991). In summary, these views suggest that the extremely high *in utero* mortality of XO offspring could be attributable either to X chromosome imprinting or to maternal environmental influences.

Observations from transgenic animals

Spontaneous mutations rather than drug-induced or experimentally induced lesions are recognised as providing valuable model systems in many fields of study. However, the possibility of generating mutant strains of animal has also arisen with advances in molecular biology and its application, enabling the production of transgenic animals. Provided that attention is directed at a single-copy gene, it should be possible to replace such a gene by homologous recombination of defined DNA into the site of the original gene. Not fully in this context but nonetheless of relevance, Behringer *et al.* (1990) were able to produce transgenic mice chronically expressing human AMH. As would be anticipated, the females showed no Müllerian duct derivatives and the germ cell population in such transgenic ovaries was markedly depleted after birth; by 2 weeks *post partum*, no oocytes remained. AMH had apparently acted adversely upon the foetal ovary, in some manner destroying the XX germ cells and thereby causing ovarian aplasia (Behringer *et al.*, 1990). As noted in Chapters 4 and 5, untimely exposure of a female foetus to AMH, as in these transgenic mice, can provoke virilisation of the ovary from both a morphological and functional standpoint. McLaren (1990) has suggested that one possible rôle for AMH in the foetal testis could be to wipe out any oocytes that had succeeded in entering meiosis despite the customary inhibiting influence of testis cords. In fact, cord-like structures resembling seminiferous tubules were also seen containing Sertoli cells in these transgenic mice, but there was a failure of the Sertoli cells to survive or to induce Leydig cell differentiation, suggesting that they were probably not fully competent. The seminiferous tubule-like structures degenerated and were undetectable in adult females. In transgenic male mice showing abnormally high AMH expression, affected animals showed external genital feminisation, abnormal Wolffian duct development and cryptorchidism. Speculation that AMH may be involved in the regulation of testicular descent has already been noted (Chapter 4). These transgenic male mice suffered a loss of testicular germ cells, suggesting that extra-physiological levels of AMH can also influence the developmental fate of XY germ cells.

There is a further observation in transgenic mice that appears of direct relevance to gonadal perturbations. In transgenic females homozygous for the deleted α-inhibin allele, the ovaries develop stromal tumours and, in one instance, seminiferous-like tubules had formed (Matzuk *et al.*, 1992). These inhibin-deficient mice had initially developed normally, but every animal eventually developed mixed or incompletely differentiated gonadal stromal tumours either unilaterally or bilaterally. This transgenic model therefore served to demonstrate that inhibin acts as a critical negative regulator of stromal cell proliferation in the gonad. Further studies are needed to clarify the extent to which differentiation of Sertoli-like cells underlies the wayward development towards a testis.

In XX mice sex reversed by transfer of a male-determining sequence into the newly fertilised egg, such transgenic males are sterile. Germ cells are prevented from progressing beyond the stage of prospermatogonia and degenerate, and the testes are significantly smaller than in fertile males (Koopman *et al.*, 1991). Sterility reflects the presence of two X chromosomes, and had previously been documented in sex-reversed mice (XX *Sxr*) which were male due to the presence of Y-derived sequences on one of their X chromosomes (see above). However, genes other than those transferred would seem essential for functional spermatogenesis.

Another transgenic model in mice that has a defective reproductive system is one in which a viral oncogene, the simian virus 40 large tumour (T) antigen, is expressed, driven by the promoter of the human GnRH gene (Radovick *et al.*, 1991). This modification results in *central hypogonadism* due to an arrest in GnRH neuronal migration during development and tumour formation along the migratory pathway. A particular value of this transgenic mouse would be as a model for the anomaly of hypogonadotrophic hypogonadism in humans, generally referred to as Kallmann's syndrome (Chapter 7). Even so, the point is made that agenesis of the olfactory bulbs, such as that seen in some cases of Kallmann's syndrome, may represent only one disorder in a spectrum of disorders of GnRH neuronal migration that result in central hypogonadism, so other models may need to be generated. An example would be the recent one involving immortalised hypothalamic GnRH neurons by means of genetically targeted tumorigenesis (Mellon *et al.*, 1990).

Other modifications or defects in the reproductive tissues of genetically manipulated mice have been noted in the review of Wilmut, Hooper & Simons (1991).

Concluding remarks

The various types of animal described in this chapter serve to illustrate some of the diverse anomalies that may arise in the reproductive system due to spontaneous modifications or specific errors in the genome. In addition to their rather special value for physiologists interested in the function of organs under a range of conditions, mutant animals or those with major chromosome anomalies can provide geneticists with powerful models with which to clarify gene function in the normal genome. Now that the era of molecular genetics is firmly upon us, one can anticipate an ever-increasing application of either targeted gene deletions or the transgenesis approach – i.e. actual gene insertion – with which to confirm precisely the nature of the modifications that generate these interesting and often bizarre animals, and indeed to generate other anomalous forms of animal.

As mutations arise spontaneously but can also be created by such technology, one can anticipate some heated debate as to what is physiological and what must be regarded as extraordinary. Perhaps a prudent starting point would therefore be to concentrate molecular analyses on 'classical' anomalies, such as the well-known W mutant mouse in which the testes demonstrate germ cell aplasia. This condition stems from disruption of migration and proliferation of the primordial germ cells (Mintz, 1957; Mintz & Russell, 1957), but the precise nature of the molecular lesion(s) remains to be specified.

And, in the genetic analysis of sexual anomalies, it is not only the modification or absence of genes that needs special consideration but also the possibility of an inappropriate timing of specific programmes. This is well illustrated in Burgoyne's views on circumstances in which ovarian determination may pre-empt testis formation in genetic males.

References

Arey, L. B. & Greene, R. R. (1957). A true hermaphroditic rat with ovotestes and evidences of double hormonal stimulation. *Anatomical Record*, **127**, 457–8.

Attardi, B. & Ohno, S. (1974). Cytosol androgen receptor from kidney of normal and testicular feminized (*Tfm*) mice. *Cell*, **2**, 205–12.

Bamber, S., Iddon, C.A., Charlton, H. & Ward, B. J. (1980). Transplantation of the gonads of hypogonadal (*hpg*) mice. *Journal of Reproduction and Fertility*, **58**, 249–52.

Bardin, C. W., Bullock, L., Schneider, G., Allison, J. E. & Stanley, A. J. (1970). Pseudohermaphrodite rat: end organ insensitivity to testosterone. *Science*, **167**, 1136–7.

Bardin, C. W., Bullock, L. P., Sherins, R. J., Mowszowicz, I. & Blackburn, W. R. (1973). Androgen metabolism and mechanism of action in male pseudohermaphroditism: a study of testicular feminisation. *Recent Progress in Hormone Research*, **29**, 65–109.

Beamer, W. G., Whitten, W. K. & Eicher, E. M. (1978). Spontaneous sex mosaicism in BALB/cWt mice. In *Gatlinburg Symposium on Genetic Mosaics and Chimeras in Mammals*, ed. L. B. Russell, pp. 195–208. New York: Plenum Press.

Behringer, R. R., Cate, R. L., Froelick, G. J., Palmiter, R. D. & Brinster, R. L. (1990). Abnormal sexual development in transgenic mice chronically expressing Müllerian inhibiting substance. *Nature (London)*, **345**, 167–70.

Biddle, F. G., Eales, B. A. & Nishioka, Y. (1991). A DNA polymorphism from five inbred strains of the mouse identifies a functional class of *domesticus*-type Y chromosome that produces the same phenotypic distribution of gonadal hermaphrodites. *Genome*, **34**, 96–104.

Biddle, F. G. & Nishioka, Y. (1988). Assays of testis development in the mouse

distinguish three classes of *domesticus*-type Y chromosome. *Genome*, **30**, 870–8.
Bishop, C. E., Roberts, C., Michot, J. L., Nagamine, C. M., Winking, H., Guenet, J. L. & Weith, A. (1989). A molecular analysis of the mouse Y chromosome. In *Evolutionary Mechanisms in Sex Determination*, ed. S. S. Wachtel, pp. 79–90. Boca Raton, Florida: CRC Press.
Bishop, C. E., Weith, A., Mattei, M. G. & Roberts, C. (1988). Molecular aspects of sex determination in mice: an alternative model for the origin of the *Sxr* region. *Philosophical Transactions of the Royal Society of London, Series B*, **322**, 119–24.
Bradbury, M. W. (1987). Testes of XX↔XY chimeric mice develop from fetal ovotestes. *Developmental Genetics*, **8**, 207–18.
Bradbury, J. T. & Bunge, R. G. (1958). Oocytes in seminiferous tubules. *Fertility and Sterility*, **9**, 18–25.
Bullock, L. P. & Bardin, C. W. (1974). Androgen receptors in mouse kidney: a study of male, female and androgen-insensitive (*Tfm*/Y) mice. *Endocrinology*, **94**, 746–56.
Burgoyne, P. S. (1978). The role of the sex chromosomes in mammalian germ cell differentiation. *Annales de Biologie Animale, Biochimie et Biophysique*, **18**, 317–25.
Burgoyne, P. S. (1982). Genetic homology and crossing over in the X and Y chromosomes of mammals. *Human Genetics*, **61**, 85–90.
Burgoyne, P. S. (1988). Role of mammalian Y chromosome in sex determination. *Philosophical Transactions of the Royal Society of London, Series B*, **322**, 63–72.
Burgoyne, P. S. (1992). Y chromosome function in mammalian development. *Advances in Developmental Biology*, **1**, 1–29.
Burgoyne, P. S. & Baker, T. G. (1981). Oocyte depletion in XO mice and their XX sibs from 12 to 200 days *post partum*. *Journal of Reproduction and Fertility*, **61**, 207–12.
Burgoyne, P. S. & Baker, T. G. (1984). Meiotic pairing and gametogenic failure. In *Controlling Events in Meiosis*, ed. C. W. Evans & H. G. Dickinson, pp. 349–62. Cambridge: Company of Biologists.
Burgoyne, P. S. & Baker, T. G. (1985). Perinatal oocyte loss in XO mice and its implications for the aetiology of gonadal dysgenesis in XO women. *Journal of Reproduction and Fertility*, **75**, 633–45.
Burgoyne, P. S. & Mahadevaiah, S. K. (1993). Unpaired sex chromosomes and gametogenic failure. In *Chromosomes Today*, ed. A. T. Sumner & A. C. Chandley, vol. 11, pp. 243–63. London: Chapman & Hall.
Burgoyne, P. S., Mahadevaiah, S. K., Sutcliffe, M. J. & Palmer, S. J. (1992*b*). Fertility in mice requires X-Y pairing and a Y-chromosomal 'spermiogenesis' gene mapping to the long arm. *Cell*, **71**, 391–8.
Burgoyne, P. S. & Palmer, S. J. (1991). The genetics of XY sex reversal in the mouse and other mammals. *Seminars in Developmental Biology*, **2**, 277–84.
Burgoyne, P. S., Levy, E. R. & McLaren, A. (1986). Spermatogenic failure in male mice lacking H-Y antigen. *Nature (London)*, **320**, 170–2.
Burgoyne, P. S., Sutcliffe, M. J. & Mahadevaiah, S. K. (1992*a*). The role of unpaired sex chromosomes in spermatogenic failure. *Andrologia*, **24**, 17–20.
Burrill, M. W., Greene, R. R. & Ivy, A. C. (1941). A case of spontaneous intersexuality in the rat. *Anatomical Record*, **81**, 99–117.
Byskov, A. G. (1978). The meiosis inducing interaction between germ cells and

rete cells in the foetal mouse gonad. *Annales de Biologie Animale, Biochimie et Biophysique*, **18**, 327–34.

Capel, B., Rasberry, C., Dyson, J., Bishop, C. E., Simpson, E., Vivian, N., Lovell-Badge, R., Rastan, C. & Cattanach, B. M. (1993). Deletion of Y chromosome sequences located outside the testis determining region can cause XY female sex reversal. *Nature Genetics*, **5**, 301–7.

Cattanach, B. M. (1962). XO mice. *Genetics Research (Cambridge)*, **3**, 487–90.

Cattanach, B. M. (1975). Sex reversal in the mouse and other mammals. In *The Early Development of Mammals*, ed. M. Balls & A. Wild, pp. 305–17. Cambridge: Cambridge University Press.

Cattanach, B. M. (1987). Sex-reversed mice and sex determination. *Annals of the New York Academy of Sciences*, **513**, 27–39.

Cattanach, B. M., Evans, E. P., Burtenshaw, M. D. & Barlow, J. (1982). Male, female and intersex development in mice of identical chromosome constitution. *Nature (London)*, **300**, 445–6.

Cattanach, B. M., Iddon, C. A., Charlton, H. M., Chiappa, S. A. & Fink, G. (1977). Gonadotrophin-releasing hormone deficiency in a mutant mouse with hypogonadism. *Nature (London)*, **269**, 338–40.

Cattanach, B. M., Pollard, C. E. & Hawkes, S. G. (1971). Sex-reversed mice: XX and XO males. *Cytogenetics*, **10**, 318–37.

Chung, K. W. (1974). A morphological and histological study of Sertoli cells in normal and XX sex-reversed mice. *American Journal of Anatomy*, **139**, 369–88.

Danforth, C. H. (1927). A gynandromorph mouse. *Anatomical Record*, **35**, 32 (Abstract).

Eicher, E. (1982). Primary sex determining genes in mice. In *Prospects for Sexing Mammalian Sperm*, ed. R. P. Amann & G. E. Seidel, pp. 121–35. Colorado: Colorado University Press.

Eicher, E. M. & Washburn, L. L. (1983). Inherited sex reversal in mice: identification of a new primary sex-determining gene. *Journal of Experimental Zoology*, **228**, 297–304.

Eicher, E. M. & Washburn, L. L. (1986). Genetic control of primary sex determination in mice. *Annual Review of Genetics*, **20**, 327–60.

Eicher, E., Washburn, L. L., Whitney, J. B. & Morrow, K. E. (1982). *Mus poschiavinus* Y chromosome in the C57BL/6J murine genome causes sex reversal. *Science*, **217**, 535–7.

Evans, E. P., Burtenshaw, M. D. & Cattanach, B. M. (1982). Meiotic crossing-over between the X and Y chromosomes of male mice carrying the sex-reversing (*Sxr*) factor. *Nature (London)*, **300**, 443–5.

Fekete, E. (1937). A case of lateral hermaphroditism in *Mus musculus*. *Anatomical Record*, **69**, 151–2.

Gehring, U. & Tomkins, G. M. (1974). Characterisation of a hormone receptor defect in the androgen-insensitivity mutant. *Cell*, **3**, 59–64.

Goldstein, J. L. & Wilson, J. D. (1972). Studies on the pathogenesis of the pseudohermaphroditism in the mouse with testicular feminisation. *Journal of Clinical Investigation*, **51**, 1647–58.

Greene, R. R., Burrill, M. W. & Ivy, A. C. (1939). Experimental intersexuality. The effect of antenatal androgens on sexual development of female rats. *American Journal of Anatomy*, **65**, 415–69.

Greep, R. O. (1942). Two hermaphroditic rats. *Anatomical Record*, **83**, 121–8.

Harris, G. W. (1972). Humours and hormones: the Dale lecture, 1971. *Journal of Endocrinology*, **53**, i–xxiii.

Hollander, W. F., Gowen, J. W. & Stadler, J. (1956). A study of 25 gynandromorphic mice of the Bagg albino strain. *Anatomical Record*, **124**, 223–43.

Holmes, R. S. & Jones, J. T. (1979). Lactate dehydrogenase C_4 in male sex accessory glands of normal mice and in testes of sex-reversed mice. *Journal of Experimental Zoology*, **207**, 43–8.

Hunt, P. A. (1991). Survival of XO mouse fetuses: effect of parental origin of the X chromosome or uterine environment? *Development*, **111**, 1137–41.

Iddon, C. A., Charlton, H. M. & Fink, G. (1980). Gonadotrophin release in hypogonadal (*hpg*) and normal mice after electrical stimulation of the median eminence or injection of luteinising hormone releasing hormone. *Journal of Endocrinology*, **85**, 105–10.

Koopman, P., Gubbay, J., Vivian, N., Goodfellow, P. & Lovell-Badge, R. (1991). Male development of chromosomally female mice transgenic for *Sry*. *Nature (London)*, **351**, 117–21.

Krieger, D. T., Perlow, M. J., Gibson, M. J., Davies, T. F., Zimmerman, E. A., Ferin, M. & Charlton, H. M. (1982). Brain grafts reverse hypogonadism of gonadotropin releasing hormone deficiency. *Nature (London)*, **298**, 468–71.

LeBarr, D. K. & Blecher, S. R. (1986). Epididymides of sex-reversed XX mice lack the Initial Segment. *Developmental Genetics*, **7**, 109–16.

Livne, I., Silverman, A-J. & Gibson, M. J. (1992). Reversal of reproductive deficiency in the *hpg* male mouse by neonatal androgenization. *Biology of Reproduction*, **47**, 561–7.

Lovell-Badge, R. & Robertson, E. (1990). XY female mice resulting from a heritable mutation in the primary testis-determining gene, *Tdy*. *Development*, **109**, 635–46.

Lyon, M. F. (1969). A true hermaphrodite mouse presumed to be an XO/XY mosaic. *Cytogenetics*, **8**, 326–31.

Lyon, M. F., Cattanach, B. M. & Charlton, H. M. (1981). Genes affecting sex differentiation in mammals. In *Mechanisms of Sex Differentiation in Animals and Man*, ed. C. R. Austin & R. G. Edwards, pp. 329–386. New York: Academic Press.

Lyon, M. F. & Glenister, P. H. (1980). Reduced reproductive performance in androgen-resistant *Tfm/Tfm* female mice. *Proceedings of the Royal Society of London, Series B*, **208**, 1–12.

Lyon, M. F. & Hawker, S. G. (1973). Reproductive lifespan in irradiated and unirradiated chromosomally XO mice. *Genetics Research (Cambridge)*, **21**, 185–94.

Lyon, M. F. & Hawkes, S. G. (1970). X-linked gene for testicular feminization in the mouse. *Nature (London)*, **227**, 1217–19.

MacIntyre, M. N. (1956). Effect of the testis on ovarian differentiation in heterosexual embryonic rat gonad transplants. *Anatomical Record*, **124**, 27–46.

MacIntyre, M. N., Hunter, J. E. & Morgan, A. H. (1960). Spatial limits of activity of foetal gonadal inductors in the rat. *Anatomical Record*, **138**, 137–47.

McLaren, A. (1980). Oocytes in the testis. *Nature (London)*, **283**, 688–9.

McLaren, A. (1981). The fate of germ cells in the testis of fetal *Sex-reversed* mice. *Journal of Reproduction and Fertility*, **61**, 461–7.

McLaren, A. (1983). Sex reversal in the mouse. *Differentiation*, **23** (Supplement), 93–8.

McLaren, A. (1988). Somatic and germ-cell sex in mammals. *Philosophical Transactions of the Royal Society of London, Series B*, **322**, 3–9.

McLaren, A. (1990). Of MIS and the mouse. *Nature (London)*, **345**, 111.

McLaren, A. (1991). Sex determination in mammals. *Oxford Reviews of Reproductive Biology*, **13**, 1–33.

McLaren, A. & Monk, M. (1982). Fertile females produced by inactivation of an X chromosome of '*sex-reversed*' mice. *Nature (London)*, **300**, 446–8.

McLaren, A., Simpson, E., Tomonari, K., Chandler, P. & Hogg, H. (1984). Male sexual differentiation in mice lacking H-Y antigen. *Nature (London)*, **312**, 552–5.

Mahadevaiah, S. K., Lovell-Badge, R. & Burgoyne, P. S. (1993). *Tdy*-negative XY, XXY and XYY female mice: breeding data and synaptonemal complex analysis. *Journal of Reproduction and Fertility*, **97**, 151–60.

Mason, A. J., Hayflick, J. S., Zoeller, R. T., Young, W. S., Philips, H. S., Nikolics, K. & Seeburg, P. H. (1986*a*). A deletion truncating the gonadotropin-releasing hormone gene is responsible for hypogonadism in the *hpg* mouse. *Science*, **234**, 1366–71.

Mason, A. J., Pitts, S. L., Nikolics, K., Szonyi, E., Wilcox, J. N., Seeburg, P. H. & Stewart, T. A. (1986*b*). The hypogonadal mouse: reproductive functions restored by gene therapy. *Science*, **234**, 1372–8.

Matzuk, M. M., Finegold, M. J., Su, J. G. J., Hsueh, A. J. W. & Bradley, A. (1992). α-inhibin is a tumour-suppressor gene with gonadal specificity in mice. *Nature (London)*, **360**, 313–19.

Mellon, P. L., Windle, J. J., Goldsmith, P. C., Padula, C. A., Roberts, J. L. & Weiner, R. I. (1990). Immortalization of hypothalamic GnRH neurons by genetically targeted tumorigenesis, *Neuron*, **5**, 1–10.

Merchant-Larios, H. & Taketo, T. (1991). Testicular differentiation in mammals under normal and experimental conditions. *Journal of Electron Microscopy Technique*, **19**, 158–71.

Miklos, G. L. G. (1974). Sex chromosome pairing and male fertility. *Cytogenetic and Cell Genetics*, **13**, 558–77.

Mintz, B. (1957). Embryological development of primordial germ cells in the mouse: influence of a new mutation, W^{j}. *Journal of Embryology and Experimental Morphology*, **5**, 396–403.

Mintz, B. & Russell, E. S. (1957). Gene-induced embryological modifications of primordial germ cells in the mouse. *Journal of Experimental Zoology*, **134**, 207–37.

Mittwoch, U. (1989). Sex differentiation in mammals and tempo of growth: probabilities *vs* switches. *Journal of Theoretical Biology*, **137**, 445–55.

Mittwoch, U. (1992). Sex determination and sex reversal: genotype, phenotype, dogma and semantics. *Human Genetics*, **89**, 467–9.

Mittwoch, U. & Buehr, M. L. (1973). Gonadal growth in sex reversed mice. *Differentiation*, **1**, 219–24.

Nagamine, C. M., Taketo, T. & Koo, G. C. (1987). Morphological development of the mouse gonad in *tda*-1XY sex reversal. *Differentiation*, **33**, 214–22.

Ohno, S. & Lyon, M. F. (1970). X-linked testicular feminization in the mouse as a non-inducible regulatory mutation of the Jacob-Monod type. *Clinical Genetics*, **1**, 121–7.

Palmer, S. J. & Burgoyne, P. S. (1991). The *Mus musculus domesticus Tdy* allele acts later than the *Mus musculus musculus Tdy* allele: a basis for XY sex-reversal in C57BL/6-Y^{POS} mice. *Development*, **113**, 709–14.

Radovick, S., Wray, S., Lee, E. *et al.* (1991). Migratory arrest of gonadotropin-releasing hormone neurons in transgenic mice. *Proceedings of the National Academy of Sciences, USA*, **88**, 3402–6.
Scott, J. E. & Blecher, S. R. (1987). β-glucuronidase activity is present in the microscopic epididymis of the *Tfm*/Y mouse. *Developmental Genetics*, **8**, 11–15.
Short, R. V. (1978). Sex determination and differentiation of the mammalian gonad. *International Journal of Andrology, Supplement*, **2**, 21–8.
Silverman, A. J., Zimmerman, E. A., Gibson, M. J., Perlow, M. J., Charlton, H. M., Kokoris, G. J. & Krieger, D. T. (1985). Implantation of normal fetal preoptic area into hypogonadal mutant mice: temporal relationship of the growth of gonadotropin-releasing hormone neurons and the development of the pituitary/testicular axis. *Neuroscience*, **16**, 69–84.
Simpson, E. M. & Page, D. C. (1991). An interstitial deletion in mouse Y chromosomal DNA created a transcribed *Zfy* fusion gene. *Genomics*, **11**, 601–8.
Singh, L. & Jones, K. W. (1982). Sex-reversal in the mouse (*Mus musculus*) is caused by a recurrent non-reciprocal crossover involving the X and an aberrant Y chromosome. *Cell*, **28**, 205–16.
Speed, R. M. (1986). Oocyte development in XO foetuses of man and mouse: the possible role of heterologous X-chromosome pairing in germ cell survival. *Chromosoma (Berlin)*, **94**, 115–24.
Sutcliffe, M. J. & Burgoyne, P. S. (1989). Analyses of the testes of H-Y negative XOSxr^{b} mice suggests that the spermatogenesis gene (*Spy*) acts during the differentiation of the A spermatogonia. *Development*, **107**, 373–80.
Sutcliffe, M. J., Darling, S. M. & Burgoyne, P. S. (1991). Spermatogenesis in XY, XYSxr^{a} and XOSxr^{a} mice: a quantitative analysis of spermatogenesis throughout puberty. *Molecular Reproduction and Development*, **30**, 81–9.
Takeda, H., Suzuki, M., Lasnitzki, I. & Mizuno, T. (1987). Visualization of X-chromosome inactivation mosaicism of *Tfm* gene in X^{Tfm}/X^{+} heterozygous female mice. *Journal of Endocrinology*, **114**, 125–9.
Taketo-Hosotani, T., Merchant-Larios, H., Thau, R. B. & Koide, S. S. (1985). Testicular cell differentiation in fetal mouse ovaries following transplantation into adult male mice. *Journal of Experimental Zoology*, **236**, 229–37.
Taketo-Hosotani, T., Nishioka, Y., Nagamine, C. M. Villalpando, I. & Merchant-Larios, H. (1989). Development and fertility of ovaries in the B6.Y^{DOM} sex-reversed female mouse. *Development*, **107**, 95–105.
Taketo, T., Saeed, J., Nishioka, Y. & Donahoe, P. K. (1991). Delay of testicular differentiation in the B6.Y^{DOM} ovotestis demonstrated by immunocytochemical staining for Müllerian inhibiting substance. *Developmental Biology*, **146**, 386–95.
Tarkowski, A. K. (1961). Mouse chimaeras developed from fused eggs. *Nature (London)*, **190**, 857–60.
Tarkowski, A. K. (1964). True hermaphroditism in chimaeric mice. *Journal of Embryology and Experimental Morphology*, **12**, 735–57.
Turner, C. D. (1964). Special mechanisms in anomalies of sex differentiation. *American Journal of Obstetrics and Gynecology*, **90**, 1208–26.
Ward, H. B., McLaren, A. & Baker, T. G. (1987). Gonadal development in T16H/X*Sxr* hermaphrodite mice. *Journal of Reproduction and Fertility*, **81**, 295–300.

Welshons, W. J. & Russell, L. B. (1959). The Y-chromosome as the bearer of male determining factors in the mouse. *Proceedings of the National Academy of Sciences, USA*, **45**, 560–6.
Whitten, W. K., Beamer, W. G. & Byskov, A. G. (1979). The morphology of foetal gonads of spontaneous mouse hermaphrodites. *Journal of Embryology and Experimental Morphology*, **52**, 63–78.
Wilmut, I., Hooper, M. L. & Simons, J. P. (1991). Genetic manipulation of mammals and its application in reproductive biology. *Journal of Reproduction and Fertility*, **92**, 245–79.

7

Abnormal sexual development in man

Est-ce un jeune homme ? est-ce une femme,
Une déesse, ou bien un dieu ?
L'amour, ayant peur d'être infâme,
Hésite et suspend son aveu.

Dans sa pose malicieuse,
Elle s'étend, le dos tourné
Devant la foule curieuse,
Sur son coussin capitonné.

Pour faire sa beauté maudite,
Chaque sexe apporta son don.
Tout homme dit : C'est Aphrodite !
Toute femme : C'est Cupidon !

(Gautier, 1852)

Introduction

Diverse disorders of sexual development exist in man, many of them having been recognised since ancient times and indeed some of them conferring on their possessors positions of dubious privilege in society (Fig. 7.1). Whilst scientific description of the disorders has tended to focus primarily on the condition of the gonads, and may find expression in the external genitalia and non-reproductive tissues, the disorders in fact stem from genetic anomalies involving the chromosome complement or perhaps even individual genes. Such anomalies were first critically analysed in the 1950s with the development of reliable methods for karyotyping and, with the ever increasing sophistication of techniques culminating in those at the molecular level, so the precision of diagnosis has increased. One outcome of the cytogenetic or molecular genetic approach has been the revelation that sexual disorders are frequently a consequence of errors at meiosis in the maternal or paternal germ cell line. Another is the realisation that errors in man usually have their counterparts in laboratory rodents and sometimes also in the large farm animals (see Chapters 5 and 6).

This chapter attempts to summarise key observations on some well-known disorders of sexual development, in each case making reference to

Fig. 7.1. Depiction in ancient oriental art of one form of intersexuality in man. (Courtesy of Professors P. Malet and P. Popescu.)

the underlying genetic lesion or the current hypothesis concerning such. Because of the very considerable practical and emotional significance of these conditions in man, a large number of prominent texts exist, many of course having a strong clinical orientation. The reader may wish to consult the volumes by Overzier (1963), Armstrong & Marshall (1964), Jirasek (1971), Van Niekerk (1974), Simpson (1976), Yen & Jaffe (1978), Edwards (1980) and Austin & Edwards (1981), with its extensive chapters by Polani (1981*a*, *b*). A useful historical flavour can be found in the review of Turner (1964). More recent statements are referred to throughout this chapter.

A crucial finding underlying much that is presented in the following pages is that of a strictly homologous region at the tip of the short arms of the X and Y chromosomes. Such so-called pseudoautosomal DNA sequences show only a partial sex linkage and can be exchanged between X and Y chromosomes during male meiosis. Indeed, the pseudoautosomal region has been calculated to undergo a recombination rate – during male meiosis – about 20 times higher than the mean value for the human genome (Weissenbach & Rouyer, 1990). Moreover, there is now evidence for a second, smaller pseudoautosomal region permitting pairing also at the end of the long arm (Fig. 7.2; reviewed by Rappold, 1993). And, clearly of relevance is the finding that the human X chromosome undergoes non-disjunction much more frequently than the autosomes during spermatogenesis, the increase being restricted to non-disjunction of the XY bivalent at the first male meiotic division (Jacobs *et al.*, 1988; Jacobs, 1992).

Sex reversal

The phenomenon of so-called sex reversal leading to the development of testes in individuals of apparently XX sex chromosome constitution has now been extensively documented and reviewed (e.g. de la Chapelle, 1972; de la Chapelle *et al.*, 1977, 1990; de la Chapelle, Koo & Wachtel, 1978; Lyon, Cattanach & Charlton, 1981; Ferguson-Smith & Affara, 1988), even though the incidence of this condition may be only about 1 in 20000 newborn males (Vergnaud *et al.*, 1986). Such XX males invariably have small testes devoid of germ cells and thus, as is the case for sex-reversed mice (Chapter 6), they are sterile, this being in some way associated with the presence of two X chromosomes (Table 7.1). There are a number of resemblances to patients with the XXY Klinefelter's syndrome, except that sex-reversed individuals are usually shorter in stature than normal XY males whereas those with Klinefelter's syndrome are generally taller. Gynaecomastia is a characteristic feature and the penis may be small, but surface virilisation including pubic and axillary hair is usually normal. Cytogenetic studies on blood lymphocytes invariably reveal a 46,XX karyotype. These sex-reversed males are not mentally impaired.

Several examples of familial XX males – and indeed XX true hermaphrodites – have been described (Berger *et al.*, 1970; de la Chapelle *et al.*, 1977), and cytogenetic evidence for mosaicism and Y-translocations involving an X chromosome has been obtained in isolated cases (Miro *et al.*, 1978; Evans *et al.*, 1979). In these early reports, two apparently demonstrated an autosomal dominant inheritance, like the Sxr mouse, but another showed

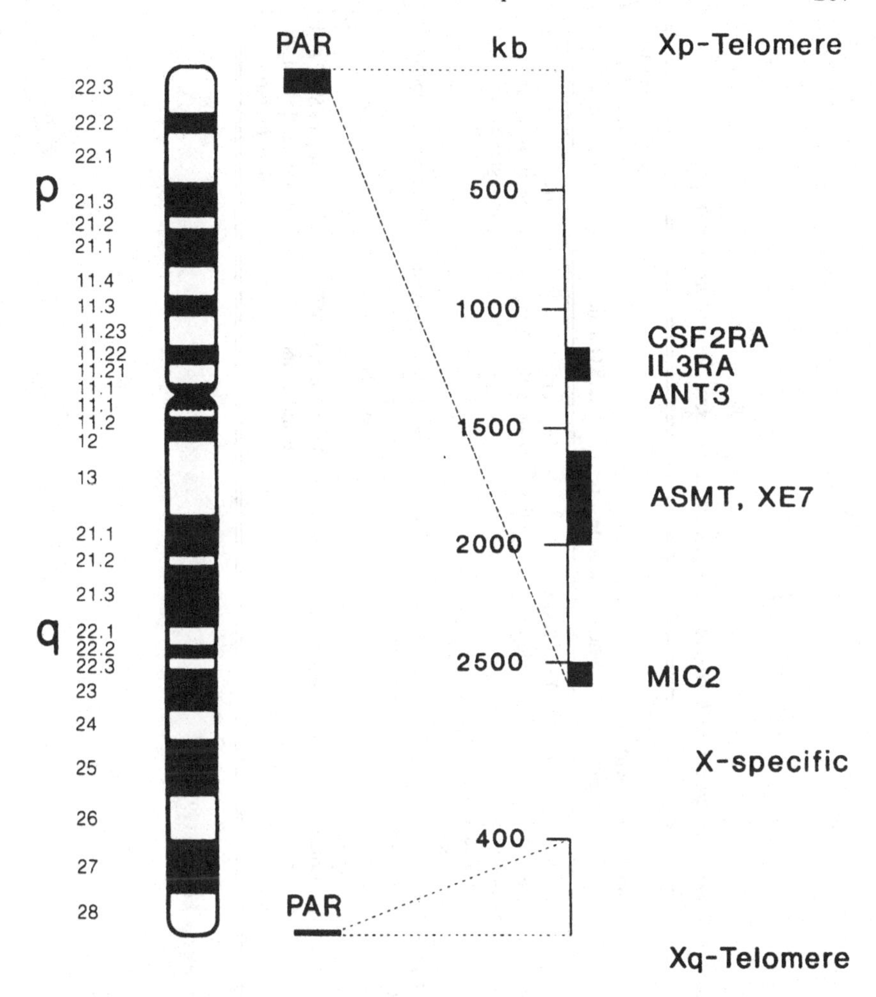

Fig. 7.2. Schematic representation of the human X chromosome to highlight the two pseudoautosomal regions (PAR) of differing size located at opposite ends of the chromosome. (Courtesy of Dr G. A. Rappold.)

evidence of an autosomal recessive inheritance, similar to that seen in goats (see Chapter 5). In the family with autosomal recessive inheritance, the mother of the affected individuals was H-Y antigen positive, and the father had somewhat elevated H-Y antigen titres (Lyon *et al.*, 1981).

Ferguson-Smith (1966) proposed that an abnormal cross-over event was responsible for sex reversal in humans, accounting for both XX males and XY females. In other words, an accidental recombination between segments of the X and Y chromosomes during meiosis in the father would

Table 7.1. *Some generalised findings associated with various abnormalities of sex determination in man*

	46,XX male	46,XX male with genital ambiguities	46,XX true hermaphrodite	45,XO male	46,XY female with pure gonadal dysgenesis
Incidence of live births	1 per 20000	1 per 20000	1 per 20000	Very rare	Rare
Clinical examination of genitalia	Apparently normal male	Male with ambiguities: hypospadia, micropenis	Male with ambiguities	Male with ambiguities	Female without ambiguities
Morphology of gonads	Testis with Sertoli and Leydig cells but no germ cells	Male with ambiguities: hypospadia, micropenis	Both testicular and ovarian tissue	Testis with Sertoli and Leydig cells, but no germ cells	Ovarian dysgenesis
Proposed genetic aetiology	Terminal X→Y transfer	Autosomal mutation since no Y-related sequences	Autosomal mutation	Translocation of Y to an autosome	Mutation in *SRY* and autosomal 'loss of function' mutation

Note:
Modified from the review of Cotinot, McElreavey & Fellous (1993).

occasionally allow transfer of Y-linked sequences to the paternal X chromosome. In support of this proposition, and expanding the original report of Madan & Walker (1974), detailed measurements of the X chromosomes by Evans *et al.* (1979) revealed a significant difference in size, with the short arm of one of the X chromosomes being slightly elongated in 70% of patients. In this abnormal situation, the entire pseudoautosomal region plus a portion of the Y-chromosome-specific region including *TDF* from the Y are transferred, leading to testis differentiation in supposedly XX males (Andersson, Page & de la Chapelle, 1986; de la Chapelle *et al.*, 1990). Conversely, loss of the same sequences by X–Y interchange allows female differentiation in a small proportion of individuals with XY gonadal dysgenesis (Ferguson-Smith & Affara, 1988). Segregation analysis of the pseudoautosomal loci in these patients demonstrated the validity of Ferguson-Smith's (1966) hypothesis, and showed the existence of sex inversions unrelated to the *TDF* locus.

Advances in molecular genetics have permitted the demonstration that DNA from a majority (approximately 85%) of all XX males studied has Y-chromosome-specific sequences including *ZFY* (Petit *et al.*, 1987; de la Chapelle *et al.*, 1990) or lacking *ZFY* (Palmer *et al.*, 1989). In other words, a testis-determining sequence is situated on the paternally derived X chromosome due to the proposed abnormal events during meiosis. This demonstration followed from the mapping and cloning of the distal part of the short arm of the human Y chromosome (Guellaen *et al.*, 1984; Page, de la Chapelle & Weissenbach, 1985; Vergnaud *et al.*, 1986). In fact, in all nine cases of XX males known to have Y-specific sequences, these have been located at the proximal tip of one of the two X chromosomes (Ferguson-Smith & Affara, 1988). These last authors also refer to several XX males in which no Y-specific sequences could be demonstrated, these individuals frequently possessing abnormalities of the external genitalia (hypospadia), in contrast to the situation in Y-positive XX men. In the latter, the amount of Y-specific sequences present is extremely variable (Ferguson-Smith *et al.*, 1990; Ferguson-Smith, 1991).

In a summary of several large series of XX males who had been screened for Y DNA, 59 of a total of 86 XX males (69%) were reported to be positive for Y-chromosomal DNA (de la Chapelle, 1987). These patients were referred to as Y(+)XX males. The 31% in which no Y-chromosome-specific DNA could be demonstrated were referred to as Y(−)XX males. Their aetiology might be explained on the basis of putative genes, downstream from *TDF*, that can trigger testis determination when they undergo a mutation. Both 'gain of function' and 'loss of function' mutations in other genes in the sex determination pathway may cause sex reversal (Ogata *et al.*, 1992). Nothing substantial is yet known about these putative genes, but their phenotypic effect is slightly different from *TDF*. As an alternative explanation for Y(−)XX males, the condition of mosaicism has been proposed, with a prevalent XX lineage but also a scarce or hidden lineage containing a Y chromosome (de la Chapelle *et al.*, 1990). Two such 46,XX/47,XXY mosaic patients were described in detail by de la Chapelle *et al.* (1990) and, in one, it required DNA hybridisation and cytogenetic studies to demonstrate the existence of an XXY cell line.

On the basis of their elegant transplantation studies of XX foetal gonads in mice, Taketo *et al.* (1993) proposed a novel aetiology for sex reversal producing human XX males. Such males and also true human hermaphrodites lacking the *SRY* sequence might first have embarked upon ovarian differentiation, but a loss of oocytes at early stages of gonadal differentiation could have permitted the subsequent formation of a testis. But, despite the discussion on transdifferentiation in Chapter 3, it is by no means clear why death of oocytes and a consequent loss of ovarian tissue should enable or indeed prompt subsequent formation of Sertoli-like cells. There is tentative evidence from grafting experiments for a regulatory involvement of the mesonephros, and also evidence from a mouse gonadal graft model that interactions with the host tissue are required for biochemical differentiation of Sertoli cells and organisation of seminiferous tubules (Taketo, 1991).

As to sex reversal giving rise to XY females, this can be reasoned to result from failure of the testis determination or differentiation pathways due to mutations. However, instead of a consequent elaboration of functional ovaries in such cases, only dysgenic – 'streak' – ovaries arise in these XY females. Women with 46,XY pure (complete) gonadal dysgenesis generally become tall since there is no secretion of follicular oestrogen to produce epiphyseal fusion in the long bones. The gonads are at risk of developing gonadoblastomas (Ferguson-Smith, 1991). As would be anticipated in this oestrogen-deficient state, patients show no signs of breast development, have very scanty pubic and axillary hair, and are amenorrhoeic. Such individuals were thought to be the human equivalent of wood lemmings so that, although they have a male karyotype (XY), the indifferent gonads are not transformed into testes, and the patients develop as phenotypic females (see Fredga, 1988). Here, a deletion in the rearranged X is thought to inactivate the male-determining pathway. In contrast to the wood lemmings, however, which are fully fertile, their germ cells do not survive, so the ovaries are rendered non-functional and degenerate into streak gonads. The condition was initially presumed to be due to an X-linked gene acting to suppress the male-determining sequence on the Y chromosome (German *et al.*, 1978), thereby resulting in failure of the testis determination or differentiation pathway (Short, 1982).

Some XY females presenting with gonadal dysgenesis have lost the sex-determining region from the Y chromosome by terminal exchange between the sex chromosomes (Affara *et al.*, 1987; Ferguson-Smith, Affara & Magenis, 1987; Levilliers *et al.*, 1989) or by other deletions (Berta *et al.*, 1990; Page *et al.*, 1990). There is sound evidence that mutations in the *SRY* open reading frame are also involved in the process in at least a proportion – perhaps 10–15% – of instances. For example, Berta *et al.* (1990) found a *de novo* point mutation (a single base-pair change) in the *SRY* gene of one of 11 XY female patients, and were able to infer from this instance of sex reversal that *SRY* is required for male sex determination. Mutant *SRY* encoded by XY females loses its particular DNA binding activity (Harley *et al.*, 1992), which implies that this activity is necessary for testicular development. And Jäger *et al* (1990*a*) also recorded a *de novo* frameshift mutation in the *SRY* gene due to a four base-pair deletion and associated with sex reversal in one of 12 human XY females (see Chapter 2). By implication, these *de novo* mutations were not present in the fathers. However, Jäger *et al.* (1991, 1992) have since reported an inherited mutation in *SRY* detected in one of their total of 15 XY females studied, in which a sex-reversed XY female shares the same amino acid substitution with her XY male relatives. McElreavey *et al.* (1991) reported a family with a single base-pair change within the putative DNA binding motif of *SRY* carried in two generations, the family including patients with pure gonadal dysgenesis. Vilain *et al.* (1992) expanded this report of a single base-pair substitution in the *SRY* open reading frame, resulting in a conservative amino acid change, which was found in five members of the family who each had a Y chromosome. In examination of 25 cases of human XY females with pure gonadal dysgenesis for mutations on the short arm of the Y chromosome, McElreavey *et al.* (1992) described a patient in whom the *SRY* open reading frame was intact but whose Y chromosome had a deletion that commenced 1.7 kilobases 5′ to the open reading frame and extended some 25–50 kilobases further. The deletion may have removed upstream exons of *SRY* and/or transcriptional regulatory motifs, either situation resulting in lack of testicular development. It should perhaps be noted here that a distinction is drawn between pure gonadal dysgenesis and mixed (partial) gonadal dysgenesis, the latter being analysed in detail by Robboy *et al.* (1982). Patients with partial gonadal dysgenesis have some testicular tissue, frequently seen as groups of Sertoli or Sertoli-like cells, but meticulous histological examination of biopsy specimens may be needed to distinguish the two categories of dysgenesis.

One further observation on an XY woman of particular interest was reported by Scherer *et al.* (1989). The patient had an apparently normal Y chromosome, but there was a tandem duplication of part of the short arm of the X containing the *ZFX* gene and which was active. Thus arose the suggestion that an abnormal dosage of genes in the *ZFX* region could in some manner have been responsible for the sex reversal, perhaps by suppressing function of the testis-determining gene carried on the Y chromosome. Prompted by this report, Burgoyne & Palmer (1992) commented that it clearly calls into question the assumption that all mutations causing XY sex reversal in man are influencing genes in the testis-determining pathway.

As a contrary and seemingly extremely important finding, Ogata *et al.* (1992), studying a case of infant sex reversal, produced evidence strongly suggesting that impaired testis formation and resultant female development had occurred in the presence of *SRY*. On the basis of a penetrating molecular analysis, the conclusion was drawn that the sex reversal was a direct or indirect consequence of having two active copies of the distal part of the short arm of the X chromosome (Xp). If such a causal relationship in fact exists, then this would imply that one or more genes subject to X inactivation and involved in testis formation will be found in the Xp distal region, and that two active copies of the gene(s) hinder the process of testis determination or differentiation (Ogata *et al.*, 1992). This interpretation would also seem to apply to four other non-mosaic sex-reversed patients with a 46,Y,Xp+ karyotype. In essence, the new hypothesis states that patients with only one active copy of the gene(s) (e.g. 47,XXY or 48,XXXY) masculinise as normal 46,XY males, whereas those with two active copies of the gene(s) (e.g. 46,Y,dup(Xp) and 46,X,Yp+), demonstrate sex reversal. Indeed, some *SRY*-positive XY females could be explained by cryptic duplications of the gene(s) proposed above (Ogata *et al.*, 1992).

Looking to the future, instances of sex reversal that can be shown to be independent of *SRY* anomalies should facilitate the identification of other genes located upstream or downstream in the sex determination pathway. As noted in Chapter 6, three examples of *Sry* 'silencing' by a position effect have been reported in XY female mice (Capel *et al.*, 1993).

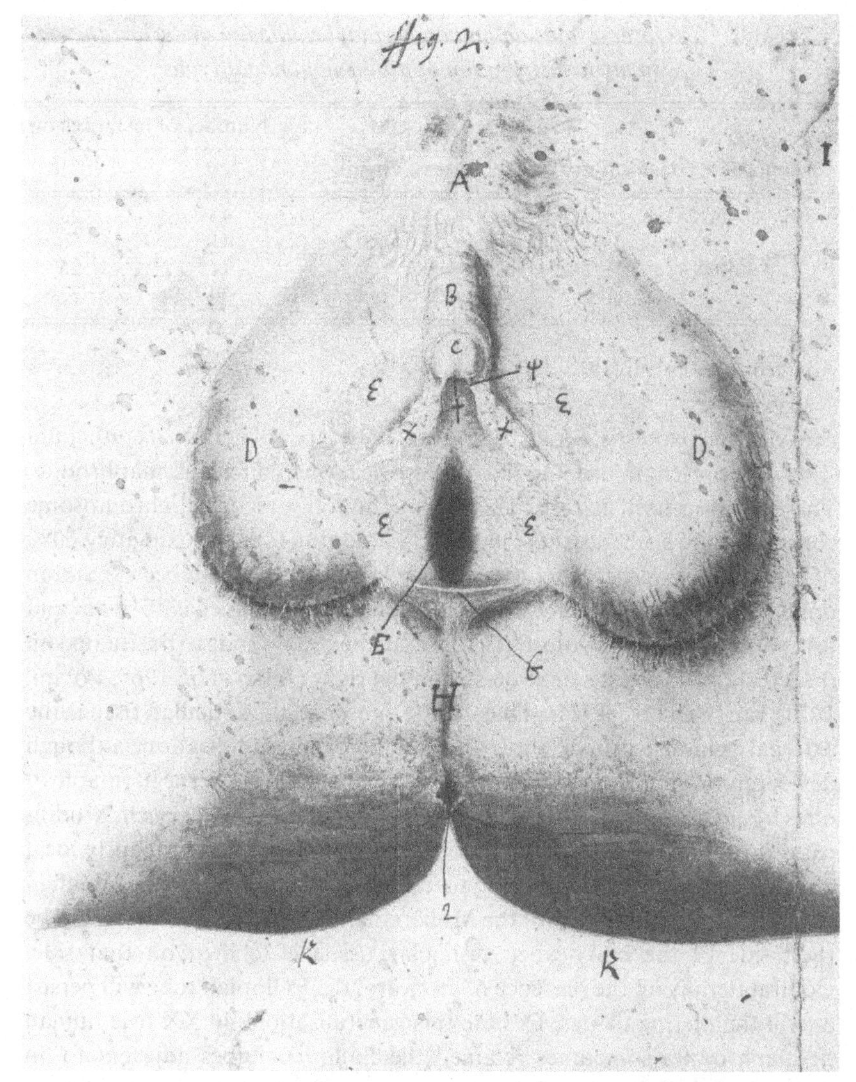

Fig. 7.3. One of the remarkable unpublished drawings from the Douglas collection (dated 1714) showing the external genitalia of an individual referred to as an hermaphrodite. The specimen is almost certainly not a true hermaphrodite. (Courtesy of Glasgow University Library.)

Hermaphrodites or intersexes

Patients identified as true hermaphrodites (Fig. 7.3) have both testicular and ovarian tissues, although the distribution of these two gonadal tissues between sides is variable. There may be any combination of the following:

Table 7.2. *Observations on human hermaphrodites to underline the asymmetric distribution of differing gonadal types*

Chromosome constitution	Nature of gonad	Number of instances on left	right
46,XX	Ovary	43	28
	Ovotestis or testis	41	56
46,XY	Testis	14	25
	Ovotestis or ovary	34	23

Note:
Adapted from Polani (1970).

ovary, testis, ovotestis. Strictly speaking, both types of germinal epithelium should be present in order to justify classification as a true hermaphrodite. The condition itself is rare. The most commonly recorded chromosome constitution in such patients is 46,XX, accounting for approximately 60% of reported instances. By contrast, XY chromosomes have been found in only 15% of hermaphrodites. When the gonads are distinct as an ovary and testis or an ovary and ovotestis (Table 7.2), the ovary tends to be situated on the left side and a testis or ovotestis on the right (Josso *et al.* 1965; Polani, 1970; van Niekerk, 1974). Due to the presence of testicular tissue, the external genitalia usually show some signs of masculinisation, although development of the breasts may also be noted at puberty. True hermaphrodites frequently possess perineal hypospadias and a vaginal pouch. Modifications to the internal genitalia represent in part the predominantly local nature of hormone action during foetal life. Development of the Wolffian duct system and inhibition of the Müllerian ducts will have occurred on the right side of the body when testicular tissue is located on that side. Contralaterally, in the presence of an ovary, the Fallopian tube will persist as will the uterine tissues. Despite this generalisation, 46,XX true human hermaphrodites sometimes retained the Fallopian tubes adjacent to an ovotestis, especially when the proportion of testicular tissue was small (van Niekerk & Retief, 1981).

There was much discussion in the early 1960s as to whether hermaphrodites could have arisen as chromosomal mosaics. Evidence in support of this possibility came from a child examined at laparotomy by Gartler, Waxman & Giblett (1962). There was an ovary on the left side and an ovotestis on the right. Biopsies of gonadal tissue revealed exclusively XX cells in the ovary but both XX and XY cells in the ovotestis. Within the ovotestis, there was a preponderance of XY cells in the testicular portion and of XX cells in the ovarian portion. A mosaic chromosome constitution was also revealed in skin and peripheral blood. Gartler *et al.* (1962) proposed that the mosaic could have resulted from a double fertilisation of the oocyte, two haploid sets of chromosomes being introduced by the penetrating spermatozoa. A similar explanation was invoked by Josso *et al.* (1965) to explain XX/XY mosaicism in a 20-year-old patient with a left ovary and right ovotestis. As a caution, it is worth noting that karyotyping of the somatic tissues of several other human hermaphrodites failed to reveal mosaicism, patients apparently being either wholly female or wholly male with regard to their sex chromosome constitution (Polani, 1962; Overzier, 1963) although, by contrast, Gilgenkrantz (1987) produced such evidence for two true hermaphrodites. Of course, hermaphroditism in the presence of an XX karyotype seemingly contradicts the requirement for a Y chromosome or portion thereof to enable differentiation of testicular tissue. Traditional explanations to be found in the literature, as for example in Eicher & Washburn (1983), included:

1 the undetected loss of a Y chromosome after initiation of testicular development, or undetected chromosomal mosaicism such as XO/XY or XX/XXY, or XX/XY chimaerism;
2 the translocation of testicular determinants from the Y to an X chromosome or to an autosome;
3 a single gene mutation.

Ferguson-Smith & Affara (1988) postulated that in sporadic or familial XX true hermaphrodites, one of the gonad-determining gene loci escapes X-inactivation because of mutation or chromosomal rearrangement, thereby resulting in mosaicism for testis- and ovary-determining cell lines in somatic cells. Y-negative XX males belong to the same clinical spectrum as XX true hermaphrodites, and gonadal dysgenesis in some XY females may be due to sporadic or familial mutations of gonad-determining gene loci. Although X–Y interchange with random inactivation of the interchanged X could well account for the gonadal findings (Ferguson-Smith, 1966), Y-specific sequences had not yet been found in XX true hermaphrodites (Vergnaud *et al.*, 1986; de la Chapelle, 1987; Ferguson-Smith & Affara, 1988). Indeed, a series of DNA probes from Yp and specifically from the testis-determining region failed to reveal a hybridisation signal in most instances of hermaphroditism examined (Abbas *et al.*, 1990; Wolf, Schempp & Scherer, 1992). However, some exceptions to this situation were noted using probes from the pseudoautosomal boundary on the Y chromosome or the immediately adjoining region (Palmer *et al.*, 1989;

Jäger *et al.*, 1990*b*). But there is a class of XX true hermaphrodites that shows inheritance of the Y–pseudoautosomal boundary and *SRY* but not *ZFY*. There is also an instance of an XX true hermaphrodite with ambiguous genitalia, bilateral ovotestes, and short arm of the X chromosome bearing a visible deletion (Berkovitz *et al.*, 1992). A large fragment of the Y chromosome short arm was present and included *SRY*, *RPS4Y* and *ZFY*. Preferential inactivation of the deleted X chromosome has been invoked, with the X-inactivation spreading into the Y chromosome fragment.

Intersexuality can be defined here as the possession, at birth, of both male and female characteristics in the gonads and/or in the internal or external genitalia (Jacobs, 1969). The incidence of intersexual individuals in the population must be low. Nonetheless, since the availability of suitable cytogenetic techniques in the late 1950s, a very large variety of abnormal sex chromosome constitutions has been reported in man. Many, but not all of these, have been shown to be associated with varying degrees of abnormality of primary and secondary sex characters.

The most frequently described chromosome abnormality associated with intersexual development is mosaicism, where one cell line has 45 chromosomes and an XO sex chromosome constitution and the other 46 chromosomes and an XY sex chromosome constitution, or more rarely 47 and an XYY sex chromosome constitution (Jacobs, 1969). The phenotype of such individuals will depend on the time after fertilisation when the error giving rise to the two cell lines occurred, and the frequency and distribution of the two types of cells in the embryonic gonads. In fact, almost every degree of intersexual development has been found to be associated with an XO/XY constitution. For example, in a mosaic intersex child there was a small ovary on the left side with stroma but no follicles; this was presumed to be composed largely of XO cells. On the contralateral side (the right), there was a histologically normal testis, presumed to be composed of XY cells. Another type of mosaicism expressed as an intersex condition is when one cell line is that of a normal female whilst the other also has a Y chromosome, e.g. XX/XXY mosaics or XX/XY/XXY mosaics. An intersex phenotype may also arise from an XX/XY chromosome complement (Jacobs, 1969, 1972), and specific clinical findings are recorded by Jacobs (1969).

Turning to instances of pseudohermaphroditism, this condition in males (46,XY) is discussed in the next section under the heading of testicular feminisation. Most cases of human female pseudohermaphroditism are due to congenital adrenal hyperplasia, but in this condition the time of onset of abnormal steroidogenesis by the foetal adrenal gland is not known. However, in a case reported by Mürset *et al.* (1970), female pseudohermaphroditism resulted from a maternal adrenal tumour, which persisted throughout the pregnancy. The urethra and labio-scrotal folds were completely masculinised, but Müllerian duct derivatives were normal.

As a small final point in this section, perhaps of some relevance to the topic of intersexuality, most marmoset monkeys have twin pregnancies and, as a consequence, many marmosets are somatic chimaeras, although germ cells of one type rarely survive in the gonads of the opposite sex. But, if fusion of the chorions and development of vascular anastomoses between heterosexual siblings is thought of as one origin of intersexuality, then it is worth noting that this situation is found frequently in marmosets and yet freemartins apparently do not occur (Hill, 1926; Wislocki, 1932; Biggers, 1968; McLaren, 1976; Ohno, 1979). Likewise in man, normal sexual development is found in twins of opposite sex in the presence of XX/XY chimaerism.

Testicular feminisation – androgen insensitivity

The syndrome of testicular feminisation, in which a chromosomally normal 46,XY male has the outward appearance of a female, has been recognised in various species for many years (see Chapter 6). Contributions dealing with this syndrome in man are numerous, and serve to illustrate the close analytical dependence of clinical problems upon cytogenetics (e.g. Jacobs *et al.*, 1959; Polani, 1970). Individuals with female external genitalia are revealed internally to have small testes devoid of germ cells, perhaps reflecting their cryptorchidism. The testes may be fully abdominal, or palpable in the inguinal canal or labial regions: the latter degree of testicular descent is found in 80% of instances (Short, 1982). There is a blind-ending vagina but never more than a rudimentary uterus and Fallopian tubes, implying suppression of Müllerian duct development due to anti-Müllerian hormone (AMH) secretion by the foetal testes (see Chapters 3 and 4). Primary amenorrhoea is invariable. Concentrations of testosterone in peripheral blood may achieve the normal range, but there are no external signs of virilisation. Pubic, facial and axillary hair are usually scanty or absent. Normal breast development at puberty may result from oestrogens produced directly from the breast itself, from peripheral aromatisation of testosterone, or from increased oestrogen secretion by the testis (Macdonald *et al.*, 1979). Individuals with the syndrome tend to be taller than average, and may excel at sport (Ferguson-Smith, 1991).

As to the cause of testicular feminisation, the primary defect was shown to be a profound insensitivity of the target organs to testosterone (George & Wilson, 1988; Griffin & Wilson, 1989), this androgen insensitivity syndrome being the most common form of male pseudohermaphroditism in man. In fact, all androgen-mediated components of male development within the Wolffian duct system and external genitalia are compromised. One demonstrable consequence of the defect is that growth of pubic hair cannot be induced by administering androgens. More specifically, the syndrome is now known to be due to an X-linked gene defect (*TFM*) which causes an absence or an instability of the cytoplasmic androgen receptor protein in cells throughout the body. Diverse tissue responses depend upon this protein, testosterone combining with the receptor to activate genes concerned in male differentiation (Ohno, 1979). In the absence of this receptor, neither testosterone nor its biologically active metabolite, 5α-dihydrotestosterone, can exert any influence. By contrast, the cytoplasmic oestrogen receptor protein is unaffected, so this would explain why patients with the syndrome are very definitely female in appearance.

Despite the relatively straightforward portrayal of the syndrome above, it is now more clearly appreciated on the basis of intensive laboratory and clinical studies that androgen resistance falls into two broad categories: those involving defects in the androgen receptor itself and those involving a deficiency in the 5α-reductase enzyme (Griffin, 1992; Wilson, 1992). A complication is that testosterone and dihydrotestosterone interact with the same receptor to produce distinct physiological effects, although dihydrotestosterone has a greater receptor binding affinity than testosterone. Whereas differentiation of the Wolffian ducts during embryogenesis is not dependent upon formation of dihydrotestosterone (see Chapter 4), virilisation of the urogenital sinus and external genitalia together with sexual maturation at puberty do require the influence of dihydrotestosterone. A clinical necessity for demonstrating these distinctions was that of a 5α-reductase deficiency model in which the respective rôles of testosterone and dihydrotestosterone could be clarified. Such models have been extensively exploited in the studies of Wilson (reviewed by Wilson *et al.*, 1983; Griffin & Wilson, 1989; Wilson, 1992).

Clinical studies have revealed a spectrum of phenotypic abnormalities in these genetic males. As a consequence, specialist texts may subdivide testicular feminisation into the 'complete' type with no detectable signs of virilisation, and an 'incomplete' type with partial virilisation including the presence of axillary and pubic hair (Griffin & Wilson, 1980; George & Wilson, 1988). Deficiency of the androgen receptor is only partial in the second type (see Keenan *et al.*, 1974; Griffin & Wilson, 1980). At the root of these classifications involving the absence of receptor function, qualitatively abnormal receptors or defects in receptor amount (and indeed of further subdivisions), there are seemingly different gene mutations, although it was initially uncertain whether such putative mutations were restricted to a single locus. Notwithstanding this important qualification, there would appear to be X-linkage of the defect in all instances (Griffin & Wilson, 1980; Lyon *et al.*, 1981).

Because the conversion of testosterone into dihydrotestosterone by steroid 5α-reductase is a key step in androgen action, the loss of a gene underlying this conversion is relevant here: male pseudohermaphrodites result, that is 46,XY males with male internal urogenital tracts but female external genitalia with a blind-ending vagina. However, it is worth stressing that 5α-reductase deficiency is a rare autosomal recessive disorder. Andersson *et al.* (1991) report on two related individuals having male pseudohermaphroditism caused by 5α-reductase deficiency in which there is deletion of a second of two genes normally involved in the enzyme expression, thereby compromising development of genital tissues. Andersson *et al.* (1991) emphasise the fact that there are at least two functional 5α-reductases in man, and point out the relationship of these isoenzymes with hormone-mediated events in male sexual differentiation. Further studies at the molecular level will undoubtedly shed light on other (rare) mutations that act to diminish the effectiveness of virilisation during development of male embryos and during their subsequent postnatal life. Wilson (1992) has underlined the fact that a functional classification underestimates the complexity of the mutations, and he further points out that major gene deletions and/or rearrangements, single amino acid substitutions and premature termination codons can all cause functional abnormalities of variable severity. Indeed, complete androgen insensitivity has been shown to result from a single point mutation in the DNA binding domain of the androgen receptor gene, the two siblings concerned having receptor-positive androgen resistance (Mowszowicz *et al.*, 1993). A further example here is that of a mutation within exon 2 of the androgen receptor gene (Lumbroso *et al.*, 1993).

Following on from this brief discussion of testicular feminisation in man, outstanding questions that still call for responses are:

1 Precisely how potent is testosterone when it cannot be converted to dihydrotestosterone?
2 In what manner is androgen action amplified by means of conversion to

dihydrotestosterone? That is to say, how does 5α-reductase sharpen up the specificity of androgen action?
3 Is the ratio of 5β- to 5α-reductase activity critical for regulating the pathway?
4 What are the intra-cellular mechanisms whereby the levels of androgen receptor are controlled?

The central and critical rôle of 5α-reductase in development and maintenance of the prostate gland will undoubtedly act to promote and sustain interest in this exciting sphere of research.

Turner's syndrome – streak gonads

One of the more prominent and interesting genetic disorders in women is the 45,XO condition of dysgenesis that characteristically produces Turner's syndrome, with the absence of one X chromosome per cell (Ford *et al.*, 1959). In fact, this can be viewed as the most common chromosome abnormality in our own species, although the incidence drops dramatically from about 1 in 100 at conception to 1 in 10000 females at birth. Some 98% or 99% of embryos or foetuses with an XO chromosome constitution will die during embryonic development and be aborted (Bishop, 1972; Short, 1982). Observations on human XO foetuses indicate that they are severely retarded and have low-for-dates weight. The classical somatic abnormalities or stigmata representative of Turner's syndrome include short stature, skin folds over the neck ('webbing'), birth marks on the skin, narrowing of the aorta, and abnormalities of the renal and lymph systems – the last resulting in oedema. Indeed, the short stature and possibly abnormal bone growth may, to some degree, be a reflection of the cardio-vascular and lymphatic abnormalities. Although the syndrome is named after the clinical report of Turner (1938), Turner himself of course had no knowledge of the underlying chromosome abnormality and indeed suspected pituitary malfunction. He did mention, however, that patients with similar phenotypic features had been described in the nineteenth and early twentieth centuries, and even cited one report of 1762 – based on a description at autopsy.

Individuals with the XO condition usually have conspicuously reduced sexual development, expressed internally through an accelerated rate of oocyte degeneration, a juvenile reproductive tract, and a premature onset of sterility. In such XO women, primary amenorrhoea is the most frequent consequence, affecting 97% of cases (Simpson, 1976), although menstruation (Weiss, 1971) and even pregnancy have been reported (Dewhurst,

1978; King, Magenis & Bennett, 1978). Indeed, Dewhurst gives a full analysis of pregnancy in five patients, with details of the offspring or foetus, and King *et al.* report on a sixth. The ovaries of XO human foetuses aborted spontaneously during the first 10 weeks appear normal, with an almost characteristic complement of germ cells (Singh & Carr, 1966). However, deficiencies of germinal-follicular tissue are noted in the ovaries of mid-term XO foetuses aborted as a sequel to amniocentesis and the expected sequence of primordial follicle formation rarely occurred. As a consequence of an enhanced rate of atresia during the second half of pregnancy, there are few if any oocytes left in the foetus by birth, the ovaries of XO infants being severely affected (Carr, Haggar & Hart, 1968; Bove, 1970; Weiss, 1971). The loss of XO germ cells in the human XO gonad occurs throughout the period when oogonia are embarking upon meiosis (Carr, 1972). On the basis of their observations in XO mice, Burgoyne & Baker (1985) were convinced that a considerable loss of oocytes must take place at the pachytene stage in XO human ovaries as a consequence of the presence of an unpaired X chromosome, that is of an X dosage deficiency in the oocytes. Both X chromosomes are active in XX germ cells throughout oogenesis and both are seemingly required from the beginning of meiosis to promote the survival and function of oocytes. However, in at least three XO human foetuses, Speed (1986) found that arrest in oogenesis can occur at the very earliest stage of meiotic prophase before chromosome pairing is even established. In the monosomic condition for the X chromosome, the ovaries degenerate into streaks or rudiments of fibrous connective tissue generally referred to as streak gonads (Simpson, 1976), resulting in a stroma unable to generate follicles or to secrete hormones. And, although affected individuals are born with morphologically normal female external genitalia, there is little or no development of secondary sexual characteristics in the absence of ovarian hormone synthesis.

Despite the strong emphasis in the two preceding paragraphs on abnormalities, it should be recalled that Turner's syndrome is not an all-or-nothing condition as regards ovarian oocytes. In rare instances, a few oocytes may apparently survive until the time of puberty, thereby permitting the formation of functional ovaries, and thus a tiny proportion of women presumed to have Turner's syndrome can be fertile. The invariable assumption in such cases is said to be that the surviving oocytes would be of XX sex chromosome constitution (Burgoyne & Baker, 1981), but these authors reason that there may occasionally be XO oocytes remaining at puberty. They note that the chances of a successful pregnancy in XO or XO mosaic mothers are low, some 33 of 48 reported pregnancies ending in spontaneous abortion, stillbirth, neonatal death or congenital abnormality (King *et al.*, 1978). This would conform with the anticipated outcome for XO oocytes, for any XO foetuses arising would almost certainly abort.

As to the aetiology of Turner's syndrome, the fertilising spermatozoon is apparently the culprit on most occasions, being devoid of a sex chromosome seemingly due to an error in meiosis or perhaps during the preceding mitotic division of the seminiferous epithelium, or even, as suggested by studies in the mouse, immediately after sperm entry into the egg (for review, see Bond & Chandley, 1983). Certainly, in about 80% of Turner's patients, the single X chromosome is derived from the mother. This observation prompted Short (1982) to comment that perhaps the inactive second X chromosome in female somatic cells is not as inactive as imagined. The observations of Krauss *et al.* (1987) may have some relevance here, for an interstitial deletion of the long arm of the X chromosome was expressed in menstrual irregularities and then in premature ovarian failure. Normal ovarian function requires structural integrity in the critical region of the long arm of the X chromosome (Fig. 7.4; Chandley, 1984).

Again, not without relevance to the diverse clinical reports, the earlier studies of such patients that examined for the Barr body (sex chromatin; see Chapter 2) noted that 80% of the nuclear chromatin patterns were of the male type (Barr, 1959; Ford *et al.*, 1959). This led to the prediction that at least some of the cases would be found to have 45 chromosomes, the sex chromosome being a single X (Polani, Lessof & Bishop, 1956) – a prediction of course now accepted as fact in most instances. However, a small proportion of cases may be mosaics, with sex chromosome constitutions of 45,XO/46,XX (Ford *et al.*, 1959) or even 45,XO/47,XXX, such mosaics perhaps affording one explanation for the wider range of clinical symptoms and discussed by Dewhurst & Lucas (1971), Hassold (1980) and Hassold *et al.* (1980).

Most discussions of Turner's syndrome tend to conclude with the observation that whilst XO females are almost always sterile in humans, this is certainly not the case in all mammals. For example, XO mice are phenotypically normal and apparently fertile (see Chapter 6), although fertility is reduced, there is an accelerated rate of atresia of oocytes, and a premature end to reproductive life (Lyon *et al.*, 1981). Differences in the time of onset and rate of atresia may represent the principal distinction here between mouse and man – in other words a precocious and much more severe ovarian failure in XO women. Relevant to this broader perspective, normal females in the creeping vole, *Microtus oregonii*, have an XO constitution and males the classical XY, but the oogonia are genetically XX

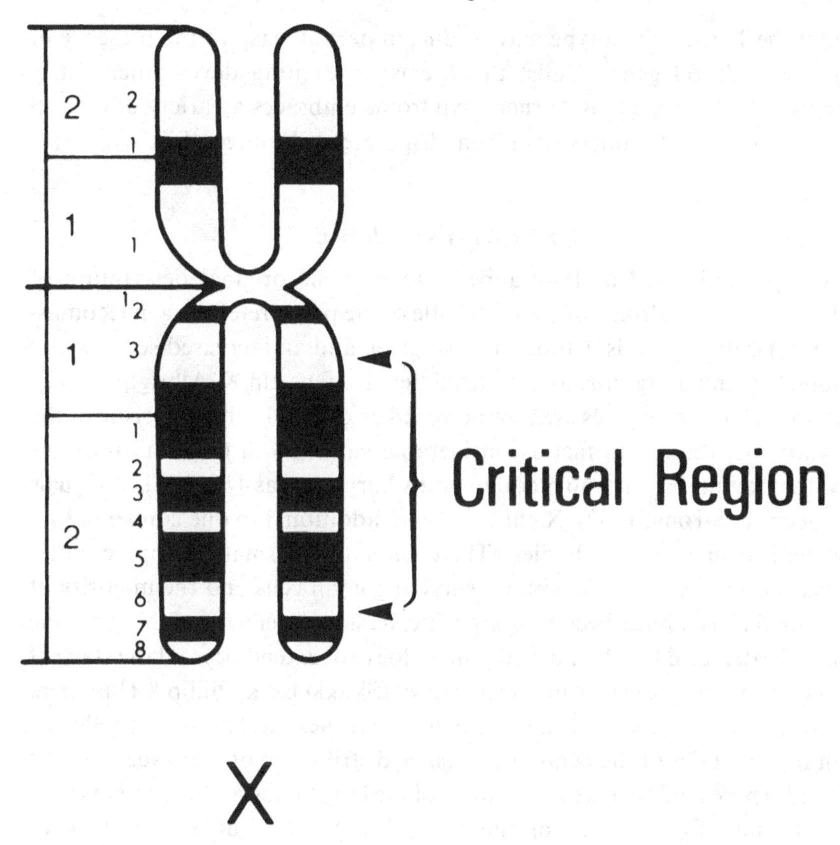

Fig. 7.4. Diagrammatic representation of the human X chromosome to show the critical region of the long arm. (Courtesy of Dr Ann Chandley.)

and spermatogonia YO due to preferential non-disjunction in the primordial germ cells (Ohno, 1967).

Fortunately, the present paragraphs can conclude at a molecular level, for there is now a suggestion that ribosomal protein S4 (RPS4) deficiency may play a rôle in generating Turner's syndrome (Watanabe *et al.*, 1993). Human X- and Y-encoded ribosomal proteins, RPS4X and RPS4Y, are interchangeable. The evidence can be summarised as follows. Some XY females have the somatic features of Turner's syndrome. XY females with such features consistently lack a 90 kilobase region located between *SRY* and *ZFY*. This 90 kilobase region contains a gene, *RPS4Y*, which satisfies the predictions for a Turner's gene: an X-linked homologue escapes X-inactivation (Watanabe *et al.*, 1993). The argument would therefore be that normal human development requires at least two *RPS4* genes per cell and that the Turner phenotype may be due, in part at least, to the presence of just one *RPS4* gene. Whilst this seems an exciting development, it is important to recall that Turner's syndrome embraces a variety of clinical anomalies that doubtless stem from disparate molecular defects.

Klinefelter's syndrome

A paper published in 1942 appears to give the original description of Klinefelter's syndrome in men. The title of the paper refers to gynaecomastia, aspermatogenesis without a-Leydigism, and an increased secretion of follicle stimulating hormone (Klinefelter, Reifenstein & Albright, 1942). Although these features are now accepted as generally characteristic of the syndrome, there is in fact a considerable variation in the clinical observations on these men who predominantly karyotype as 47,XXY individuals (Jacobs & Strong, 1959). X chromosomes additional to one condense into heterochromatic Barr bodies. There tend to be small, atrophic testes usually, but not always, with no surviving germ cells and the majority of seminiferous tubules becoming atrophic. Residual elements of the germ line may be detected but their development does not extend beyond the stage of first spermatocyte except in rare instances (Skakkebaek, Philip & Hammen, 1969). The testes secrete low levels of testosterone, and this may be reflected in the small size of the penis and a sparse distribution of hair except on the head. There is often increased growth of the long bones so that patients tend to be tall. Enlargement of the breasts may occur at a late stage of development. The condition affects 1 in every 800–1000 newborn males, and the patients are not unlike eunuchs from various points of view (Short, 1982). A mild form of mental deficiency is usually present but this is increased markedly in the presence of extra sex chromosomes, such as 48,XXXY. Boys with Klinefelter's syndrome may have delayed speech development (Ratcliffe, 1982).

Because the diagnosis of individuals with Klinefelter's syndrome is frequently made at or after puberty, the extent and manner of prepuberal expression of the syndrome are worth noting. Birth weights are lower in 47,XXY infants when compared with 46,XY controls and, in summarising the results on two such series, six of 65 newborn XXY infants had undescended testes (Ratcliffe, Axworthy & Ginsborg, 1979; Robinson *et al.*, 1979). In a subsequent paper, the incidence of incomplete descent of one or both testes in newborn boys with XXY Klinefelter's syndrome was given as 6.3% as opposed to 0.87% in XY controls (Ratcliffe, 1982). And in prepuberal boys with the syndrome, the complement of spermatogonia was conspicuously reduced (Ferguson-Smith, 1959), perhaps reflecting the finding of abnormal germ cells before birth (Coerdt *et al.*, 1985). Poor development of the genitalia is recorded in a significant minority of prepuberal boys, and major somatic abnormalities in non-reproductive tissues such as kidney agenesis, cleft palate and inguinal hernia may be a feature of XXY children, possibly achieving an incidence as high as 18%.

The question arises as to whether karyotypes other than 47,XXY occasionally found in Klinefelter's syndrome (e.g. 46,XY, 48, XXYY and various mosaics) could be correlated with germ cell survival. Foss & Lewis (1971) made a detailed study of four unusual cases of Klinefelter's syndrome; these were drawn from 466 patients attending an infertility clinic, of which 15 (3.2%) were classified initially as having a karyotype of 47,XXY in cultures of leucocytes. The four unusual patients within these 15 had motile spermatozoa in the ejaculate on more than one occasion (in 21 of 46 ejaculates). On further investigation, one of the four was revealed as a 46,XY/47,XXY mosaic in both leucocyte and fibroblast cultures; the others remained with the initial diagnosis. Testicular histology in biopsies from these patients did not look normal for mature, healthy individuals, for there were varying degrees of tubular fibrosis (hyaline sclerosis and atrophy of seminiferous tubules) associated with Leydig cell hyperplasia. In only one instance was the whole range of spermatogenic stages seen within one tubule. This was previously noted by Warburg (1963) in a Klinefelter's patient found to be a 46,XY/47,XXY mosaic and having an extremely low sperm count. Short (1982) comments on occasional seminiferous tubules in XXY mice in which spermatogenesis is taking place, noting that these particular germ cells have reverted to a normal male XY karyotype by loss of the surplus X. He suggests that a similar situation may exist in certain individuals having Klinefelter's syndrome.

As to the origin of the extra sex chromosome(s) in Klinefelter's patients, Jacobs (1990) investigated 111 males with a 47,XXY chromosome complement. Using a molecular probe, she deduced which parent had contributed the extra chromosome, and the cell division during which the error occurred. Some 49% of cases were paternal in origin, having both an X and a Y chromosome from the father as a result of an error in the first paternal meiotic division. And 51% were maternal in origin, the error having occurred in 72% of the cases at the first maternal meiotic division and in 28% at the second maternal meiotic division. Errors could not be demonstrated at an early post-zygotic division although, in the preceding report of Jacobs *et al.* (1988), the proportion of post-zygotic mitotic error was recorded as 3%. In the light of the above proportions, it is difficult to interpret the significance of the traditional – if contended – assertion that the incidence of Klinefelter's syndrome increases with increasing age of the mother (see Jacobs *et al.*, 1988).

As already noted, the XXY anomaly somehow affects spermatogenesis in patients with Klinefelter's syndrome. Although germ cells are usually found in the testes before puberty, their number may be reduced. In general, the loss of germ cells from the testis in Klinefelter's syndrome occurs much later than the loss of germ cells from the ovary in the XO anomaly in man (Carr *et al.*, 1968). For reasons not well understood, XXY germ cells are unable to survive. As for XO, there may be a dosage phenomenon, germ cells with two X chromosomes failing in a testicular environment.

Congenital adrenal hyperplasia; Kallmann's syndrome

Congenital adrenal hyperplasia is an endocrine disturbance ultimately involving androgens that may strongly influence the sexual phenotype in our own species and has thus been referred to as the adreno-genital syndrome. As noted by Cook (1992), it may first have been described by de Graaf (1672). Due to a genetic defect, one or more enzymes of the adrenal cortex such as 21-hydroxylase fail to be synthesised so that the biosynthetic pathway responsible for generating steroid hormones of the group including cortisol and aldosterone cannot be fully effective (reviewed by Kelnar, 1993). The function of endocrine glands such as the adrenal cortex is under the influence of feedback control. In the absence of an adequate feedback of corticosterone, cortisol or aldosterone upon the hypothalamus, the releasing factor (CRF) for the anterior pituitary trophic hormone (ACTH) is secreted in excess in an attempt, in turn, to generate more target gland secretion. However, the enzyme defect precludes an appropriate response whereas the pathway for steroid sex hormones in the adrenal cortex may be much enhanced and associated with hypertrophy and hyperplasia. The resultant androgen secretion may achieve titres sufficient to cause extensive masculinisation of the external genitalia of baby girls (Short, 1982), for the foetal adrenal glands commence to function during the third month of gestation. Such masculinisation may be noted as a clitoris enlarged into a penis-like structure and the vulval labiae fused to give a presumptive scrotum. But the timing of excessive androgen secretion does not promote retention of the Wolffian duct derivatives in female foetuses nor does it prevent development of the ovaries and transformation of the Müllerian ducts into Fallopian tubes and uterus (Turner, 1964). The true phenotypic state requires to be restored by means of plastic surgery whilst replacement therapy with adrenal hormones – corticosteroids – compensates for the cortisol deficiency.

Kallmann's syndrome, named after the report of Kallmann, Schoenfeld & Barrera (1944), is a condition with major consequences for the reproductive system, affecting 1 in 10000 males but 5–7 times fewer females (Hardelin *et al.*, 1992). Known also as olfactogenital dysplasia, and frequently associated with an arrest of development of the anterior part of the brain, it is a genetic disorder of both the gonadal and olfactory functions. As to the former, Kallmann's syndrome results in a high incidence of cryptorchidism in males and, in females, amenorrhoea associated with the persistence of primary follicles or the death of follicles at their early stages of growth. It is characterised by low levels of gonadotrophins and oestrogens, small ovaries, and primordial follicles clustered next to a thin tunica albuginea (Goldenberg *et al.*, 1976). A few follicles may occasionally begin to grow. Early studies concluded that the defect resided in the hypothalamic–pituitary axis rather than in the ovary, because many patients exhibiting the syndrome could be induced to ovulate in response to gonadotrophins. More recent molecular investigations have pinpointed the defect and modified the preceding conclusion.

The characteristic hypogonadotrophic hypogonadism is a consequence of defective gonadotrophin releasing hormone (GnRH) secretion whereas the inability to smell (anosmia) is due to the absence or hypoplasia of olfactory bulbs and tracts. Why should these two conditions be linked? Immunocytochemical studies in mice indicated that neurons producing GnRH share a common embryonic origin and migration pathway to the hypothalamus with the vomeronasal and terminalis nerves, suggesting that a human cell migration defect specifically affecting such neurons in the olfactory placode of the developing brain could be involved in the syndrome (Schwanzel-Fukuda & Pfaff, 1989; Schwanzel-Fukuda, Bick & Pfaff, 1989; Wray, Grant & Gainer, 1989). This is now known to be the case, and a gene has been isolated that maps to the Kallmann's syndrome critical region on the distal part of the short arm of the human X chromosome; it escapes X-inactivation (Franco *et al.*, 1991). This particular gene (*KALIG-1*) shares homology with molecules involved in cell adhesion and axonal pathfinding, suggesting that a molecular defect in this gene – a partial or complete deletion – causes the neuronal migration defect underlying Kallmann's syndrome. A contribution from pioneer cells (a form of glial cell) from the olfactory placode has been invoked, since they leave a neural cell adhesion molecule (NCAM) trail or scaffold which is then followed by vomeronasal and GnRH neurons (Pfaff & Schwanzel-Fukuda, 1993).

However, it is accepted that more than one gene will be involved in the different genetic forms of the syndrome, for segregation analysis has revealed three modes of transmission: X-linked, autosomal recessive and autosomal dominant (Hardelin *et al.*, 1992). It is now also appreciated that the *KAL* gene responsible for X-linked Kallmann's syndrome, the most frequent form, has a rôle in other neuronal pathways and in kidney organogenesis (Hardelin *et al.*, 1992, 1993*a*). A recent proposition is that the *KAL* protein is an extracellular matrix component with possible antiprotease and adhesion functions (Petit, 1993), a conclusion in line with that of Rugarli *et al.* (1993) who suggest a signal molecule required for neuronal targeting throughout life. The *KAL* gene consists of 14 exons spanning 120–200 kilobases which correlate with the distribution of domains in the predicted protein (Castillo *et al.*, 1992).

Although large deletions have been noted in the terminal part of the X chromosome short arm in some patients with Kallmann's syndrome, specifically in the Xp22.3 region, only two such deletions were found in a search of 20 unrelated males affected with isolated X-linked Kallmann's syndrome (Hardelin *et al.*, 1993*a*). On the basis of Southern blot analysis, it was therefore concluded that large deletions are uncommon in patients affected with Kallmann's syndrome alone. Both deletions in this study encompassed the entire *KAL* gene responsible for X-linked Kallmann's syndrome, and both patients exhibited other clinical anomalies such as unilateral renal aplasia and unilateral absence of a vas deferens (see Chapter 9). More subtle molecular defects have been recorded such as three different base transitions, all leading to a stop codon, and one single-base deletion responsible for a frameshift (Hardelin *et al.*, 1992). Overall, there is conclusive evidence for heterogeneity in the mutations responsible for X-linked Kallmann's syndrome (Hardelin *et al.*, 1993*b*). Indeed, hybridisation of genomic DNA with a *KAL* cDNA probe revealed a homologous locus on the Y chromosome. The locus was mapped to Yq11, a region which shares extensive homology with sequences on Xp22.3 (Castillo *et al.*, 1992).

Concluding remarks

The diverse clinical conditions described in this chapter serve to emphasise the fact that successful completion of processes as fundamental as human sex determination and differentiation cannot be taken for granted, and that errors of differentiation and development should always be anticipated when a population of sufficient size is under consideration. The mammalian genome is by no means as stable as many clinicians may hitherto have imagined, for somatic anomalies can be demonstrated to be a consequence of gross genetic errors or arising from more precise lesions in the form of deletions or specific point mutations. As an example, deletions and translocations occur at a very high frequency in the Xp22.3 region of the human chromosome, a segment immediately proximal to the pseudoautosomal boundary which is considered to represent a deletion 'hot spot' (Lee *et al.*, 1993). The range of astonishingly powerful techniques afforded by modern molecular biology means that bizarre clinical conditions should be susceptible to clear-cut analysis and interpretation, even when extremely subtle defects are being sought. Mutations leading to an aromatase deficiency in an adult female are a good example, for they can result in sexual infantilism and polycystic ovaries (Ito *et al.*, 1993).

Inspiration for much of the molecular work on anomalies of human sexual differentiation came from the studies of Fellous and Weissenbach at the Pasteur Institute in Paris, de la Chapelle in Helsinki, Wolf and colleagues in Freiburg, Ferguson-Smith & Affara in Cambridge, Page in Boston and Goodfellow and colleagues in London. Indeed, the observations in XX males and XY females, notably the presence of a sex-determining Y DNA sequence in many XX males (e.g. Guellaen *et al.*, 1984; Page *et al.*, 1985) and, conversely, deletion of the sex-determining sequence on the Y chromosome in certain XY females, were a major step in mapping the position of *TDF* on the human Y chromosome. Of course, the question then arises as to whether abnormal genetic exchange between the X and Y chromosomes in male meiosis may itself be under the influence of some form of genetic predisposition. Irrespective of the answer to this particular thorny question, one is forced to remark that the intellectual insights gained by molecular analysis have not yet provided the physician with a ready means of mitigating clinical problems, and this remains especially so in the emotive context of infertility and sterility. In fact, approaches such as the transplantation of foetal ovarian tissue to women with dysgenic gonads (Gosden, 1992) may offer more hope in the short term than a specific molecular therapy. In infertile or oligospermic males, further studies on the recently discovered gene *AZF* (azoospermia factor) and its protein product that regulates spermatogenesis may eventually lead to a fruitful treatment (see Ma *et al.*, 1993).

At the time of writing, genes in the Xq/Yq pseudoautosomal region have not been identified, but they may well have a crucial bearing on some of the clinical anomalies discussed in this chapter.

References

Abbas, N. E., Toublanc, J. E., Boucekkine, C., Toublanc, M., Affara, N. A., Job, J. C. & Fellous, M. (1990). A possible common origin of 'Y-negative' human XX males and XX true hermaphrodites. *Human Genetics*, **84**, 356–60.

Affara, N. A., Ferguson-Smith, M. A., Magenis, R. E., Tolmie, J. L., Boyd, E., Cooke, A., Jamieson, D., Kwok, K., Mitchell, M. & Snadden, L. (1987). Mapping the testis determinants by an analysis of Y-specific sequences in males with apparent XX and XO karyotypes and females with XY karyotypes. *Nucleic Acids Research*, **15**, 7325–42.

Andersson, S., Berman, D. M., Jenkins, E. P. & Russell, D. W. (1991). Deletion of steroid 5α-reductase 2 gene in male pseudohermaphroditism. *Nature (London)*, **354**, 159–61.

Andersson, M., Page, D. C. & de la Chapelle, A. (1986). Chromosome Y-specific DNA is transferred to the short arm of X chromosome in human XX males. *Science*, **233**, 786–8.

Armstrong, C. N. & Marshall, A. J. (1964). *Intersexuality in Vertebrates Including Man*. London: Academic Press.

Austin, C. R. & Edwards, R. G. (ed.) (1981). *Mechanisms of Sex Differentiation in Animals and Man*. London: Academic Press.

Barr, M. L. (1959). Sex chromatin and phenotype in man. *Science*, **130**, 679–85.

Berger, R., Abonyi, D., Nodot, A., Vialatte, J. & LeJeune, J. (1970). Hermaphrodisme vrai et 'garçon XX' dans une fratrie. *Revue Européenne d'Études de Clinique et Biologie*, **15**, 330–3.

Berkovitz, G. D., Fechner, P. Y., Marcantonio, S. M., Bland, G.,Stetten, G., Goodfellow, P. N., Smith, K. D. & Migeon, C. J. (1992). The role of the sex-determining region of the Y chromosome (SRY) in the etiology of 46,XX true hermaphroditism. *Human Genetics*, **88**, 411–16.

Berta, P., Hawkins, J. R., Sinclair, A. H., Taylor, A., Griffiths, B. L., Goodfellow, P. N. & Fellous, M. (1990). Genetic evidence equating *SRY* and the testis-determining factor. *Nature (London)*, **348**, 448–50.

Biggers, J. D. (1968). Aspects of intersexuality in domestic mammals. *Proceedings 6th International Congress on Animal Reproduction and Artificial Insemination*, Paris, **2**, 841–70.

Bishop, C. E., Weith, A., Mattei, M. G. & Roberts, C. (1988). Molecular aspects of sex determination in mice: an alternative model for the origin of the *Sxr* region. *Philosophical Transactions of the Royal Society of London, Series B*, **322**, 119–24.

Bishop, M. W. H. (1972). Genetically determined abnormalities of the reproductive system. *Journal of Reproduction and Fertility, Supplement*, **15**, 51–78.

Bond, D. J. & Chandley, A. C. (1983). *Aneuploidy*. Oxford: Oxford University Press.

Bove, K. E. (1970). Gonadal dysgenesis in a newborn with XO karyotype. *American Journal of Diseases of Children*, **120**, 363–6.

Burgoyne, P. S. & Baker, T. G. (1981). Oocyte depletion in XO mice and their XX sibs from 12 to 200 days *post partum*. *Journal of Reproduction and Fertility*, **61**, 207–12.

Burgoyne, P. S. & Baker, T. G. (1985). Perinatal oocyte loss in XO mice and its implications for the aetiology of gonadal dysgenesis in XO women. *Journal of Reproduction and Fertility*, **75**, 633–45.

Burgoyne, P. S. & Palmer, S. J. (1992). Cellular basis of sex determination and sex reversal in mammals. In *Gonadal Development and Function*, ed. S. G. Hillier. Serono Symposium Publication No. 94, pp. 17–29. New York: Raven Press.

Capel, B., Rasberry, C., Dyson, J., Bishop, C. E., Simpson, E., Vivian, N., Lovell-Badge, R., Rastan, S. & Cattanach, B. M. (1993). Deletion of Y chromosome sequences located outside the testis determining region can cause XY female sex reversal. *Nature Genetics*, **5**, 301–7.

Carr, D. H. (1972). Cytogenetic aspects of induced and spontaneous abortions. *Clinics in Obstetrics and Gynaecology*, **15**, 203–19.

Carr, D. H., Haggar, R. A. & Hart, A. G. (1968). Germ cells in the ovaries of XO female infants. *American Journal of Clinical Pathology*, **49**, 521–6.

Castillo, I. del, Cohen-Salmon, M., Blanchard, S., Lutfalla, G. & Petit, C. (1992). Structure of the X-linked Kallmann syndrome gene and its homologous pseudogene on the Y chromosome. *Nature Genetics*, **2**, 305–10.

Chandley, A. C. (1984). Infertility and chromosome abnormality. *Oxford Reviews of Reproductive Biology*, **6**, 1–46.

Coerdt, W., Rehder, H., Gausman, I. & Johannisson, R. (1985). Quantitative histology of human fetal testes in chromosomal disease. *Pediatric Pathology*, **3**, 245–9.

Cook, B. (1992). *Contributions of the Hunter Brothers to Our Understanding of Reproduction*. Glasgow: University of Glasgow Press.

Cotinot, C., McElreavey, K. & Fellous, M. (1993). Sex determination: genetic control. In *Reproduction in Mammals and Man*, ed. C. Thibault, M. C. Levasseur & R. H. F. Hunter, pp. 213–26. Paris: Ellipses.

De Graaf, R. (1672). *De Mulierum Organis Generationi Inservientibus Tractatus Novus*. Leyden.

de la Chapelle, A. (1972). Analytic review: nature and origin of males with XX sex chromosomes. *American Journal of Human Genetics*, **24**, 71–105.

de la Chapelle, A. (1987). The Y-chromosomal and autosomal testis-determining genes. *Development*, **101**, Supplement, 33–8.

de la Chapelle, A., Hästbacka, J., Korhonen, T. & Mäenpää, J. (1990). The etiology of XX sex reversal. *Reproduction, Nutrition, Développement*, Supplement **1**, 39s–49s.

de la Chapelle, A., Koo, G. C. & Wachtel, S. S. (1978). Recessive sex-determining genes in human XX male syndrome. *Cell*, **15**, 837–42.

de la Chapelle, A., Schröder, J., Murros, J. & Tallqvist, G. (1977). Two XX males in one family and additional observations bearing on the aetiology of XX males. *Clinical Genetics*, **11**, 91–106.

Dewhurst, J. (1978). Fertility in 47,XXX and 45,X patients. *Journal of Medical Genetics*, **15**, 132–5.

Dewhurst, C. J. & Lucas, M. (1971). Gonadal dysgenesis and chromosome mosaicism. *British Journal of Hospital Medicine*, **6**, 807–12.

Edwards, R. G. (1980). *Conception in the Human Female*. London & New York: Academic Press.

Eicher, E. M. & Washburn, L. L. (1983). Inherited sex reversal in mice: identification of a new primary sex-determining gene. *Journal of Experimental Zoology*, **228**, 297–304.

Evans, H. J., Buckton, K. E., Spowart, G. & Carothers, A. D. (1979). Heteromorphic X chromosomes in 46,XX males: evidence for the involvement of X-Y interchange. *Human Genetics*, **49**, 11–31.

Ferguson-Smith, M. A. (1959). The prepubertal testicular lesion in chromatin-positive Klinefelter's syndrome (primary micro-orchidism) as seen in mentally handicapped children. *Lancet*, **i**, 219–22.

Ferguson-Smith, M. A. (1965). Karyotype–phenotype correlations in gonadal dysgenesis and their bearing on the pathogenesis of malformations. *Journal of Medical Genetics*, **2**, 142–55.

Ferguson-Smith, M. A. (1966). X–Y chromosomal interchange in the aetiology of true hermaphroditism and of XX Klinefelter's syndrome. *Lancet*, **ii**, 475–6.

Ferguson-Smith, M. A. (1991). Genotype–phenotype correlations in individuals with disorders of sex determination and development including Turner's syndrome. *Seminars in Developmental Biology*, **2**, 265–76.

Ferguson-Smith, M. A., Affara, N. A. & Magenis, R. E. (1987). Ordering of Y-specific sequences by deletion mapping and analysis of X–Y interchange males and females. *Development*, **101** (Supplement), 41–50.

Ferguson-Smith, M. A. & Affara, N. A. (1988). Accidental X–Y recombination and the aetiology of XX males and true hermaphrodites. *Philosophical Transactions of the Royal Society of London, Series B*, **322**, 133–44.

Ferguson-Smith, M. A., Cooke, A., Affara, N. A., Boyd, E. & Tolmie, J. L. (1990). Genotype–phenotype correlations in XX males and their bearing on current theories of sex determination. *Human Genetics*, **84**, 198–202.

Ford, C. E., Jones, K. W., Polani, P. E., de Almeida, J. C. & Briggs, J. H. (1959). A sex-chromosome anomaly in a case of gonadal dysgenesis (Turner's syndrome). *Lancet*, **i**, 711–13.

Foss, G. L. & Lewis, F. J. W. (1971). A study of four cases with Klinefelter's syndrome, showing motile spermatozoa in their ejaculates. *Journal of Reproduction and Fertility*, **25**, 401–8.

Franco, B., Guioli, S., Pragliola, A. *et al.* (1991). A gene deleted in Kallmann's syndrome shares homology with neural cell adhesion and axonal path-finding molecules. *Nature (London)*, **353**, 529–36.

Fredga, K. (1988). Aberrant chromosomal sex-determining mechanisms in mammals, with special reference to species with XY females. *Philosophical Transactions of the Royal Society of London, Series B*, **322**, 83–95.

Gartler, S. M., Waxman, S. H. & Giblett, E. (1962). An XX/XY human hermaphrodite resulting from double fertilisation. *Proceedings of the National Academy of Sciences, USA*, **48**, 332–5.

Gautier, T. (1852). *Emaux et Camées*. Paris.

George, F. W. & Wilson, J. D. (1988). Sex determination and differentiation. In *The Physiology of Reproduction*, ed. E. Knobil & J. Neill *et al.*, pp. 3–26. New York: Raven Press.

German, J., Simpson, J. L., Chaganti, R. S. K., Summitt, R. L., Reid, L. B. & Merkatz, I. R. (1978). Genetically determined sex-reversal in 46,XY humans. *Science*, **202**, 53–6.

Gilgenkrantz, S. (1987). Hermaphrodisme vrai et double fécondation. *Journal de Génétique Humaine*, **35**, 105–18.

Goldenberg, R. L., Powell, R. D., Rosen, S. W., Marshall, J. R. & Ross, G. T. (1976). Ovarian morphology in women with anosmia and hypogonadotrophic hypogonadism. *American Journal of Obstetrics and Gynecology*, **126**, 91–4.

Gosden, R. G. (1992). Transplantation of fetal germ cells. *Journal of Assisted Reproduction and Genetics*, **9**, 118–23.

Griffin, J. E. (1992). Androgen resistance – the clinical and molecular spectrum. *New England Journal of Medicine*, **326**, 611–18.

Griffin, J. E. & Wilson, J. D. (1980). The syndromes of androgen resistance. *New England Journal of Medicine*, **302**, 198–209.

Griffin, J. E. & Wilson, J. D. (1989). The androgen resistance syndromes: 5α-reductase deficiency, testicular feminization, and related syndromes. In *The Metabolic Basis of Inherited Disease*, 6th edn, ed. C. R. Scriver, A. L. Beaudet, W. S. Sly & D. Valle, pp. 1919–44. New York: McGraw-Hill.

Guellaen, G., Casanova, M., Bishop, C., Geldwerth, D., Andre, G., Fellous, M. & Weissenbach, J. (1984). Human XX males with Y single-copy DNA fragments. *Nature (London)*, **307**, 172–3.

Hardelin, J. P., Levilliers, J., del Castillo, I., Cohen-Salmon, M., Legouis, R., Blanchard, S., Compain, S., Bouloux, P., Kirk, J., Moraine, C., Chaussain, J. L., Weissenbach, J. & Petit, C. (1992). X chromosome-linked Kallmann syndrome: stop mutations validate the candidate gene. *Proceedings of the National Academy of Sciences, USA*, **89**, 8190–4.

Hardelin, J. P., Levilliers, J., Young, J., Pholsena, M., Legouis, R., Kirk, J., Bouloux, P., Petit, C. & Schaison, G. (1993*a*). Xp22.3 deletions in isolated familial Kallmann's syndrome. *Journal of Clinical Endocrinology and Metabolism*, **76**, 827–31.

Hardelin, J. P., Levilliers, J., Blanchard, S., Carel, J. C., Leutenegger, M., Bertelletto, J. P. P., Bouloux, P. & Petit, C. (1993*b*). Heterogeneity in the mutations responsible for X chromosome-linked Kallmann syndrome. *Human Molecular Genetics*, **2**, 373–7.

Harley, V. R., Jackson, D. I., Hextall, P. J., Hawkins, J. R., Berkovitz, G. D., Sockanathan, S., Lovell-Badge, R., Goodfellow, P. N. (1992). DNA binding activity of recombinant *SRY* from normal males and XY females. *Science*, **255**, 453–6.

Hassold, T. J. (1980). A cytogenetic study of repeated spontaneous abortions. *American Journal of Human Genetics*, **32**, 723–30.

Hassold, T., Chen, N., Funkhouser, J., Jooss, T., Manuel, B., Matsuura, J., Matsuyama, A., Wilson, C., Yamane, J. A. & Jacobs, P. A. (1980). A cytogenetic study of 1000 spontaneous abortions. *Annals of Human Genetics*, **44**, 151–78.

Hill, J. P. (1926). Development of *Hapale jacchus*. *Journal of Anatomy*, **60**, 486–7.

Ito, Y., Fisher, C. R., Conte, F. A., Grumbach, M. M. & Simpson, E. R. (1993). Molecular basis of aromatase deficiency in an adult female with sexual infantilism and polycystic ovaries. *Proceedings of the National Academy of Sciences, USA*, **90**, 11673–7.

Jacobs, P. A. (1966). Abnormalities of the sex chromosomes in man. *Advances in Reproductive Physiology*, **1**, 61–91.

Jacobs, P. A. (1969). The chromosome basis of some types of intersexuality in man. *Journal of Reproduction and Fertility, Supplement*, **7**, 73–8.

Jacobs, P. A. (1972). Chromosome abnormalities and fertility in man. In *The Genetics of the Spermatozoon*, ed. R. A. Beatty & S. Gluecksohn-Waelsch, pp. 346–58. Copenhagen: Bogtrykkeriet Forum.

Jacobs, P. A. (1990). The role of chromosome abnormalities in reproductive failure. *Reproduction, Nutrition, Dévelopement*, Supplement, **1**, 63–74.

Jacobs, P. A. (1992). The chromosome complement of human gametes. *Oxford Reviews of Reproductive Biology*, **14**, 47–72.

Jacobs, P. A., Baikie, A. G., Court Brown, W. M., Forrest, H., Roy, J. R., Stewart, J. S. S. & Lennox, B. (1959). Chromosomal sex in the syndrome of testicular feminisation. *Lancet*, **ii**, 591–2.

Jacobs, P. A., Hassold, T. J., Whittington, E., Butler, G., Collyer, S., Keston, M.

& Lee, M. (1988). Klinefelter's syndrome: an analysis of the origin of the additional sex chromosome using molecular probes. *Annals of Human Genetics*, **52**, 93–109.

Jacobs, P. A. & Strong, J. A. (1959). A case of human intersexuality having a possible XXY sex-determining mechanism. *Nature (London)*, **183**, 302–3.

Jäger, R. J., Anvret, M., Hall, K. & Scherer, G. (1990*a*). A human XY female with a frame shift mutation in the candidate testis-determining gene *SRY*. *Nature (London)*, **348**, 452–4.

Jäger, R. J., Ebensperger, C., Fraccaro, M. & Scherer, G. (1990*b*). A *ZFY*-negative 46,XX true hermaphrodite is positive for the Y pseudoautosomal boundary. *Human Genetics*, **85**, 666–8.

Jäger, R. J., Harley, V. R., Pfeiffer, R. A., Goodfellow, P. N. & Scherer, G. (1992). A familial mutation in the testis-determining gene *SRY* shared by both sexes. *Human Genetics*, **90**, 350–5.

Jäger, R. J., Pfeiffer, R. A. & Scherer, G. (1991). A familial amino acid substitution in *SRY* can lead to conditional XY sex inversion. *American Journal of Human Genetics*, **49**, Supplement,219.

Jirasek, J. E. (1971). *Development of the Genital System and Male Pseudohermaphroditism*. Baltimore, Maryland: Johns Hopkins University Press.

Josso, N., de Grouchy, J., Auvert, J., Nezelov, C., Jayle, M. F., Moullec, J., Frézal, J., de Casaubon, A. & Lamy, M. (1965). True hermaphroditism with XX/XY mosaicism, probably due to double fertilisation of the ovum. *Journal of Clinical Endocrinology and Metabolism*, **25**, 114–26.

Kallmann, F. J., Schoenfeld, W. A. & Barrera, S. E. (1944). The genetic aspects of primary eunuchoidism. *American Journal of Mental Deficiency*, **48**, 203–36.

Keenan, B. S., Meyer, W. J., Hadjian, A. J., Jones, H. W. & Migeon, C. J. (1974). Syndrome of androgen insensitivity in man: absence of 5α-dihydrotestosterone binding protein in skin fibroblasts. *Journal of Clinical Endocrinology and Metabolism*, **38**, 1143–6.

Kelnar, C. J. H. (1993). Congenital adrenal hyperplasia (CAH) – the place for prenatal treatment and neonatal screening. *Early Human Development*, **35**, 81–90.

King, C. R., Magenis, E. & Bennett, S. (1978). Pregnancy and the Turner syndrome. *Obstetrics and Gynaeology*, **52**, 617–24.

Klinefelter, H. F. Jr., Reifenstein, E. C. Jr. & Albright, F. (1942). Syndrome characterized by gynecomastia, aspermatogenesis without A-Leydigism, and increased excretion of follicle stimulating hormone. *Journal of Clinical Endocrinology and Metabolism*, **2**, 615–27.

Krauss, C. M., Turksoy, N., Atkins, L., McLaughlin, C., Brown, L. G. & Page, D. C. (1987). Familial premature ovarian failure due to an interstitial deletion of the long arm of the X chromosome. *New England Journal of Medicine*, **317**, 125–31.

Lee, W. C., Ferrero, G. B., Chinault, A. C., Yen, P. H. & Ballabio, A. (1993). A yeast artificial chromosome contig linking the steroid sulfatase and Kallmann syndrome loci on the human X chromosome short arm. *Genomics*, **18**, 1–6.

Levilliers, J., Quack, B., Weissenbach, J. & Petit, C. (1989). Exchange of terminal portions of X- and Y-chromosomal short arms in human XY females. *Proceedings of the National Academy of Sciences, USA*, **86**, 2296–300.

Lumbroso, S., Lobaccaro, J. M., Belon, C., Martin, D., Chaussain, J. L. &

Sultan, C. (1993). A new mutation within the DNA-binding domain of the androgen receptor gene in a family with complete androgen insensitivity syndrome. *Fertility and Sterility*, **60**, 814–19.

Lyon, M. F., Cattanach, B. M. & Charlton, H. M. (1981). Genes affecting sex differentiation in mammals. In *Mechanisms of Sex Differentiation in Animals and Man*, ed. C. R. Austin & R. G. Edwards, pp. 329–85. London: Academic Press.

MacDonald, P. C., Madden, J. D., Brenner, P. F., Wilson, J. D. & Siiteri, P. K. (1979). Origin of estrogen in normal men and in women with testicular feminization. *Journal of Clinical Endocrinology and Metabolism*, **49**, 905–16.

McElreavey, K., Vilain, E., Abbas, N., Brauner, R., Nihoul-Fekete, C., Richaud, F., Rappaport, R., Raymond, J. P., Toublanc, J. E. & Fellous, M. (1991). Conditional mutation in the testis-determining region suggests a second sex-determining gene in humans. *American Journal of Human Genetics*, **49**, Supplement, 413 (abstract).

McElreavey, K., Vilain, E., Abbas, N., Costa, J. M., Souleyreau, N., Kucheria, K., Boucekkine, C., Thibaud, E., Brauner, R., Flamant, F. & Fellous, M. (1992). XY sex reversal associated with a deletion 5′ to the *SRY* 'HMG box' in the testis-determining region. *Proceedings of the National Academy of Sciences, USA*, **89**, 11016–20.

McLaren, A. (1976). *Mammalian Chimaeras*. Cambridge & London: Cambridge University Press.

Ma, K., Inglis, J. D., Sharkey, A., Bickmore, W. A., Hill, R. E., Prosser, E. J., Speed, R. M., Thomson, E. J., Jobling, M., Taylor, K., Wolfe, J., Cooke, H. J., Hargreave, T. B. & Chandley, A. C. (1993). A Y chromosome gene family with RNA-binding protein homology: candidates for the azoospermia factor AZF controlling human spermatogenesis. *Cell*, **75**, 1287–95.

Madan, K. & Walker, S. (1974). Possible evidence for Xp+ in an XX male. *Lancet*, **i**, 1223.

Miró, R., Caballin, M. R., Marina, S. & Egozcue, J. (1978). Mosaicism in XX males. *Human Genetics*, **45**, 103–6.

Mowszowicz, I., Lee, H. J., Chen, H. T., Mestayer, C., Portois, M. C., Cabrol, S., Mauvais-Jarvis, P. & Chang, C. (1993). A point mutation in the second zinc finger of the DNA-binding domain of the androgen receptor gene causes complete androgen insensitivity in two siblings with receptor-positive androgen resistance. *Molecular Endocrinology*, **7**, 861–9.

Mürset, G., Zachmann, M., Prader, A., Fischer, J. & Labhart, A. (1970). Male external genitalia of a girl caused by a virilizing adrenal tumour. *Acta Endocrinologia (Copenhagen)*, **65**, 627–38.

Ogata, T., Hawkins, J. R., Taylor, A., Matsuo, N., Hata, J. & Goodfellow, P. N. (1992). Sex reversal in a child with a 46XYp+ karyotype: support for the existence of a gene(s), located in distal Xp, involved in testis formation. *Journal of Medical Genetics*, **29**, 226–30.

Ohno, S. (1967). *Sex Chromosomes and Sex-Linked Genes*. Berlin: Springer-Verlag.

Ohno, S. (1979). *Major Sex-Determining Genes*. Berlin: Springer-Verlag.

Overzier, C., ed. (1963). *Intersexuality*. New York: Academic Press.

Page, D. C., de la Chapelle, A. & Weissenbach, J. (1985). Chromosome Y-specific DNA in related human XX males. *Nature (London)*, **315**, 224–6.

Page, D. C., Fisher, E. M. C., McGillivray, B. & Brown, L. G. (1990). Additional deletion in sex-determining region of human Y chromosome

resolves paradox of X,t(Y;22) female. *Nature (London)*, **346**, 279–81.
Palmer, M. S., Sinclair, A. H., Berta, P., Ellis, N. A., Goodfellow, P. N., Abbas, N. E. & Fellous, M. (1989). Genetic evidence that *ZFY* is not the testis-determining factor. *Nature (London)*, **342**, 937–9.
Petit, C. (1993). Molecular basis of the X-chromosome-linked Kallmann's syndrome. *Trends in Endocrinology and Metabolism*, **4**, 8–13.
Petit, C., de la Chapelle, A., Levilliers, J., Castillo, S., Noel, B. & Weissenbach, J. (1987). An abnormal terminal X–Y interchange accounts for most but not all cases of human XX maleness. *Cell*, **49**, 595–602.
Pfaff, D. W. & Schwanzel-Fukuda, M. (1993). Migration of GnRH neurons from olfactory placode to basal forebrain. *Journal ofReproduction and Fertility*, Abstract Series No. **11**, 5, S5.
Polani, P. E. (1962). Sex chromosome anomalies in man. In *Chromosomes in Medicine*, ed. J. L. Hamerton. Little Club Clinics in Developmental Medicine, No. 5, pp. 74–133. London: Heinemann Medical.
Polani, P. E. (1970). Hormonal and clinical aspects of hermaphroditism and the testicular feminizing syndrome in man. *Philosophical Transactions of the Royal Society of London, Series B*, **259**, 187–204.
Polani, P. E. (1981*a*). Abnormal sex development in man. I. Anomalies of sex-determining mechanisms. In *Mechanisms of Sex Differentiation in Animals and Man*, ed. C. R. Austin & R. G. Edwards, pp. 465–547. London: Academic Press.
Polani, P. E. (1981*b*). Abnormal sex development in man. II. Anomalies of sex-differentiating mechanisms. In *Mechanisms of Sex Differentiation in Animals and Man*, ed. C. R. Austin & R. G. Edwards, pp. 549–90. London: Academic Press.
Polani, P. E., Lessof, M. H. & Bishop, P. M. F. (1956). Colour-blindness in 'ovarian agenesis' (gonadal dysplasia). *Lancet*, **ii**, 118–20.
Rappold, G. A. (1993). The pseudoautosomal regions of the human sex chromosomes. *Human Genetics*, **92**, 315–24.
Ratcliffe, S. G. (1982). The sexual development of boys with the chromosome constitution 47,XXY (Klinefelter's syndrome). *Clinics in Endocrinology and Metabolism*, **11**, 703–16.
Ratcliffe, S. G., Axworthy, D. & Ginsborg, A. (1979). The Edinburgh study of growth and development of children with sex chromosome abnormalities. *Birth Defects*, **15**, 243–60.
Robboy, S. J., Miller, T., Donahoe, P. K., Jahre, C., Welch, W. R., Haseltine, F. P., Miller, W. A., Atkins, L. & Crawford, J. D. (1982). Dysgenesis of testicular and streak gonads in the syndrome of mixed gonadal dysgenesis: perspective derived from a clinicopathologic analysis of twenty-one cases. *Human Pathology*, **13**, 700–16.
Robinson, A., Lubs, H. A., Nielsen, J. & Sørensen, K. (1979). Summary of clinical findings: profiles of children with 47XXY,47XXX and 47XYY karyotypes. *Birth Defects*, **15**, 261–6.
Rugarli, E. I., Lutz, B., Kuratani, S. C., Wawersik, S., Borsani, G., Ballabio, A. & Eichele, G. (1993). Expression pattern of the Kallmann syndrome gene in the olfactory system suggests a role in neuronal targeting. *Nature Genetics*, **4**, 19–26.
Scherer, G., Schempp, W., Baccichetti, C., Lenzini, E., Bricarelli, F. D., Carbone, L. D. L. & Wolf, U. (1989). Dupliction of an Xp segment that includes the *ZFX* locus causes sex inversion in man. *Human Genetics*, **81**, 291–4.

Schwanzel-Fukuda, M. & Pfaff, D. W. (1989). Origin of luteinizing hormone-releasing hormone neurons. *Nature (London)*, **338**, 161–4.

Schwanzel-Fukuda, M., Bick, D. & Pfaff, D. W. (1989). Luteinizing hormone-releasing hormone (LHRH)-expressing cells do not migrate normally in an inherited hypogonadal (Kallmann) syndrome. *Molecular Brain Research*, **6**, 311–26.

Short, R. V. (1982). Sex determination and differentiation. In *Reproduction in Mammals*, ed. C. R. Austin & R. V. Short, 2nd edn, vol. 2, pp. 70–113. Cambridge: Cambridge University Press.

Simpson, J. L. (1976). *Disorders of Sexual Differentiation. Etiology and Clinical Delineation.* New York & London: Academic Press.

Singh, R. P & Carr, D. H. (1966). The anatomy and histology of XO human embryos and fetuses. *Anatomical Record*, **155**, 369–84.

Skakkebaek, N. E., Philip, J. & Hammen, R. (1969). Meiotic chromosomes in Klinefelter's syndrome. *Nature (London)*, **221**, 1075–6.

Speed, R. M. (1986). Oocyte development in XO foetuses of man and mouse: the possible role of heterologous X-chromosome pairing in germ cell survival. *Chromosoma (Berlin)*, **94**, 115-24.

Taketo, T. (1991). Production of Müllerian-inhibiting substance (MIS) and sulfated glycoprotein-2 (SGP-2) associated with testicular differentiation in the XX mouse gonadal graft. *Annals of the New York Academy of Sciences*, **637**, 74–89.

Taketo, T., Saeed, J., Manganaro, T., Takahashi, M. & Donahoe, P. K. (1993). Müllerian inhibiting substance production associated with loss of oocytes and testicular differentiation in the transplanted mouse XX gonadal primordium. *Biology of Reproduction*, **49**, 13–23.

Turner, C. D. (1964). Special mechanisms in anomalies of sex differentiation. *American Journal of Obstetrics and Gynecology*, **90**, 1208–26.

Turner, H. H. (1938). A syndrome of infantilism, congenital webbed neck, and cubitus valgus. *Endocrinology*, **23**, 566–74.

Van Niekerk, W. A. (1974). *True Hermaphroditism: Clinical, Morphologic and Cytogenetic Aspects.* Hagerstown, Maryland: Harper & Row.

Van Niekerk, W. A. & Retief, A. E. (1981). The gonads of human true hermaphrodites. *Human Genetics*, **58**, 117–22.

Vergnaud, G., Page, D. C., Simmler, M-C., Brown, L., Rouyer, F., Noel, B., Botstein, D., de la Chapelle, A. & Weissenbach, J. (1986). A deletion map of the human Y chromosome based on DNA hybridization. *American Journal of Human Genetics*, **38**, 109–24.

Vilain, E., McElreavey, K., Jaubert, F., Raymond, J-P., Richaud, F. & Fellous, M. (1992). Familial case with sequence variant in the testis-determining region associated with two sex phenotypes. *American Journal of Human Genetics*, **50**, 1008–11.

Warburg, E. (1963). A fertile patient with Klinefelter's syndrome. *Acta Endocrinologica (Copenhagen)*, **43**, 12–26.

Watanabe, M., Zinn, A. R., Page, D. C. & Nishimoto, T. (1993). Functional equivalence of human X- and Y-encoded isoforms of ribosomal protein S4 consistent with a role in Turner syndrome. *Nature Genetics*, **4**, 268–71.

Weiss, L. (1971) Additional evidence of gradual loss of germ cells in the pathogenesis of streak ovaries in Turner's syndrome. *Journal of Medical Genetics*, **8**, 540–4.

Weissenbach, J. & Rouyer, F. (1990). Chromosome Y et détermination du sexe. *Reproduction, Nutrition, Développement*, Supplement **1**, 27–38.

Welshons, W. J. & Russell, L. B. (1959). The Y-chromosome as the bearer of male determining factors in the mouse. *Proceedings of the National Academy of Sciences, USA*, **45**, 560–6.

Wilson, J. D. (1992). Syndromes of androgen resistance. *Biology of Reproduction*, **46**, 168–73.

Wilson, J. D., Griffin, J. E., Leshin, M. & MacDonald, P. C. (1983). The androgen resistance syndromes: 5-alpha reductase deficiency, testicular feminization, and related disorders. In *The Metabolic Basis of Inherited Disease*, ed. J. B. Stanbury, J. B. Wyngaarden *et al.*, 5th edn, pp. 1001–26. New York: McGraw Hill.

Wislocki, G. (1932). Placentation in the marmoset (*Oedipomidas geoffroyi*), with remarks on twinning in monkeys. *Anatomical Record*, **52**, 381–99.

Wolf, U., Schempp, W. & Scherer, G. (1992). Molecular biology of the human Y chromosome. *Reviews of Physiology, Biochemistry and Pharmacology*, **121**, 147–213.

Wray, S., Grant, P. & Gainer, H. (1989). Evidence that cells expressing luteinizing hormone-releasing hormone mRNA in the mouse are derived from progenitor cells in the olfactory placode. *Proceedings of the National Academy of Sciences, USA*, **86**, 8132–6.

Yen, S. C. & Jaffe, R. B. (1978). *Reproductive Endocrinology: Physiology, Pathophysiology and Clinical Management*. Philadelphia & London: W. B. Saunders.

8

Sexual differentiation in chimaeras

Introduction

In Greek mythology, a chimaera was a monster bearing a lion's head, goat's body and serpent's tail (Fig. 8.1). In the present chapter, however, chimaeras will be regarded as composite animals containing genetically different cell populations derived from more than one zygote, and primary chimaeras are those in which the genetically different cell populations have coexisted from a very early stage of embryogenesis or from fertilisation itself. In a mosaic, by contrast, the cells are derived from a single zygote lineage (Mintz, 1974). Artificially generated chimaeras have been used quite extensively in experimental embryology since the 1960s (e.g. Tarkowski, 1961, 1964; Mintz, 1962, 1964, 1965*a*, *b*), and especially during the 1970s and 1980s to examine somatic cell lineages (McLaren, 1976; Gardner, 1978, 1982; Gardner & Rossant, 1979; Beddington, 1982; Le Douarin & McLaren, 1984; Bradbury, 1987), not least with the focus on malignancy and on possible ways of controlling the growth of tumours (Brinster, 1974; Illmensee & Mintz, 1976; Hardy *et al.*, 1990). The approach of bringing cell lineages together from two or more different individuals to produce a viable conceptus has permitted analysis of a number of steps in the process of differentiation such as the time of allocation of commitment, and has shed light on the origin and fate of individual cell lines (Gardner, 1985). The experimental generation of chimaeras has also enabled interactions between germ cells and the surrounding soma to be analysed under a variety of conditions, and has led to conclusions concerning differentiation and organogenesis, and survival of oogonia in testicular tissue or spermatogonia in ovarian tissue. Whilst work relevant to the present chapter has thereby focused on the potential for development of XX germ cells in XY somatic tissues, or vice versa, the full implications of such interactions should certainly be viewed more widely than simply the rôle of the

Fig. 8.1. Etruscan chimaera assuming a classical posture and depicting its composite parts – those of a lion's head, goat's body and serpent's tail. (Reproduced from *Cambridge Ancient History*, courtesy of Cambridge University Library.)

respective sex chromosome constitutions. Despite this qualification, the chapter will concentrate on cells within reproductive tissues, and also on the extent to which hermaphrodites are formed or can be detected. Most experimental studies have been conducted in mice, although perhaps the first recorded approach to producing chimaeras by aggregation was in rats (Nicholas & Hall, 1942) and there is now a useful body of work in domestic farm animals (see Ruffing *et al.*, 1993).

Experimental generation of chimaeras

Techniques for the generation of chimaeras vary in detail, and were reviewed in the volume edited by Le Douarin & McLaren (1984). A classical approach involves the aggregation of synchronous embryos, that is fusing entire embryos at an early stage of cleavage such as the 8-cell stage, to produce conceptuses or offspring (Tarkowski, 1961; Mintz, 1962, 1964; McLaren & Bowman, 1969). To achieve aggregation, pairs of young morulae are denuded of the protective zona pellucida and cultured together until reorganisation into a single integrated embryo is achieved containing twice the usual number of cells for a morula. This is then transplanted to the uterus of a pseudopregnant recipient mouse that serves as a foster mother. Exploitation of the aggregation method is generally restricted to closely synchronous embryos of the same species before the major phases of tissue differentiation are initiated and before the formation of tight junctions between blastomeres commences at about the 16-cell stage prior to organisation of the blastocyst. A detailed description of aggregation methods has been given by Wood *et al.* (1993), but with the specific objective of incorporating embryonal stem cells into the chimaera (Table 8.1). This is simpler than the alternative approach to producing mammalian chimaeras of injecting somatic cells, such as those of the inner cell mass of one embryo, into the blastocyst cavity of a second embryo (Gardner, 1968, 1971; Gardner & Lyon, 1971). Although technically more demanding, this approach offers greater versatility than the aggregation of morulae, for specific cell types or pieces of tissue rather than whole embryos can be introduced into the blastocoele of a different genotype during an *in vitro* manipulation. Indeed, a single inner cell mass cell can be introduced into the blastocoele of a recipient and successfully generate chimaeras as, for example, for coat colour in mice. More dramatically still, Gardner (1982) noted that a single cell so injected into a blastocyst could produce cell lines composing fully half of a newborn mouse. But, as a word of caution, Gardner (1978) emphasised that the routine production of mammalian chimaeras is far more difficult than the procedural descriptions might lead one to suppose. Within this caveat, there is the view for the injection model that the cells so introduced may in some manner be protected by the surrounding trophoblast, thereby frequently promoting a greater incidence of integration than found in the aggregation approach. In fact, injected embryonal stem cells have been shown to contribute extensively to both somatic tissues and the germ line of chimaeras (Bradley *et al.*, 1984), although the genetic combination of recipient blastocyst and embryonal stem cell line is important here (Kelley *et al.*, 1993).

As inferred above, artificially produced chimaeras provide an elegant means of tackling questions of developmental biology, i.e. of differentiation and dedifferentiation, especially when mixing cell lines of known embryonic origin that can be identified subsequently by means of suitable genetic markers. Whilst such cell markers tended originally to be morphological in nature, they have become far more sophisticated and powerful thanks to histochemical approaches and then to developments in molecular biology; the procedure of DNA hybridisation *in situ* is now much favoured

Table 8.1. *To illustrate the relative effectiveness of two quite distinct techniques for generating germline chimaeras in mice*

Technique for constructing chimaera	Embryos transplanted (no.)	Foetuses born (no.)	Chimaeric offspring				Offspring transmitting germ line	
			Males		Females			
			No.	(%)	No.	(%)	No.	(%)
Embryonic stem cell aggregation with a morula	321	89	31	(9.6)	1	(0.3)	11	(3.4)
Embryonic stem cell injection into blastocoele	631	250	72	(11.4)	23	(3.6)	24	(3.8)

Note:
Adapted from Wood *et al.* (1993), who present further details on the origin of the donor embryonic stem cells.
NB: Percentages are expressed as a proportion of the number of embryos transplanted into pseudopregnant recipients.

(see Patek *et al.*, 1991). A proportion of the offspring resulting from aggregation or injection techniques will contain cells derived from both embryos in many or all organs and tissues. So the approach can enable definition of cell lineages in the normal embryo, or prescribe the anatomical focus of action of lethal genes (Gardner, 1982). Nonetheless, not all *in vitro* manipulations performed to generate chimaeras will result in tetraparental offspring, for there may be loss of one of the cell populations during differentiation and development. In mice, at least, the incidence of such cell loss depends upon the genetic strains under combination. A further disadvantage of this experimental approach is that conclusions derived from experimentally produced chimaeras may not correspond precisely with the allocation of cell lineages during undisturbed development.

Quite apart from extensive work in laboratory rodents, which includes formation of chimaeras between rat and mouse embryos to investigate interspecific development (Gardner & Johnson, 1973; Zeilmaker, 1973), between *Mus musculus* and *Mus caroli* (Rossant & Frels, 1980; Rossant & Chapman, 1983; Rossant *et al.*, 1983), and between mouse and bank vole (Mystkowska, 1975), chimaeras have also been generated in sheep (e.g. Pighills, Hancock & Hall, 1968; Tucker, Moor & Rowson, 1974; Tucker, Dain & Moor, 1978; Fehilly, Willadsen & Tucker, 1984*a*, *b*), cows (Summers, Shelton & Bell, 1983; Brem, Tenhumberg & Kräusslich, 1984; Picard, King & Betteridge, 1986; Picard *et al.*, 1990) and pigs (Kashiwazaki *et al.*, 1992). The experimental approach in sheep was either by means of aggregation (Pighills *et al.*, 1968) or by transplanting a blastomere from an early cleavage stage embryo to a morula still within the zona pellucida (Willadsen & Fehilly, 1983; Polge, 1985). Tucker *et al.* (1978) reported two chimaeric rams able to produce functional spermatozoa, one of which with an XY↔XY constitution produced progeny from both the XY components. The somewhat bizarre sheep–goat chimaera has also been produced by an aggregation technique (Fehilly *et al.*, 1984*a*; Meinecke-Tillmann & Meinecke, 1984; Ruffing *et al.*, 1993; Table 8.2) or by inner cell mass transplantation (Polzin *et al.*, 1987; Roth *et al.*, 1989), coupled with a subsequent sealing of the zona pellucida by an agar embedding technique (Willadsen & Polge, 1981). In cattle, one *Bos taurus–Bos indicus* chimaeric calf was produced by Summers *et al.* (1983) using the blastocyst injection technique whilst Brem *et al.* (1984) succeeded in aggregating morulae within the same species. Picard *et al.* (1990) used the approach of aggregating inner cell masses with early morulae, and were able to determine the extent to which aggregation was affected by the age and morphological differentiation of the inner cell mass. In pigs, injection of inner cell mass

Table 8.2. *To illustrate the incidence of pregnancy and proportion of sheep–goat chimaeric embryos reaching term when transferred to the two species of host*

Species of recipient	Recipients (no.)	Embryos transferred (no.)	Offspring at term (no.)	No. of recipients with		No. of offspring	
				early pregnancy failure	term pregnancy	live	stillborn
Sheep	52	118	28 (24%)	0	24 (46%)	25	1
Goat	61	133	21 (16%)	15	14 (23%)	16	5

Notes:
Modified from Ruffing *et al.* (1993).
The sex ratio of the chimaeras was 2 male : 1 female.

cells into blastocysts generated at least two chimaeras amongst 11 piglets arising from the manipulation and subsequent transplantation (Kashiwazaki *et al.*, 1992). Overall, in these experiments with domestic farm animals, the notion underlying the formation of chimaeras has been that of incorporating desirable traits from one breed into another, in reality producing new strains or breeds of animal. But, very much in this context and as a consequence of advances in molecular biology, the introduction of specific gene constructs into a pronucleate egg is more likely to be the way forward in future experiments.

Spontaneously arising chimaeras

Whilst micromanipulation of very young embryos is an elegant approach to producing novel genetic combinations, the fact should not be overlooked that chimaeras do arise spontaneously in humans, laboratory and farm animals and can provide invaluable experimental material, especially if identified at an early stage of development. Double fertilisation of an oocyte with two maternal nuclei or the spontaneous fusion of zygotes in cases of multiple ovulation can produce the condition. Again, chimaeras may arise by the exchange of cells between foetuses, for example between the placental circulations of dizygotic twins. Such chimaerism may be limited to an incorporation of blood cells (see Chapter 5). In spontaneous chimaeras, just as in those generated artificially, it needs to be appreciated that a loss of certain cell lines may have occurred by the time the condition is recognised.

In instances of spontaneous XX↔XY chimaerism in man (reviewed by Tippett, 1984), the somatic cells are of two types: those with an XY complement are testis determining and those with an XX complement are ovary determining (Ferguson-Smith & Affara, 1988). Human twins have been described in which both offspring were found to have 46,XX and 46,XY cells in their haemopoietic tissues, and the lymphocytes of each twin were unresponsive to those of the co-twin. The two twins were developing as a normal female and male, despite the clear evidence of an admixture of two cell lines and the presence of more XY than XX lymphocytes in both of them. These chimaeras must have formed early in life for such a remarkable similarity in their stem cell lines to have been established (Edwards, 1980). Human XX↔XY twin chimaeras may be normally fertile, for at least seven females of XX↔XY human chimaeric pairs have had children (McLaren, 1984). A different form of spontaneous human chimaera was reported by Mayr, Pausch & Schnedl (1979). In this instance, chimaerism was demonstrable only in the germ line and not in blood cells or skin fibroblasts. The individual came to light due to a lack of concordance between her own blood group and those of her offspring.

Dizygotic twins with fused placentae and vascular anastomoses usually become chimaeras, as demonstrated by XX↔XY bull calves (Teplitz, Moon & Basrur, 1967). Both twins were chimaeras. By contrast in man and marmoset, blood chimaerism in heterosexual twin pairs is repeatedly stated not to be accompanied by any abnormalities of sexual development (e.g. McLaren, 1976, 1984; see Chapter 5), and human XX↔XY twin chimaeras are usually fertile (Tippett, 1984). As to goats, pockets of germ cells can occasionally be found in the sex-reversed testes of adults (Chapter 5), and may result from chimaerism which is common in this species (Short, 1972). Tortoiseshell cats – the coat colour is a patchwork of black and 'orange' hairs – can also provide examples of spontaneous chimaeras. The 'orange' gene is X-linked and most tortoiseshell cats are heterozygous females, the coat pattern arising by X chromosome inactivation resulting in functional mosaicism. Although male tortoiseshells are rare, some of these chimaeras may contain a mixture of XX and XY cells or two populations of XY cells and be fertile (Short, 1982).

Sex chromosome chimaeras

In the context of sex chromosome chimaeras, that is XX↔XY chimaeras generated by combining cleaving embryos of opposite sex, a surprising finding has been that only a very small proportion develop as hermaphrodites or intersexes with ovotestis tissue in contrast to the 50% that would have been anticipated (Mintz, 1968; McLaren, 1972; McLaren, Chandley & Kofman-Alfaro, 1972; Gardner, 1982); most develop as males. Likewise in chimaeric mice produced by embryonal stem cell injection into host blastocysts, only one of the 63 chimaeras arising (externally female) was found to be hermaphrodite with an ovotestis composed of 70–80% testicular tissue, although there were no mature spermatozoa (Hardy *et al.*, 1990). Why should the incidence of hermaphrodites be so low? Have potential hermaphrodites or intersexes been destined for early embryonic loss and died *in utero*? By way of response, Bradbury (1987) found that most foetal XX↔XY mouse chimaeras develop initially as hermaphrodites, with varying proportions of presumptive ovarian and testicular tissue, but such gonads show a subsequent transition from foetal ovotestes to postnatal testes due to regression of the ovarian portions, a process which is well under way by day 14 of gestation. This follows rather precisely the earlier speculation of Whitten, Beamer & Byskov (1979). The survival of granulosa cells is known to require the presence of oocytes, and Vigier *et al.* (1987) reported that anti-Müllerian hormone (AMH) produced by Sertoli cells kills foetal oocytes. This led Burgoyne *et al.* (1988*b*) to propose that the formation of testes in most adult XX↔XY mouse chimaeras could be due to the regression of ovarian tissue following the elimination of oocytes in foetal life by Sertoli cell action. Despite the loss of female germ cells, abundant XX somatic cells would remain in the gonads.

Many sex chromosome chimaeras develop as fertile males, having histologically normal testes, although containing a mixed population of XY and XX cells distributed throughout the body (Mintz, 1968; Mystkowska & Tarkowski, 1968). In terms of the sex ratio of chimaeric offspring, a significant excess of males has been reported in most studies (McLaren, 1975), although XX↔XY females certainly can be produced. Apparently, the genetic strains of mice used in the combination technique have an important influence on whether 'balanced' chimaeras are produced (Mullen & Whitten, 1971), with approximately equal contributions from the two constituent embryos, and on whether there is an excess of phenotypic males. In some 'unbalanced' combinations, one strain always contributing more cells than the other, the sex ratio was normal because the sex of the chimaeras tended to be that of the embryo of the dominant strain, while balanced combinations were skewed towards 3 males : 1 female (Mullen & Whitten, 1971; Gardner, 1982). In other words, in a balanced combination, XX↔XY chimaeras tended to develop as males (Whitten, 1975). In the absence of a 'timing mismatch' (see Chapter 6), XX↔XY chimaeras develop as females only if less than about 25% of the cells are from the XY component (McLaren, 1991*a*), perhaps inferring some form of quantitative molecular influence.

This situation of a small proportion (<20–25%) of XY cells in the somatic elements of the genital ridge has been examined further by Burgoyne *et al.* (1988*b*) and by Palmer & Burgoyne (1991*b*). In such chimaeras, XY cells may apparently enter differentiation along the Sertoli cell pathway but are too few in number and inappropriately distributed to promote formation of testis cords. In the absence of an influence of such testis cords, germ cells would be free to enter meiosis and the oocytes so formed would seemingly recruit the full population of supporting somatic cells to constitute follicles, including those that had initially embarked on differentiation as Sertoli cells (McLaren, 1991*a*). The molecular nature of such a putative influence of the oocytes remains unknown. XY cells would therefore be contributing to the population of follicles in a female chimaera in approximately the same proportion as their contribution to other tissues and organs (Burgoyne *et al.*, 1988*b*; Palmer & Burgoyne, 1991*b*). This stands in contrast to the Sertoli cell population in male chimaeras which is derived principally from XY supporting cells, relatively few neighbouring XX cells being recruited (see below).

Turning to the germ cell lines in chimaeric embryos, some plasticity in their early development is certainly demonstrable but this is perhaps worth setting in a broader biological context. In lower vertebrates such as amphibians, it has long been appreciated that primordial germ cells have dual potentiality in terms of their sexual destiny (Witschi, 1965). Here, the germ cells respond to environmental cues that may act to induce alterations in the predicted stem cells of the germ line. In other words, the cells appear at variance with their sex chromosome complement. However, if animals showing such transformation are bred successfully, then the genetic sex constitution remains unchanged as deduced from the aberrant sex ratio of the progeny (Mintz, 1968). As a contrast, this last author suggests that in the mouse and possibly all other placental mammals, complete and thereby functional alteration of germ cell sex phenotype does not and cannot occur.

If XX↔XY male chimaeras can be demonstrably fertile, the origin of functional germ cells in such chimaeras is important and seemingly bears on their postnatal survival. But perhaps the most exciting question is whether two genetically distinct types of gamete can be formed within the same gonad. If fertile XX↔XY male chimaeric mice are subjected to test breeding, the offspring show a normal sex ratio and invariably correspond in type with the XY component of the chimaera. As suggested at the end of the previous paragraph, functional reversal of differentiation of XX germ cells does not occur in the mouse or in several other mammals (see Chapter 6). Especially topical, therefore, is the question of what happens to XX germ cells in the gonads of XX↔XY male chimaeras. Such XX germ cells could be detected in meiotic prophase alongside normal prospermatogonia in the testes of late foetuses (Mystkowska & Tarkowski, 1970). However, McLaren *et al.* (1972) noted that whilst XX germ cells entered meiosis on schedule in spite of their testicular environment, which might have been anticipated to antagonise such development, the XX germ cells nonetheless degenerated and disappeared soon after birth. No XX germ cells were found even among primary spermatocytes in male XX↔XY chimaeras. Teplitz *et al.* (1967) also referred to the loss of XX germ cells in testicular tissue of bovine chimaeras. All the available evidence suggests that XX germ cells in sex-chromosome chimaeras are incapable of undergoing spermatogenesis in mammals (Mystkowska & Tarkowski, 1968). Therefore, functional sex reversal seems not to take place in such chimaeras. The induction of meiosis may have been prompted by an influence of the mesonephric rete region, a second X chromosome perhaps rendering a germ cell more susceptible to this putative influence (McLaren, 1981), although this latter viewpoint is not held by Burgoyne (1989; see below).

Because XX↔XY female chimaeras are generally less common than males, the opportunities for study have been infrequent. Even so, Gardner (1982) refers to the presence of a mature XY oocyte in a female chimaera, although no information was available on its potential for fertilisation and embryonic development. A frequent interpretation is that XY oocytes in XX↔XY chimaeras will probably not survive until ovulation. The overall prospects for germ cells, at least in mice, have been summarised as follows by McLaren (1987). Whether a germ cell enters the male or female pathway of development depends only on whether it is exposed to a testicular environment, rather than on a direct effect of the Y chromosome. XY germ cells can develop as oocytes, while germ cells lacking any Y chromosome DNA sequences can embark on spermatogenesis. Dosage compensation is negated, as the second X chromosome in XX germ cells is reactivated before birth and the number of X chromosomes plays an important rôle: continuation of spermatogenesis past the time of birth is not possible in the presence of two X chromosomes, while germ cells with a single X are at a disadvantage during oogenesis. The Y chromosome is thought to be involved only in the later stages of spermatogenesis. A further point to note is that XX↔XY chimaeras very seldom produce progeny from more than one component: the XY component if they are males and the XX component if they are females (McLaren, 1984).

Sex chimaerism in the testis

Chapters 2 and 3 have already made reference to the fact that Sertoli cells are the first male-specific cell type to become identifiable in the embryonic testis, indicating that it is in the supporting cell lineage that the testis-determining factor (*Tdy* or, more specifically, *Sry*) is first expressed. Accordingly, it is of considerable interest to discover the origin of this cell type in XX↔XY chimaeras, bearing in mind that both XX and XY cells can be found in all somatic tissues of the gonad (Patek *et al.*, 1991).

Several lines of evidence indicate that Sertoli cells differentiate largely from the XY component in XX↔XY chimaeras, although not exclusively so (Palmer & Burgoyne, 1991*a*; Patek *et al.*, 1991), as was first suspected and reported from studies in mice during the 1980s (Burgoyne *et al.*, 1988*b*).

In a concise review of relevant work, McLaren (1991*a*) emphasised this point by noting that Sertoli cells are predominantly XY in chromosome constitution, irrespective of the proportion of XX and XY cells in the chimaeric embryo overall. This stands in marked contrast to the seemingly random contribution of XX and XY components to all other somatic cell populations that have been examined; for example, XX cells may contribute significantly to the Leydig cell population and to the peritubular cells (Burgoyne *et al.*, 1988*b*). As to the specific quantitative contribution of XX cells within a chimaera to the formation of Sertoli cells, preliminary results were somewhat confusing but this question has now been tackled by *in situ* hybridisation in a study in which the XX component of the chimaeras carried an identifiable β-globin transgenic marker. Use of this molecular technique indicated the presence of up to 20% of XX cells in the foetal Sertoli cell population, although this proportion becomes substantially reduced by the stage of the adult testis, perhaps to less than 3% (see Palmer & Burgoyne, 1991*a*).

Even so, it is of major significance to realise that a somatic cell of XX sex chromosome constitution can be recruited and reprogrammed into a cell that resembles the classical and pivotal somatic cell of the seminiferous tubule. How can this occur? Clearly, there must be a potent inductive influence of the local cellular milieu in the embryonic gonad upon processes of differentiation in these XX cells, and analysis of the factors that impose the change by overriding the cells' intrinsic programme would be invaluable. A short-range diffusible molecule might be involved and direct contact between XX and XY cells would offer another avenue for reprogramming. Palmer & Burgoyne (1991*a*) and Burgoyne (1992) suggest the influence of a gene product in the *Sry*-initiated cascade, implying that there exists a separate and distinct step between *Sry* action and Sertoli cell differentiation. Precisely when the commitment of XX cells to follicle formation is compromised in an XX↔XY chimaera remains to be specified. In other words, just how far can differentiation of such a somatic cell proceed whilst still remaining functionally reversible rather than destined to die? McLaren (1991*a*) suggested that XY pre-Sertoli cells in the developing genital ridge of male chimaeric embryos may be able to recruit some XX cells from the immediate neighbourhood into the Sertoli cell population, but the developmental stage achieved by the XX cells before such recruitment and the nature of the recruitment process require clarification. In particular, what molecular cues are involved and what modifications render a nucleus no longer susceptible to such forms of reprogramming? Moreover, it is not yet certain whether Sertoli cells of presumptive

XX origin are fully competent in terms of the anticipated range of synthetic functions peculiar to that cell type, not least since the above-mentioned decrease in the XX Sertoli cell population between foetal and adult stages suggests a lack of viability or even that such pseudo-Sertoli cells are actively selected against (Palmer & Burgoyne, 1991*a*).

In XX↔XY chimaeras, and provided that the XY component is not from a very slow developing genetic strain, McLaren (1991*a*) indicated that the fate of the XY supporting cells – and thus the sexual phenotype of the individual – will depend on the actual size of the XY contribution. If XY supporting cells in the gonad are relatively abundant, pre-Sertoli cells will differentiate in numbers sufficient to form testis cords, germ cells will not enter meiosis, and the embryo will develop as a male. On the other hand, chimaeras with a low proportion of XY cells will develop as females. In the latter instance, XY supporting cells are presumed initially to differentiate as pre-Sertoli cells, but these remain too few in number to form testis cords. Precisely how the quantitative threshold is assessed is unknown. The germ cells therefore enter meiosis and the resulting oocytes are thought to induce the pre-Sertoli cells to transdifferentiate into follicle cells (McLaren, 1991*b*). Follicle cells of XY constitution have been detected in chimaeras (Burgoyne, Buehr & McLaren, 1988*a*). Although this is the current scheme of interpretation, a more explicit definition of the quantitative and qualitative aspects of gonadal cell programming is clearly essential. Of course, an alternative interpretation might be that when somatic cells of XY constitution are too few in number in early chimaeric embryos, a majority of these proceeds directly into the pathway of follicular cell formation rather than 'reversing' the pre-Sertoli cell status. It is relevant to recall that Sertoli and granulosa cells are believed to arise from the same stem cell line (see Chapter 3). It is also worth recalling that Singh, Matsukuma & Jones (1987) reported testis development in a mouse with only 10% of XY cells.

As noted above, whether germ cells embark on the pathway of oogenesis or spermatogenesis depends not on their own sex chromosome constitution but, rather, on whether or not they are exposed to the meiosis-inhibiting influence of testicular cords. In some contrast, however, their later development depends critically on their own sex chromosome constitution as their own gene programme begins to be expressed. Thus, in a male XX↔XY chimaera, most of the XX as well as XY germ cells enter mitotic arrest and develop as prospermatogonia within the testis cords, so a Y chromosome is not needed for diversion to the male pathway, but the XX germ cells degenerate shortly after birth as a consequence of the incorrect X dosage (McLaren, 1991*b*). It is now appreciated that normal proliferation and survival of male germ cells involves a gene on the Y chromosome (*Spy*) and which is distinct from *Tdy*. As already discussed, no XX germ cells have been found among the germ cell population undergoing spermatogenesis in postnatal life (Mystkowska & Tarkowski, 1968), although they presumably enter the genital ridges with the XY germ cells. The finding by Mystowska & Tarkowski (1970) of meiotic germ cells in the foetal testes of such chimaeras suggested that these might be XX germ cells entering meiosis autonomously according to their own developmental programme (see McLaren, 1981). However, a current interpretation is that this entry into meiosis is a consequence of an incompletely masculinised testicular soma (Burgoyne, 1989).

Sex chimaerism in the ovary

As already stated, a mixed population of XX and XY cells in the body usually leads to male development. XX↔XY female chimaeras are generally far less common than males, and until recently have been studied less intensively. However, sex chromosome chimaerism in phenotypic females was noted by Mintz (1969), Mystkowska & Tarkowski (1970), McLaren *et al.* (1972) and various subsequent authors. Reports in the late 1980s, based on observations in female chimaeras, indicated that XY as well as XX cells can contribute to the follicle population (Burgoyne, 1988; McLaren, 1988; Palmer & Burgoyne, 1991*b*), but only chimaeras with a minority of XY cells develop as fertile females. Burgoyne *et al.* (1988*a*) suggested that XY cells can be recruited to form follicle cells in XX↔XY chimaeras when there is a developmental mismatch between the two components, such that an ovary-determining signal produced by the XX component pre-empts the testis-determining action of the Y.

As a more recent example, Patek *et al.* (1991) generated chimaeras by means of male embryonal stem cell injection into recipient blastocysts of unknown sex, and examined a total of 30 ovaries from 18 chimaeras; XY cells were detected in 11 of the 30 ovaries. The overall pattern of distribution of XY cells was similar in all ovaries, such cells usually being detected in all somatic lineages. In all XX↔XY ovaries, $<10\%$ of the granulosa cells were XY. In other words, ovarian granulosa cells were largely but not exclusively XX. About 20% of follicles in the study contained XX and XY granulosa cells. The ovarian surface epithelium was predominantly XX, but again XY cells were detected in all somatic cell lineages in chimaeric ovaries including the theca cells; the latter were predominantly or exclusively XX. Current thinking favours mesenchyme cells or the granulosa cell lineage as the source of thecal cells. McLaren (1991*a*) suggested that all XY supporting cells in the ovary contribute to the follicle cell population in female XX↔XY chimaeras. Even so, it is improbable that the sex chromosome constitution of cell types as analysed in adult chimaeras is a true reflection of their overall pattern at the time of commitment.

Even when the Y chromosome appears entirely normal, as in a female XX↔XY chimaera, the XY germ cells can enter meiosis and may give rise to an oocyte (Evans, Ford & Lyon, 1977; McLaren, 1988). This observation prompted Evans *et al.* (1977) to remark that the sex of a germ cell is not an autonomous property, but is determined by the nature of the gonad in which it finds itself. Indeed, XY as well as XX germ cells will enter meiosis before birth and form oocytes in female XX↔XY chimaeras, but few if any of the XY oocytes will survive to ovulation. Evans *et al.* (1977) have produced direct microscopical evidence on the capacity of a single XY germ cell in the ovarian tissue of a female mouse chimaera to become an oocyte. This fertile female (XX↔XY) yielded an oocyte at diakinesis after a period of culture, but the functional potential of this germ cell remained unknown. A full delineation of when and why XY oocytes are lost would be invaluable, and even more so would be precise reasons for the loss. Therefore, although the presence of an entire Y chromosome may not impede a germ cell from undergoing oogenesis, XY relative to XX germ cells in an XX↔XY ovary may be at a strong disadvantage because of the lack of a second X chromosome (McLaren, 1984).

Reproductive competence of chimaeras

As to the overall fertility of these composite animals, most sex chimaeras have reduced fertility and male chimaeras with XX↔XY testes were either sterile or less fertile than chimaeras with testes composed entirely of XY cells (see Patek *et al.*, 1991). This impaired fertility was associated with the loss of XY germ cells in atrophic seminiferous tubules (Patek *et al.*, 1991). Because this progressive lesion, found in adult testes, was correlated with a high proportion of XX Leydig cells, it suggested that XX Leydig cells are in some manner functionally defective and unable to support spermatogenesis or indeed that they are actively secreting molecules inappropriate to normal function of the seminiferous tubules. However, in part at least, impaired fertility could also be a consequence of germ cell depletion upon the death of any XX germ cells as T_1 prospermatogonia (Palmer & Burgoyne, 1991*a*).

Even so, as pointed out by Patek *et al.* (1991), it is difficult to envisage how an early loss of XX germ cells could cause a progressive disruption of spermatogenesis in adult mice 6–12 months old.

As noted above, only mouse chimaeras with a minority of XY cells can develop as fertile females but such XX↔XY females can demonstrate full reproductive potential. However, of the 348 progeny summarised by McLaren (1984) as being derived from XX↔XY females, 347 were derived from the XX component whereas only one could have come from the XY component. A possible reason for this outcome is mentioned in the previous section.

In the case of chimaeras produced by the aggregation technique in golden hamsters (Piedrahita, Gillespie & Maeda, 1992), two confirmed XX↔XY chimaeras were available for breeding studies. One of these animals bred normally, producing nine litters, whilst the other was sterile with abnormal testicular weight and morphology. In the latter animal, the spermatic cords appeared to have a normal complement of somatic cells, but there was a complete absence of germ cells. This finding in hamsters may mirror the situation in mouse chimaeras, in which a proportion of the germ cells of an XX↔XY male can develop into oocytes but will degenerate soon after birth (Mystkowska & Tarkowski, 1970). Alternatively, and reflecting the interpretation of Patek *et al.* (1991) in mice, germ cells of XX constitution situated within an embryonic testis may degenerate instead of developing into oocytes, leading to the finding of spermatic cords devoid of germ cells soon after birth. Once again, therefore, there is a strong suggestion that the presence of two functional X chromosomes in germ cells is incompatible with their survival in the testes (see McLaren, 1980).

Turning to interspecific chimaeras, the splendid sheep–goat specimens of Fehilly, Willadsen & Tucker (1984a) were produced by the aggregation method. The approach was to combine a single cell from a 4-cell embryo of one species with three cells from a comparable embryo of the other species, or to surround an 8-cell embryo of one species with the disaggregated blastomeres from three 8-cell embryos of the other (Fehilly *et al.*, 1985). With notions of potential immunological rejection in mind upon transplantation to the host uterus, an attempt was made to form the chorion of cells belonging to the recipient female (see Rossant, Mauro & Croy, 1982; Rossant *et al.*, 1983; Meinecke-Tillmann & Meinecke, 1984). Eight of 26 offspring were identified as sheep–goat chimaeras with interspecific chimaerism of the blood lymphocytes, red cells, and plasma transferrins and amylases (Fehilly *et al.*, 1985). Details of the fertility of one such interspecific chimaera were presented by Betteridge (1986). In this example, the sheep cells were male whereas the composition of the goat cells was unknown. Socially, this animal preferred the company of sheep to that of goats, and was demonstrably fertile in many matings with ewes, despite the fact that electrophoretic analysis showed its seminal plasma proteins to be derived from cells of both species.

Two further points of significance have been noted in interspecific animals. First, live-born offspring developing from chimaeric embryos were biased in terms of both genotype and phenotype (conformation) towards the species of the recipient foster mother, probably indicating some degree of reaction by the recipient against cells of another species (see Fehilly *et al.*, 1985). Second, close analysis of sheep–goat chimaeras showed that the relative proportions of lymphocytes and red cells changed as the animals aged, as previously noted in sheep–sheep chimaeras (Tucker *et al.*, 1978), probably suggesting a selective fitness of one cell type over another.

Concluding remarks

The emphasis in this chapter has been very much on questions of sexual differentiation, and in particular upon interactions between germ cells and the soma in experimental animals. Generation of tetraparental animal models in the laboratory has enabled reasonably firm conclusions to be drawn concerning the developmental potential and phenotype of germ cells of known constitution in a genetically dissimilar somatic milieu. Even so, a more rigorous testing of these conclusions would seem possible by using current techniques for the introduction of sex-determining genes into specific cell lines (see Chapter 2) or by the approach of introducing genetically modified stem cells (see Beddington, 1992). There is a suspicion, no more than that at present, that such modification of individual genetic constitutions will lead to interpretations more complex than those currently available as the subtle interactions between cell lines and individual gene products are highlighted. For both somatic and germ cells of the gonad, this remark refers to what may be regarded as the molecular tensions between expression of genetic constitution and influences of the local somatic milieu in which cells are situated, encompassing both cell surface interactions and the response to diffusible molecules. And, as should be clear from the information in this chapter, the cellular constitution of a chimaeric gonad represents a dynamic situation and is much influenced by the relative proportions of XX and XY cells. But, although transdifferentiation of cell types is possible, there appears to be a tendency towards stabilisation before the adult stage is reached, frequently involving loss of individual cell lines.

For example, supporting cells can enter or embark upon transdifferentiation but fail to survive the adventure due to an inappropriate chromosome complement and thereby an inadequate gene programme. Hence, Sertoli cells are predominantly XY and granulosa cells predominantly XX in XX↔XY chimaeras.

Still in a reproductive context, another topical example of the use of chimaeras would be for deducing the pattern of growth and direction of cell division during ovarian follicular development (Boland & Gosden, 1992). However, the reader will doubtless appreciate that application of the 'chimaera technique' to the analysis of cell lineages and destinies (Ford *et al.*, 1974; West, 1978) may be powerfully exploited in work on malignancy and development of specific tumours, be it in reproductive tissues or in other organ systems of the body. A classical starting point here would be the work of Brinster (1974) demonstrating that teratocarcinoma cells microinjected into the blastocyst cavity of mice could be incorporated to form chimaeras with an increased immunotolerance related to the donor strain. Moreover, such chimaeras showed clear phenotypic evidence of tissues colonised by cells which had formerly been malignant. Tumour cells propagated *in vitro* were thus able to participate in development of the embryo. More recent studies that highlight the scope of the technique include those of Illmensee & Mintz (1976), Papaioannou *et al.* (1978), Stewart & Mintz (1981), Rossant & McBurney (1982), Bradley *et al.* (1984), Hardy *et al.* (1990) and Ansell *et al.* (1991).

One of the most exciting observations in the above context is that of an apparent quantitative relationship between the number of injected tumour cells and the capacity of the 'host' blastocyst to regulate development. Whilst a few embryonal carcinoma cells can seemingly be integrated into the developing egg cylinder, some carcinoma cells were able to escape from the regulatory influence of the inner cell mass when too many were injected, enabling proliferation as colonies of embryonal carcinoma cells and thereby formation of tumours in neonatal chimaeras (Hardy *et al.*, 1990). Progress towards an analysis of the molecular conversations underlying this potential waywardness would seem to be of very considerable significance.

References

Ansell, J. D., Samuel, K., Whittingham, D. G., Patek, C. E., Hardy, K., Handyside, A. H., Jones, K. W., Muggleton-Harris, A. L., Taylor, A. H. & Hooper, M. L. (1991). Hypoxanthine phosphoribosyl transferase deficiency, haematopoiesis and fertility in the mouse. *Development*, **112**, 489–98.

Beddington, R. S. P. (1982). An audoradiographic analysis of tissue potency in different regions of the embryonic ectoderm during gastrulation in the mouse. *Journal of Embryology and Experimental Morphology*, **69**, 265–85.

Beddington, R. S. P. (1992). Transgenic strategies in mouse embryology and development. In *Transgenic Mice in Biology and Medicine*, ed. F. Grosveld & G. Collias. London: Academic Press.

Betteridge, K. J. (1986). Increasing productivity in farm animals. In *Reproduction in Mammals*, ed. C. R. Austin & R. V. Short, 2nd edn, Book 5, pp. 1–47. Cambridge & London: Cambridge University Press.

Boland, N. I. & Gosden, R. G. (1992). A clonal analysis of chimaeric mouse ovaries using DNA *in situ* hybridisation. *Journal of Reproduction and Fertility*, Abstract Series No. 9, p. 33.

Bradbury, M. W. (1987). Testes of XX↔XY chimeric mice develop from fetal ovotestes. *Developmental Genetics*, **8**, 207–18.

Bradley, A., Evans, M., Kaufman, M. H. & Robertson, E. (1984). Formation of germ-line chimaeras from embryo-derived teratocarcinoma cell lines. *Nature (London)*, **309**, 255–6.

Brem, G., Tenhumberg, H. & Kräusslich, H. (1984). Chimerism in cattle through microsurgical aggregation of morulae. *Theriogenology*, **22**, 609–13.

Brinster, R. L. (1974). The effect of cells transferred into the mouse blastocyst on subsequent development. *Journal of Experimental Medicine*, **140**, 1049–56.

Burgoyne, P. S. (1988). Role of mammalian Y chromosome in sex determination. *Philosophical Transactions of the Royal Society of London, Series B*, **322**, 63–72.

Burgoyne, P. S. (1989). Genetics of XX and XO sex reversal in the mouse. In *Evolutionary Mechanisms in Sex Determination*, ed. S. S. Wachtel, pp. 161–69. Boca Raton, Florida: CRC Press.

Burgoyne, P. S. (1992). Y chromosome function in mammalian development. *Advances in Developmental Biology*, **1**, 1–29.

Burgoyne, P. S., Buehr, M. & McLaren, A. (1988*a*). XY follicle cells in ovaries of XX↔XY female mouse chimaeras. *Development*, **104**, 683–88.

Burgoyne, P. S., Buehr, M., Koopman, P., Rossant, J. & McLaren, A. (1988*b*). Cell-autonomous action of the testis-determining gene: Sertoli cells are exclusively XY in XX↔XY chimaeric mouse testes. *Development*, **102**, 443–50.

Edwards, R. G. (1980). *Conception in the Human Female*. London: Academic Press.

Evans, E. P., Ford, C. E. & Lyon, M. F. (1977). Direct evidence of the capacity of the XY germ cell in the mouse to become an oocyte. *Nature (London)*, **267**, 430–1.

Fehilly, C. B., Willadsen, S. M., Dain, A. R. & Tucker, E. M. (1985). Cytogenetic and blood group studies of sheep/goat chimaeras. *Journal of Reproduction and Fertility*, **74**, 215–21.

Fehilly, C. B., Willadsen, S. M. & Tucker, E. M. (1984*a*). Interspecific chimaerism between sheep and goat. *Nature (London)*, **307**, 634–6.

Fehilly, C. B., Willadsen, S. M. & Tucker, E. M. (1984*b*). Experimental chimaerism in sheep. *Journal of Reproduction and Fertility*, **70**, 347–51.

Ferguson-Smith, M. A. & Affara, N. A. (1988). Accidental X–Y recombination and the aetiology of XX males and true hermaphrodites. *Philosophical Transactions of the Royal Society of London, Series B*, **322**, 133–44.

Ford, C. E., Evans, E. P., Burtenshaw, M. D., Clegg, H., Barnes, R. D. & Tuffrey, M. (1974). Marker chromosome analysis of tetraparental AKR↔CBA-T6 mouse chimeras. *Differentiation*, **2**, 321–33.

Gardner, R. L. (1968). Mouse chimaeras obtained by the injection of cells into the blastocyst. *Nature (London)*, **220**, 596–7.

Gardner, R. L. (1971). Manipulations on the blastocyst. *Advances in Biosciences*, **6**, 279–96.

Gardner, R. L. (1978). Production of chimaeras by injecting cells or tissue into the blastocyst. In *Methods in Mammalian Reproduction*, ed. J. C. Daniel, pp. 137–65. New York & London: Academic Press.

Gardner, R. L. (1982). Manipulation of development. In *Reproduction in Mammals*, ed. C. R. Austin & R. V. Short, 2nd edn, book 2, pp. 159–80. Cambridge: Cambridge University Press.

Gardner, R. L. (1985). Clonal analysis of early mammalian development. *Philosophical Transactions of the Royal Society of London, Series B*, **312**, 163–78.

Gardner, R. L. & Johnson, M. H. (1973). Investigation of early mammalian development using interspecific chimaeras between rat and mouse. *Nature, New Biology*, **246**, 86–9.

Gardner, R. L. & Lyon, M. (1971). X chromosome inactivation studied by injection of a single cell into the mouse blastocyst. *Nature (London)*, **231**, 385–6.

Gardner, R. L. & Rossant, J. (1979). Investigation of the fate of 4.5 day *post-coitum* mouse inner cell mass cells by blastocyst injection. *Journal of Embryology and Experimental Morphology*, **52**, 141–52.

Gordon, J. (1976). Failure of XX cells containing the sex reversed gene to produce gametes in allophenic mice. *Journal of Experimental Zoology*, **198**, 367–74.

Hardy, K., Carthew, P., Handyside, A. H. & Hooper, M. L. (1990). Extragonadal teratocarcinoma derived from embryonal stem cells in chimaeric mice. *Journal of Pathology*, **160**, 71–6.

Ilmensee, K. & Mintz, B. (1976). Totipotency and normal differentiation of single teratocarcinoma cells cloned by injection into blastocysts. *Proceedings of the National Academy of Sciences, USA*, **73**, 549–53.

Kashiwazaki, H., Nakao, H., Ohtani, S. & Nakatsuji, N. (1992). Production of chimeric pigs by the blastocyst injection method. *Veterinary Record*, **130**, 186–7.

Kelley, K. M., Johnson, T. R., Gwatkin, R. B. L., Ilan, J. & Ilan, J. (1993). Transgenic strategies in reproductive endocrinology. *Molecular Reproduction and Development*, **34**, 337–47.

Le Douarin, N. & McLaren, A. (ed.) (1984). *Chimeras in Developmental Biology*. London & New York: Academic Press.

McLaren, A. (1972). Germ cell differentiation in artificial chimaeras of mice. In *The Genetics of the Spermatozoon*, ed. R. A. Beatty & S. Gluecksohn-Waelsch, pp. 313–24. Copenhagen: Bogtrykkeriet Forum.

McLaren, A. (1975). Sex chimaerism and germ cell distribution in a series of chimaeric mice. *Journal of Embryology and Experimental Morphology*, **33**, 205–16.

McLaren, A. (1976). *Mammalian Chimaeras*. Cambridge & London: Cambridge University Press.

McLaren, A. (1980). Oocytes in the testis. *Nature (London)*, **283**, 688–9.
McLaren, A. (1981). The fate of germ cells in the testis of fetal *Sex-reversed* mice. *Journal of Reproduction and Fertility*, **61**, 461–7.
McLaren, A. (1984). Germ cell lineages. In *Chimeras in Developmental Biology*, ed. N. Le Douarin & A. McLaren, pp. 111–29. London: Academic Press.
McLaren, A. (1987). Clues from other animals and theoretical considerations. *Development*, **101**, Supplement, 3–4.
McLaren, A. (1988). Somatic and germ-cell sex in mammals. *Philosophical Transactions of the Royal Society of London, Series B*, **322**, 3–9.
McLaren, A. (1991*a*). Development of the mammalian gonad: the fate of the supporting cell lineage. *BioEssays*, **13**, 151–6.
McLaren, A. (1991*b*). Sex determination in mammals. *Oxford Reviews of Reproductive Biology*, **13**, 1–33.
McLaren, A. & Bowman, P. (1969). Mouse chimaeras derived from fusion of embryos differing by nine genetic factors. *Nature (London)*, **224**, 238–40.
McLaren, A., Chandley, A. C. & Kofman-Alfaro, S. (1972). A study of meiotic germ cells in the gonads of foetal mouse chimaeras. *Journal of Embryology and Experimental Morphology*, **27**, 515–24.
Mayr, W. R., Pausch, V. & Schnedl, W. (1979). Human chimaera detectable only by investigation of her progeny. *Nature (London)*, **277**, 210–11.
Meinecke-Tillmann, S. & Meinecke, B. (1984). Experimental chimaeras – removal of reproductive barrier between sheep and goat. *Nature (London)*, **307**, 637–8.
Mintz, B. (1962). Experimental study of the developing mammalian egg: removal of the zona pellucida. *Science*, **138**, 594–5.
Mintz, B. (1964). Formation of genetically mosaic mouse embryos, and early development of 'Lethal (t^{12}/t^{12}) – Normal' mosaics. *Journal of Experimental Zoology*, **157**, 273–92.
Mintz, B. (1965*a*). Genetic mosaicism in adult mice of quadriparental lineage. *Science*, **148**, 1232–3.
Mintz, B. (1965*b*). Experimental genetic mosaicism in the mouse. In *Preimplantation Stages of Pregnancy* Ciba Foundation Symposium, ed. M. O'Connor & G. E. W. Wolstenholme, pp. 194–207. London: J. & A. Churchill.
Mintz, B. (1968). Hermaphroditism, sex chromosomal mosaicism and germ cell selection in allophenic mice. *Journal of Animal Science*, **27**, 51–60.
Mintz, B. (1969). Developmental mechanisms found in allophenic mice with sex chromosomal and pigmentary mosaicism. *Birth Defects, Original Article Series*, **5**, 11–22.
Mintz, B. (1974). Gene control of mammalian differentiation. *Annual Review of Genetics*, **8**, 411–70.
Mullen, R. J. & Whitten, W. K. (1971). Relationship of genotype and degree of chimerism in coat colour to sex ratios and gametogenesis in chimeric mice. *Journal of Experimental Zoology*, **178**, 165–76.
Mystkowska, E. T. (1975). Development of mouse–bank vole interspecific chimaeric embryos. *Journal of Embryology and Experimental Morphology*, **33**, 731–44.
Mystkowska, E. T. & Tarkowski, A. K. (1968). Observations on CBA-p/CBA-T6T6 mouse chimaeras. *Journal of Embryology and Experimental Morphology*, **20**, 33–52.
Mystkowska, E. T. & Tarkowski, A. K. (1970). Behaviour of germ cells and sexual differentiation in late embryonic and early postnatal mouse chimeras. *Journal of Embryology and Experimental Morphology*, **23**, 395–405.

Nicholas, J. S. & Hall, B. V. (1942). Experiments on developing rats. II. The development of isolated blastomeres and fused eggs. *Journal of Experimental Zoology*, **90**, 441–59.

Palmer, S. J. & Burgoyne, P. S. (1991*a*). *In situ* analysis of fetal, pre-puberal and adult XX↔XY chimaeric mouse testes: Sertoli cells are predominantly, but not exclusively, XY. *Development*, **112**, 265–8.

Palmer, S. J. & Burgoyne, P. S. (1991*b*). XY follicle cells in the ovaries of XO/XY and XO/XY/XYY mosaic mice. *Development*, **111**, 1017–19.

Papaioannou, V. E., Gardner, R. L., McBurney, M. W., Babinet, C. & Evans, M. J. (1978). Participation of cultured teratocarcinoma cells in mouse embryogenesis. *Journal of Embryology and Experimental Morphology*, **44**, 93–104.

Patek, C. E., Kerr, J. B., Gosden, R. G., Jones, K. W., Hardy, K., Muggleton-Harris, A. L., Handyside, A. H., Whittingham, D. G. & Hooper, M. L. (1991). Sex chimaerism, fertility and sex determination in the mouse. *Development*, **113**, 311–25.

Picard, L., Chartrain, I., King, W. A. & Betteridge, K. J. (1990). Production of chimaeric bovine embryos and calves by aggregation of inner cell masses with morulae. *Molecular Reproduction and Development*, **27**, 295–304.

Picard, L., King, W. A. & Betteridge, K. J. (1986). Naissance d'une chimère bovine suite à la micromanipulation d'embryons. *Médecine Vétérinaire de Québec*, **16**, 139–40.

Piedrahita, J. A., Gillespie, L. & Maeda, N. (1992). Production of chimeric hamsters by aggregation of eight cell embryos. *Biology of Reproduction*, **47**, 347–54.

Pighills, E., Hancock, J. L. & Hall, J. G. (1968). Attempted induction of chimaerism in sheep. *Journal of Reproduction and Fertility*, **17**, 543–7.

Polge, C. (1985). Embryo manipulation and genetic engineering. *Symposium of The Zoological Society of London*, **54**, 123–35.

Polzin, V. J., Anderson, D. L., Anderson, G. B., Bondurant, R. H., Butler, J. E., Pashen, R. L. & Penedo, M. C. T. (1987). Production of sheep-goat chimeras by inner cell mass transplantation. *Journal of Animal Science*, **65**, 325–30.

Rossant, J. (1984). Somatic cell lineages in mammalian chimeras. In *Chimeras in Developmental Biology*, ed. N. Le Douarin & A. McLaren, ch. IV, pp. 89–107. London: Academic Press.

Rossant, J. & Chapman, V. M. (1983). Somatic and germline mosaicism in interspecific chimaeras between *Mus musculus* and *Mus caroli*. *Journal of Embryology and Experimental Morphology*, **73**, 193–205.

Rossant, J., Croy, B. A., Clark, D. A. & Chapman, V. M. (1983). Interspecific hybrids and chimeras in mice. *Journal of Experimental Zoology*, **228**, 223–33.

Rossant, J. & Frels, W. I. (1980). Interspecific chimeras in mammals: successful production of live chimeras between *Mus musculus* and *Mus caroli*. *Science*, **208**, 419–21.

Rossant, J. & McBurney, M. W. (1982). The developmental potential of a euploid male teratocarcinoma cell line after blastocyst injection. *Journal of Embryology and Experimental Morphology*, **70**, 99–112.

Rossant, J., Mauro, V. M. & Croy, B. A. (1982). Importance of trophoblast genotype for survival of interspecific murine chimaeras. *Journal of Embryology and Experimental Morphology*, **69**, 141–9.

Roth, T. L., Anderson, G. B., Bondurant, R. H. & Pashen, R. L. (1989). Survival of sheep X goat hybrid inner cell masses after injection into ovine embryos. *Biology of Reproduction*, **41**, 675–82.

Ruffing, N. A., Anderson, G. B., Bondurant, R. H., Currie, W. B. & Pashen, R. L. (1993). Effects of chimerism in sheep-goat concepti that developed from blastomere-aggregation embryos. *Biology of Reproduction*, **48**, 889–904.

Short, R. V. (1972), Germ cell sex. In *The Genetics of the Spermatozoon*, ed. R. A. Beatty & S. Gluecksohn-Waelsch, pp. 325–45. Copenhagen, Bogtrykkeriet Forum.

Short, R. V. (1982). Sex determination and differentiation. In *Reproduction in Mammals*, ed. C. R. Austin & R. V. Short, 2nd edn, book 2, pp. 70–113. Cambridge: Cambridge University Press.

Singh, L., Matsukuma, S. & Jones, K. W. (1987). Testis development in a mouse with 10% of XY cells. *Developmental Biology*, **122**, 287–90.

Stewart, T. A. & Mintz, B. (1981). Successive generations of mice produced from an established culture line of euploid teratocarcinoma cells. *Proceedings of the National Academy of Sciences, USA*, **78**, 6314–18.

Summers, P. M., Shelton, J. N. & Bell, K. (1983). Synthesis of primary *Bos taurus–Bos indicus* chimaeric calves. *Animal Reproduction Science*, **6**, 91–102.

Tarkowski, A. K. (1961). Mouse chimaeras developed from fused eggs. *Nature (London)*, **190**, 857–60.

Tarkowski, A. K. (1964). True hermaphroditism in chimaeric mice. *Journal of Embryology and Experimental Morphology*, **12**, 735–57.

Tarkowski, A. K. (1969). Consequences of sex chromosome chimaerism for sexual differentiation in mammals. *Annales d'Embryologie et de Morphogenèse, Supplement* **1**, 211–22.

Teplitz, R. L., Moon, Y. S. & Basrur, P. K. (1967). Further studies of chimerism in heterosexual cattle twins. *Chromosoma*, **22**, 202–9.

Tippett, P. (1984). Human chimeras. In *Chimeras in Developmental Biology*, ed. N. Le Douarin & A. McLaren, pp. 165–78. London: Academic Press.

Tucker, E. M., Dain, A. R. & Moor, R. M. (1978). Sex chromosome chimaerism and the transmission of blood group genes by tetraparental rams. *Journal of Reproduction and Fertility*, **54**, 67–73.

Tucker, E. M., Moor, R. M. & Rowson, L. E. A. (1974). Tetraparental sheep chimaeras induced by blastomere transplantation: changes in blood type with age. *Immunology*, **26**, 613–21.

Vigier, B., Watrin, F., Magre, S., Tran, D. & Josso, N. (1987). Purified bovine AMH induces a characteristic freemartin effect in fetal rat prospective ovaries exposed to it *in vitro*. *Development*, **100**, 43–55.

West, J. D. (1978). Analysis of clonal growth using chimaeras and mosaics. In *Development in Mammals*, ed. M. H. Johnson, vol. 3, pp. 413–60 Amsterdam: Elsevier.

Whitten, W. K. (1975). Chromosomal basis for hermaphroditism in mice. *Symposium of the Society for Developmental Biology*, **33**, 189–205.

Whitten, W. K., Beamer, W. G. & Byskov, A. G. (1979). The morphology of fetal gonads of spontaneous mouse hermaphrodites. *Journal of Embryology and Experimental Morphology*, **52**, 63–78.

Willadsen, S. M. & Fehilly, C. B. (1983). The developmental potential and regulatory capacity of blastomeres from two-, four- and eight-cell sheep embryos. In *Fertilization of the Human Egg in vitro*, ed. H. M. Beier & H. R. Lindner, pp. 353–7. Berlin: Springer-Verlag.

Willadsen, S. M. & Polge, C. (1981). Attempts to produce monozygotic quadruplets in cattle by blastomere separation. *Veterinary Record*, **108**, 211–13.

Witschi, E. (1965). Hormones and embryonic induction. *Archives d'Anatomie Microscopique et de Morphologie Expérimentale*, **54**, 601.
Wood, S. A., Allen, N. D., Rossant, J., Auerbach, A. & Nagy, A. (1993). Non-injection methods for the production of embryonic stem cell-embryo chimaeras. *Nature (London)*, **365**, 87–9.
Zeilmaker, G. H. (1973). Fusion of rat and mouse morulae and formation of chimaeric blastocysts. *Nature (London)*, **242**, 115–16.

9

Asymmetries in the reproductive system and their significance

> But the Kers were aye the deadliest foes
> That e'er to Englishmen were known,
> For they were all bred left-handed men,
> And fence against them there was none.
> *(Hogg, 1830)*

Introduction

To write a chapter on unilateral phenomena in terms of gonadal development and genital physiology may seem unusual in a monograph assembled under the present title, and especially so in the context of intersexuality. The chapter will accordingly be kept reasonably brief in order to avoid appearing as a diversion. Nonetheless, there is a strong feeling that if one could understand precisely why one side of the body may be privileged in terms of the rate and/or extent of organ development, then some of the frequently noted differences in the reproductive system might be more accessible to a meaningful explanation. This thought particularly concerns differentiation and development of the gonads, for inequalities in rates of growth of the two gonads may find expression in the formation of abnormal tissues, perhaps more appropriately termed anomalous tissues. At the very least, it appears that there may be a unilateral predisposition to potentially wayward development. So, whilst the pages that follow will in due course focus most attention upon the gonads themselves, differences in the form and/or function of the genital ducts – especially in female mammals – will be considered largely to reflect different inputs or influences from the two gonads. As to unilateral differences in the rate and extent of development of mammary tissues, interpretations may concern factors not related directly to secretion of ovarian hormones but rather more generally to reflect an underlying asymmetry of the mammalian body. In this broader context, the reader is referred to the discussions of Oppenheimer (1974), Neville (1976) and the volume edited by Bock & Marsh (1991).

The following lines may also appeal to readers of this chapter. In a brief book review entitled 'Extremes of symmetry', Ryder (1992) noted that symmetry is one of the most important unifying ideas by which we can make sense of the world. It is commonly used by scientists and mathematicians in contexts in which it is related to invariance under particular transformations, but it is also clearly linked to notions of beauty and aesthetics. Its importance in natural philosophy has been recognised from the time of Plato, who stated that God ever geometrises, to Paul Dirac, the predictor of antimatter, who declared that theories cannot be correct unless they exhibit mathematical beauty – an idea which may be thought to be taking the matter too far. Even so, the cultural implications of symmetry would appear to have a solid historical basis.

A complementary point of view, and one that sets the scene for this chapter, was expressed by Altmann (1992) in a review entitled 'Perturbing patterns'. He noted that one of the most fascinating cases of broken symmetry arises in the problem of the growth of biological structures. Although these systems may start by being fairly homogeneous, this homogeneity becomes unstable, thus allowing patterns and structures to emerge. This is the observation that led Alan Turing to his famous mathematical theory of morphogenesis. But more surprises await us in the way in which symmetry breaking creeps into everything. If this is accepted, then a fundamental question in the words of Morgan (1991) is how the information about left–right asymmetry is transmitted across generations in such a way as to allow structures on the left and right sides of the body to develop differently. This seemingly leads us into the realm of genetics and considerations of a specific gene or group of genes.

Asymmetry of gonadal growth

Mammalian gonads may differ significantly in size within the same individual (bilateral asymmetry), this being widely appreciated in the case of the mature male since the testes have descended from an abdominal location in a majority of species. Thus, those who have cast an alert and interested eye over classical sculpture in, for example, the Louvre, the Palais de Versailles or perhaps even in Florence or Rome can scarcely have failed to notice an apparent asymmetry in the portrayed disposition of scrotal contents. Let it be proposed straight away that this is primarily due to differences in testicular size rather than a reflection of muscular function or the dimensions of ligaments. This is born out by the extensive autopsy study of Johnson, Petty & Neaves (1984), in which the left testes weighed 10% less than the right (Table 9.1). How can we shed light on the origin of such gonadal differences?

Table 9.1. *Left–right differences in human testis weight as determined at autopsy within 24 hours of sudden, unattended death*

Age of specimen (years)	Testis weight ± SEM (grams)[a]	
	Left	Right
21–50	17.2 ± 0.5	18.9 ± 0.5
51–80	17.5 ± 0.9	19.2 ± 0.9

Note:
Adapted from Johnson *et al.* (1984).
[a] Values represent mean weights ± standard error of the mean.

In a series of papers concerned primarily with placental mammals, Mittwoch referred to differential rates of growth between the two gonads (e.g. Mittwoch, 1976*a*, *b*). Not only did the left and right gonads differ in their number of cells and number of mitoses, the right gonad usually being larger, but the right gonad in hermaphrodites also had an increased tendency to develop into a testis or an ovotestis whilst the left was more likely to develop into an ovary. Hence, there arose the generalisation that when asymmetry is present in rates of gonadal growth, the faster-growing gonad is predestined to differentiate as a testis. In fact, Mittwoch (1986) felt confident in stating that the bilateral asymmetry of the gonads in true hermaphroditism could be related to an asymmetry of growth in human foetal gonads. The right gonads were larger than the left in human foetal material and testes were larger than ovaries (Mittwoch & Mahadevaiah, 1980). Earlier studies in foetal mice had failed to demonstrate any size differences between the right and left gonads, but testes were larger than ovaries at all stages examined (Mittwoch & Buehr, 1973).

Could such differential rates of growth to some degree be an expression of patterns of migration of the primordial germ cells (Chapter 3)? For example, might primordial germ cells reach one genital ridge sooner, or in significantly greater numbers, than the other and – as a consequence – act locally to prompt more rapid growth? Or might rapid growth be more a reflection of the proliferating vascular system and the provision of substrate, together with appropriate endocrine factors such as peptide growth factors? As a quite separate question, is there some fundamental link between the asymmetry of gonadal size in mammals and the fact that one of the gonads is essentially vestigial in female birds, only the left ovary usually being functional? In most avian species, the cortex of the right side is less extensive than on the left (Witschi, 1951). Consequently, fewer germ cells

Table 9.2. *To highlight the unequal distribution of germ cells in passerine birds observed during differentiation of male and female gonads*

	Left gonad		Right gonad		
Age of embryo (hours)	Cortex	Medulla	Cortex	Medulla	Sex
100	241	128	28	39	Indifferent stage
102	1423	500	180	187	Indifferent stage
132	1536	252	25	410	Female
132	1724	2763	68	1435	Male

Note:
Modified after Witschi (1951).
Note that the number of germ cells counted in the gonad of a male embryo far exceeds the number in a female embryo of comparable age. The right-side gonad of both male and female birds exhibits the early stages of testicular differentiation whilst the left gonad differentiates into either ovary or testis.

are attracted towards this side and at the end of migration the right cortex contains scarcely over one tenth the number of those now embedded in the left (Table 9.2). The more vigorous left gonad in female birds, the heterogametic sex, contrasts strikingly with the larger right testis in mammals, in which of course the male is the heterogametic sex. Even so, one does not yet understand at all what it is that regulates these unilateral tendencies nor why full development of the right 'residual' gonad in birds can occur only after removal of the functional ovary on the left. In some manner, it would appear that the left ovary may act to suppress functional development of the right gonad which, when liberated from this inhibitory influence, differentiates into a testis which may contain spermatozoa (Benoit, 1932; Frankenhuis & Kappert, 1980).

From the point of view of school biology, the teaching used to be that all triploblastic animals acquire bilateral symmetry early in development, so that there is a median plane on either side of which the two halves of the body are more or less exact mirror images of each other. In the annelids, there is relatively little difference between the two sides, but in the molluscs and chordates there is a greater or lesser degree of secondary internal asymmetry, especially noticeable in the coiling of the gut, but often showing itself in blood vessels, gonads and other organs (Grove & Newell, 1953; Borradaile, 1955). If one now jumps to the higher mammals, handed asymmetry or handed laterality can be readily distinguished, that is consistent differences between left and right sides of the body, especially in the thoraco-abdominal organs. For example, the heart and stomach are located on the left side, the aortic loop is to the left, and there are differing

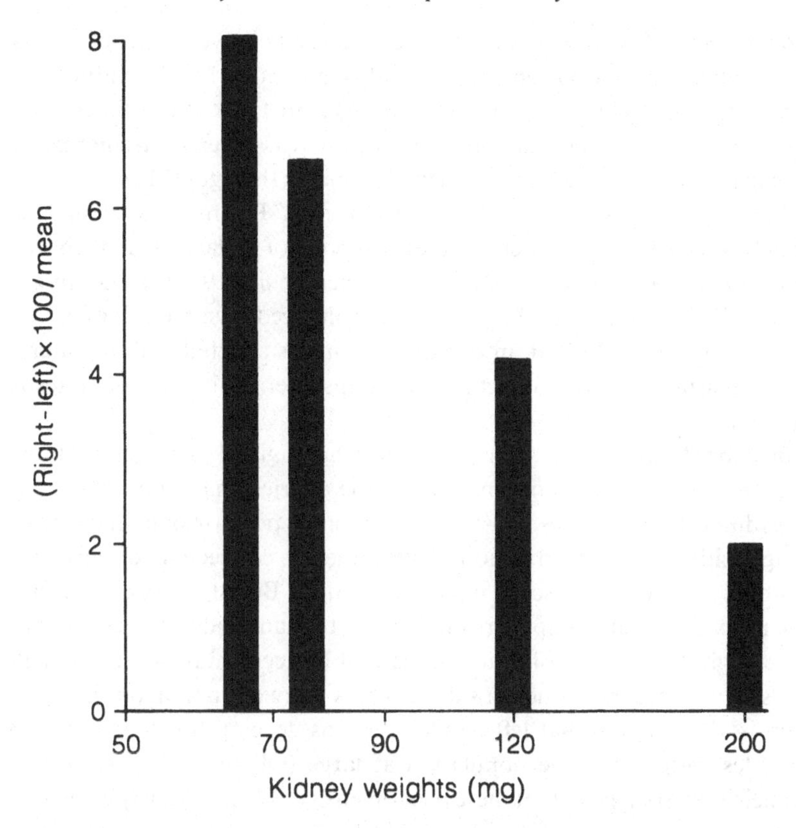

Fig. 9.1. Histogram to demonstrate that the bilateral asymmetry in favour of the right kidney in mice decreases with increasing kidney weight. Seven wild populations of mice were divided into two groups according to increasing kidney weight, and each group was further subdivided into younger and older mice. (After Mittwoch, 1992.)

numbers of lung lobes between the two sides. As to other paired organs, there may be differences between the kidneys and frequently also between the adrenal glands. Renal differences may be more conspicuous in younger mammals (Fig. 9.1) or, as in the case of the guinea-pig, in females rather than in males: the left kidney is heavier than the right in females (Shirley, 1976). Whilst there was no significant difference in size between the two kidneys in human male foetuses, an asymmetry was again found in females in favour of the left (Mittwoch & Mahadevaiah, 1980). It is worth recalling the close embryonic association between the kidneys, adrenal glands (suprarenals) and gonads, and that differences in the architecture of the blood vessels – especially the renal ones – can usually be noted between left and right sides (Fig. 9.2). In humans, the right testicular vein enters the vena cava whereas the left one enters the renal vein. And in human pathology, adrenal tumours giving primary aldosteronism are two to three times more likely to arise on the left than on the right, whereas adrenal carcinomas of Cushing's syndrome are more frequently found on the right (Munro Neville & Mackay, 1972; Cook, 1987). In instances of Kallmann's syndrome, already referred to in Chapter 7, there are reports of unilateral renal aplasia or hypoplasia (Wegenke *et al.*, 1975; Franco *et al.*, 1991), suggesting a possible link between neural disorders and enhanced asymmetries of paired abdominal organs. Indeed, in one of three males affected with unilateral renal aplasia, this was associated with ipsilateral absence of a testis (Wegenke *et al.*, 1975).

In a brief review, Berstein (1993) has highlighted the fact that the frequency of tumours in organs such as the ovaries, mammary glands or lungs differs between sides. He suggests that predisposition of paired organs to right-sided or left-sided tumour development is connected not only with anatomy but also with some internal function. Breast cancer in several generations of relatives appears on one and the same side, suggesting that left or right localisation of a tumour cannot be accidental. And, although breast cancer occurs on the left side 1.5 times more often than on the right, there is also the fact that left-handed persons develop breast cancer 2–3 times less often than the population at large (Olsson & Ingvar, 1991). Berstein (1993) suggests that these phenomena could in some way be related to the observation that left handedness may result from hyperandrogenisation during the phase of embryonic development (Geschwind & Behan, 1982). And left-handed people also reveal an increased incidence of autoimmune disease (Geschwind & Behan, 1982).

Testicular hypoplasia is a well-documented pathological condition in cattle, sheep and pigs, caused by aneuploidy or other chromosomal abnormalities (Knudsen, 1961; Hancock & Daker, 1981) and a recessive autosomal gene with partial penetrance (Laing & Young, 1956). Hypoplastic testes are invariably smaller than normal and show a diminished spermatogenic activity which, when present in both gonads, leads to reduced fertility or infertility. However, normal fertility tends to be maintained in unilaterally affected animals (Roberts, 1956). Hypoplasia stemming from the influence of a recessive autosomal gene is unilateral and generally affects the left testis (Laing & Young, 1956). Failure of testicular descent or cryptorchidism may also reveal unilateral tendencies. For example, cryptorchidism in humans often occurs unilaterally with right-sided cryptorchidism significantly outnumbering the left-sided condition

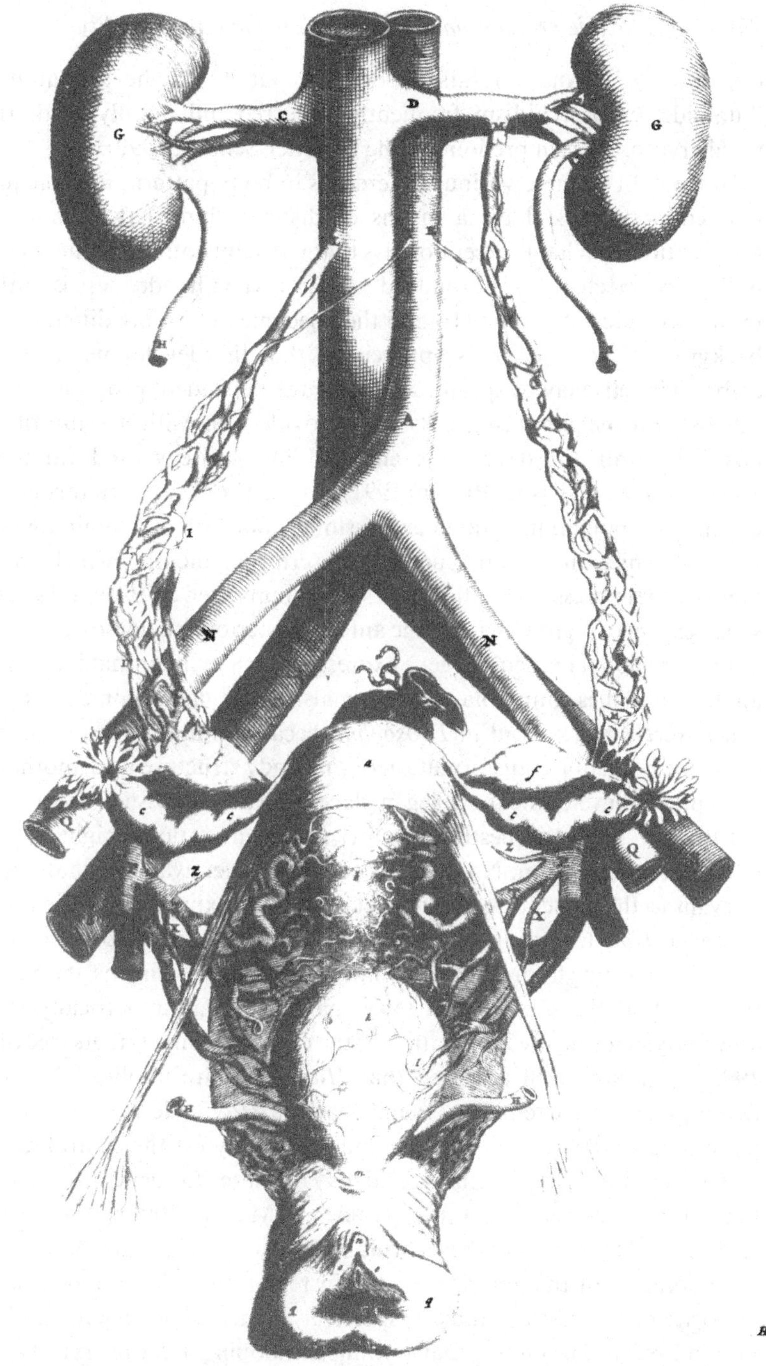

Fig. 9.2. Classical illustration from Regnier de Graaf (1672) of the kidneys and female reproductive organs. An asymmetry can be noted in the disposition of blood vessels and even in the size of the kidneys. (Courtesy of Edinburgh University Library.)

(Heyns, 1987). And, in rats exposed prenatally to the anti-androgen flutamide, cryptorchidism frequently occurred unilaterally, with right-sided cryptorchidism predominating (van der Schoot, 1992).

In the light of these various differences in body pattern, the conclusion that embryos must have a means of distinguishing left from right is unexceptional. What is exceptional is that virtually nothing is known of the underlying mechanisms nor indeed precisely when handedness is initiated in the two sides of an embryo and they become detectably different. As a background, however, it is appreciated that the developmental fate of embryonic cells may be specified by a lineage-dependent programme or by signals from neighbouring cells that provide cells with a sense of their position within the developing embryo. The strategy used varies with species, stage and tissue (Brown, 1991). At least some pattern formation is dependent upon an integrated expression of homeobox-containing genes, most of which appear to encode transcription factors, which in turn regulate the expression of other genes. As summarised by Brown (1991), the specification of segments along the antero-posterior and dorso-ventral axes requires regional homeobox gene expression, in a manner that is seemingly analogous in flies, amphibia and mammals. The so-called homeobox (*Hox*) genes were first revealed in *Drosophila* because mutations in such genes cause the homeotic transformation of one body structure into another, for example the development of a leg in place of an antenna. Such observations led Brown (1991) to question whether similar mutations might influence left–right specification. No detailed response is yet available that appears relevant to the present treatment, but abolishing expression of the homeobox gene *Hox* 1.6 produces animals lacking the fifth hindbrain rhombomere. By injecting fluorescent dyes into these animals' neurons, it is possible to show that the seventh and lower ganglia are asymmetrically shifted anteriorly, altering the innervation of the corresponding targets (see Short, 1992). It is now well accepted that *Hox* genes are implicated in brain development, a current view being that the *Hox* code is crucial for the patterning of the head as a whole and not simply for the central nervous system. The morphogen retinoic acid may regulate *Hox* gene complexes in a time- and dose-specific manner (Holder & Maden, 1993). For example, retinoic acid treatment at the gastrula stage leads to a transformation in the rhombomeres of the anterior hindbrain region in which rhombomeres 2 and 3 are respecified as 4 and 5, respectively; hence, abnormal formation of the hind-brain. The finding that asymmetric looping of the heart in *Xenopus laevis* embryos can be eliminated by inhibition of proteoglycan synthesis also strongly underlines molecular considerations (Yost, 1990).

As pointed out by Neuman-Silberberg & Schüpbach (1993), achievement of asymmetry must be dependent upon spatial localisation of at least one component in the system. For example, patterning may involve the local activation of receptors distributed uniformly in the egg membrane. Another strategy for generating asymmetry is the localisation of RNA, as in *Drosophila* and *Xenopus*. The resulting proteins therefore form gradients that give rise to the embryonic body plan by determining cell fate during early development (Wilhelm & Vale, 1993). Different threshold levels of the protein would be required for the activation of patterning genes responsible for the specification of a plane of asymmetry. However, despite these valuable concepts and indeed some elegant experimental evidence, the actual means of regulating the localisation of RNA remains unknown.

Brown & Wolpert (1990) have proposed a model and conceptual framework to stimulate further discussion and analysis of the problems of handedness. They suggested a mechanism for the conversion from asymmetries at a molecular level to the formation of right and left structures by specific tissues (Fig. 9.3). The model involves three essential steps: a process of conversion from molecular asymmetry (handedness) to cellular handedness at the multi-cellular level. This, in turn, biases a system for the random generation of asymmetry. Finally, there would be an interpretation step for individual organs. Brown & Wolpert (1990) readily confess that the effects of the mouse mutant gene, *iv*, causing situs inversus viscerum, played a key rôle in generating their model. Situs inversus in mice is the result of an autosomal recessive gene, although the condition can occur spontaneously or as the result of certain other mutations. Because embryos of homozygous *iv*/*iv* mice develop as though they have lost their sense of left and right, their asymmetry is not handed but random, indicating that, in terms of the above model, the conversion or biasing process is faulty and does not interact with the random generation of asymmetry. Half the embryos therefore develop with normal visceral situs, half with the mirror image pattern of situs inversus. There is good evidence that each organ has its asymmetry specified independently (Brown, McCarthy & Wolpert, 1990). A rôle for morphogens has been invoked, and Brown & Wolpert (1990) suggest that asymmetry could be produced by diffusion reaction, which does seem capable of subdividing an organism into separate distinct compartments and theoretically could be random. If the origins of handedness reside in a protein as a morphogenic pump, then in theory at least it could produce a consistent imbalance that biases the otherwise random part of the process (Galloway, 1990).

A further aspect of asymmetry has received attention in the major review

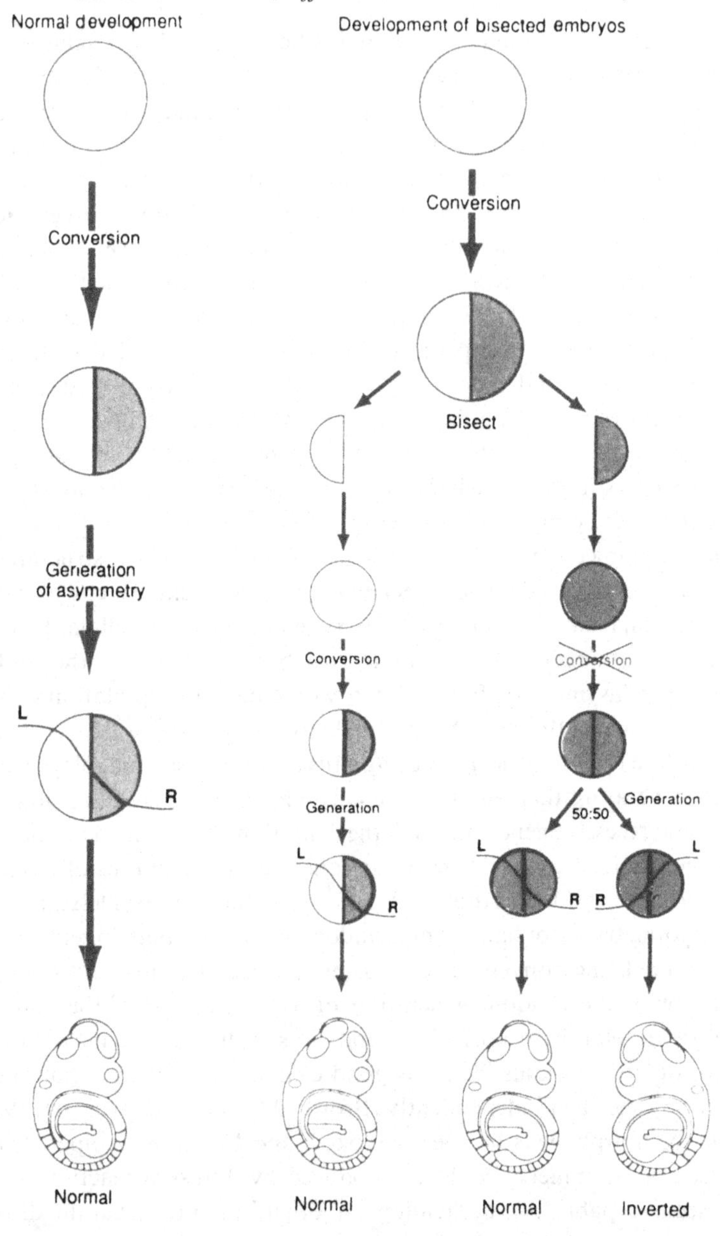

Fig. 9.3. Development of asymmetry in bisected embryos with the suggestion that *conversion* results in a stable property in right, but not left, cells. The left half of an embryo bisected after *conversion* is labile and would be able to go through *conversion* again and develop normally. The right half cannot be respecified so asymmetry is random and half the embryos develop with situs inversus. (After Brown, McCarthy & Wolpert, 1991).

of Berridge (1993) dealing with inositol trisphosphate and calcium signalling. Berridge notes that there is a hint that a phosphoinositide signalling pathway could be involved in the development of normal body asymmetry, and specifically left-right sidedness. Apparently, the situs inversus viscerum mutation of mice can be reproduced in rats if embryos are cultured with the α_1-adrenergic agonist noradrenaline (Fujinaga & Baden, 1991*a*; Fujinaga *et al.*, 1992). Because the α_1-adrenergic receptor is invariably coupled to phospholipase C, inositol trisphosphate or diacylglycerol would seem to be an important candidate for determining left–right handedness during embryonic development, at least in this particular model system. Not surprisingly, there is a critical period during development of the rat embryo when the sidedness of asymmetric body structures is determined (Fujinaga & Baden, 1991*b*).

Another experimental observation that may turn out to be of relevance comes from transgenic mice (Yokoyama *et al.*, 1993). An unforeseen insertional mutation was produced upon introduction of a tyrosinase minigene into newly fertilised eggs. In some manner, this prompted development of a body plan which was the mirror image of normal, a consistent finding in all homozygous mutants so generated. This stands in marked contrast to the previous maximum 50% incidence of situs inversus generated by various treatments of rodent embryos in culture (McCarthy, Wolpert & Brown, 1990). One interpretation of the Yokoyama mutant would be that the inserted minigene had disrupted the host genome, perhaps by activating a default pathway.

In any event, the extent to which the novel approach of Brown & Wolpert (1990) is relevant to interpreting differences between the left and right gonads is not yet certain, especially with respect to the formation of an ovary, ovotestis or testis. Undoubtedly, molecular cues are involved but because the very early embryonic gonad is bipotential (see Chapter 3), the critical factors in differentiation may be more closely dependent upon the timing of molecular cues and indeed the actual extent of a given molecular influence. As argued by Mittwoch (1976*a*, *b*, 1986), regulation of rates of gonadal growth may be the most important single factor influencing the nature of the gonad so produced, precocious development in some way prescribing formation of testicular tissue. Unilateral differences in growth rate could predestine gonadal tissue differentiation towards testis formation.

Unilateral formation of ovotestis

Reasoning from first principles, if there is a predisposition for one of the gonads to undergo anomalous forms of development when compared with its contralateral partner, then perhaps this situation could offer a clue as to a portion of the diverse mechanisms involved in sexual differentiation. In the first instance, it may be valuable to note which side, left or right, achieves predominance, although this may vary with species. It has already been seen that in domestic animal intersexes (e.g. pig, horse), the situation is usually one of an ovary on the left and a testis on the right (Chapter 5). However, the more important and fundamental question remains as to why the left–right differences occur in the first place.

For example if, during elaboration of an ovotestis, testicular tissue is frequently destined to develop within only one of the gonads of a presumptively XX female, it certainly seems pertinent to question as to why this anomaly should usually or invariably arise in the right hand gonad. And, if part of the answer concerns the differential rates of growth already mentioned, the right gonad being noted to grow faster than the left (Mittwoch, 1976*b*, 1986), then the presumed relationship between testis-determining genes such as *Tdy* (*Sry*) and those influencing gonadal growth needs to be focused upon and analysed in depth. In unilateral situations, why should testicular tissue tend to be expressed in the faster-growing gonad? As noted by Mittwoch (1992), the process of X chromosome inactivation cannot account for the observed bilateral asymmetry of gonadal differentiation in XY hermaphrodites in humans and mice. And in the earlier words of Mittwoch (1986): 'it can be regarded as certain that no theory of sex determination will be able to withstand the test of time if it cannot explain the anomaly of bilateral asymmetry'.

These comments lead to the question of whether there may be sex-related genes specifically to promote unilateral differences or, perhaps better phrased in a reproductive context, to promote unilateral susceptibilities. How might such genes function, and how might they act preferentially on only one gonad? In other words, how might a unilateral genetic influence be explained? There must be gene(s) to programme or prepare the way for differences between the left and right sides in rates of growth. As noted above, the data on human alternate hermaphrodites suggest that the right side of the body promotes or at least facilitates testicular differentiation owing to an inherent enhancement of growth (Mittwoch, 1986). Furthermore, the distribution of ovotestes resembles that of testes in being biased towards the right (van Niekerk & Retief, 1981). And, when there is a unilateral ovotestis, a contralateral ovary is two and a half times as common as a contralateral testis. Turning from humans to intersex pigs, not only is testicular tissue noted predominantly in the right gonad, but testicular structures on the right are conspicuously larger than ovarian structures on the left (Hunter, Cook & Baker, 1985). However, when both gonads have developed as abdominal testes in XX intersex animals, these structures greatly exceed the size of normal mature ovaries, despite the absence of a sperm-producing epithelium within the tubules (Hunter, Baker & Cook, 1982). In instances of unilateral cryptorchidism in intersex pigs, the scrotal gonad was found twice as frequently on the right-hand than left-hand side and it was always a testis-like structure, not an ovotestis.

In the case of hermaphrodite mice described by Eicher & Washburn (1983) in which ovotestes represented the predominant gonadal type, the combination of left testis/right ovotestis occurred more frequently than the reverse combination. By contrast, the biased asymmetry found by Ward, McLaren & Baker (1987) in sex-reversed hermaphrodite mice (T16/X*Sxr*) was in the form of a testis on the left side and an ovary on the right. In other combinations of hermaphrodite mice not sex reversed but with a *domesticus*-type Y chromosome, the gonadal disposition was predominantly a left testis and right ovotestis (Biddle, Eales & Nishioka, 1991). At first sight, the mouse might appear somewhat of an exception in the context of left–right tendencies, for there is also the observation of Billington (1965) that the left testis is almost invariably heavier than the right testis in this species. But Mittwoch (1992) notes that in adult mice, the right testis is the heavier and she then considers the possibility of a reversal in asymmetry with increasing size of the gonad, relating this to rates of development and gonadal size in mice and men.

Inequalities of ovarian function

Moving on from comments concerning rates of gonadal growth and perturbations in gonadal differentiation, observations have been made repeatedly in mature animals on ovarian activity as monitored by the number of corpora lutea or the number of eggs shed at ovulation, usually termed the ovulation rate, during spontaneous oestrous or menstrual cycles.

In laboratory species, there is long-standing evidence for differences in ovulation rate between the two ovaries. The right ovary invariably shows a slight superiority in the number of eggs released when compared with the left in mice (Hollander & Strong, 1950; Runner, 1951; McLaren & Michie,

1959; Falconer *et al.*, 1961; McLaren, 1963; Wiebold & Becker, 1987), although the difference seldom attains statistical significance beyond $P<0.05$. The right–left predominance is reversed in instances of situs inversus in mice. There is also a strong indication in rats of greater ovulatory activity in the right ovary under conditions of spontaneous ovulation (Barr, Hensh & Brent, 1970; Buchanan, 1974). As a more extreme example, the mountain viscacha, *Legidium peruanum*, a South American rodent, invariably ovulates only from the right ovary. However, the left ovary can be demonstrated to have retained physiological competence if a unilateral ovariectomy is performed on the right (Pearson, 1949) so, once again, there is a suggestion in the intact animal of a form of gonadal dominance.

Whether the pattern of innervation by the vagus can contribute to these left–right differences in rodents has been given frequent consideration, with the suggestion that such innervation may somehow act to influence the number of mature follicles in the two ovaries (Burden *et al.*, 1986). The results of experiments involving vagotomy and/or unilateral ovariectomy have been adduced to support this point of view (Burden *et al.*, 1986; Chavez, Cruz & Dominguez, 1987), and such experiments have been set in an endocrine perspective (Cook, 1987). Even so, an outstanding question remains as to why such apparent differences in nervous input to the two gonads have evolved.

Species of Chiroptera such as the black mastiff bat, *Molossus ater*, demonstrate the extreme form of gonadal asymmetry with essentially a complete dominance by the right ovary (Wimsatt, 1975, 1979; Rasweiler, 1988). For example, in 85 of 86 animals examined from a laboratory breeding colony, the right ovary was significantly larger than the left (Rasweiler, 1988). Furthermore, although primordial follicles were demonstrable in both ovaries, Graafian follicles and corpora lutea were found only in the right ovary, together with a prominent intra-ovarian vascular network. Rasweiler suggested that the relatively poor blood supply to the left ovary might have acted to limit its development, whilst further possibilities included a reduced availability of gonadotrophin receptors. Nearly all other members of the bat family, Molossidae, so far examined have revealed asymmetries similar to those described above. The California leaf-nosed bat, *Macrotus californicus*, exhibits a distinct dominance by the right ovary (Bleier & Ehteshami, 1981), as do two British horseshoe bats, species of *Rhinolphus* (Harrison Mathews, 1937), and a Japanese one (Oh, Mori & Uchida, 1985). Despite this form of dominance by the right ovary, there are species of bat in which the left ovary is the functional one (see Wimsatt, 1975).

Amongst marine mammals, Sørensen (1992) has recorded expressions of ovarian asymmetry in the small-toothed whale, *Phocoena phocoena* (the harbour porpoise). Whereas no significant differences were found between left and right ovarian weights in the immature animal, the left ovary was significantly larger than the right after puberty ($P < 0.05$). And, although primordial follicles were present in both ovaries, Graafian follicles, corpora lutea and corpora albicantia were noted only in the left ovary. In pregnant animals, and as a reflection of such asymmetric ovarian activity, it is no surprise that conceptuses were always located in the left Fallopian tube or uterine horn. No differences could be distinguished in the intra-ovarian vascular supply, and yet it must be assumed that gonadotrophic influences were in some manner privileged in the left ovary.

Turning to farm animals, and to ruminants in particular, left–right inequalities in ovarian function are well documented, with a majority of ovulations occurring in the right ovary. One of the earliest reports for cattle was that of Casida *et al.* (1935) in spontaneously cyclic heifers in which the right ovary was more active than the left, an observation endorsed in many subsequent reports. For example, 55–65% of ovulations and pregnancies in cattle and also in sheep are found on the right side (Casida, Woody & Pope, 1966; Reimers *et al.*, 1973; Salisbury, Van Demark & Lodge, 1978; McDonald, 1980). In sheep, such a distribution of ovulations between the two gonads has been noted repeatedly (McKenzie & Terrill, 1937; Henning, 1939; Hunter, 1959; Wheeler, 1978). Exceptionally, there may be a preponderance as high as 77% of ovulations from the right ovary (Hunter & Nichol, 1986). The strongly asymmetric disposition in this last experiment may have been accentuated by local factors, not least the orientation of ewes to specific external sources of stimulation such as artificial light.

A conventional explanation for the left–right differences in cattle and sheep concerns the location of the rumen on the left side of the abdominal cavity. The suggestion is that this organ may in some manner act to reduce blood flow to the left ovary, and thereby the supply of gonadotrophic hormones. In such circumstances, an enhanced ovarian activity would be anticipated on the side opposite the rumen, that is on the right (see McDonald, 1980). Certainly, location of a foetus in the right-hand portion of the bipartite uterus could be advantageous in the face of a distended rumen. Nonetheless, because asymmetry extends widely through mammalian groups, a primary involvement of the rumen cannot logically be invoked as a fundamental cause of gonadal asymmetry in cattle and sheep.

In primates, perhaps the best available evidence concerning asymmetry is for our own species. Prolonged follow-up of women with normal menstrual

Table 9.3. *To illustrate the side and sequence of ovulation in 16 women with normal menstrual cycles*

No. of cycles (no. of women)	90 (16)
No. of follicles[a]	97
Right ovary (%)	62 (64)
Left ovary (%)	35 (36)
Alternate ovulations[b]	
No. of women (%)	5 (31)
No. of cycles (%)	22 (24)
Unilateral ovulations[b]	
No. of women (%)	4 (25)
No. of cycles (%)	19 (21)

Notes:
Modified from Potashnik *et al.* (1987). 'Alternate ovulations' indicates ovulation from opposite ovaries in succeeding cycles whereas 'unilateral ovulations' indicates the same ovary in succeeding cycles.
[a] Difference between left and right: $P < 0.01$.
[b] Observed in at least four consecutive cycles.

cycles has shown that ovulation occurs in the right ovary twice as frequently as in the left (Potashnik *et al.*, 1987; Table 9.3).

At a biochemical level, left–right differences have also been reported for porcine corpora lutea in the respective ovaries. Such differences include values for gonadotrophin receptors, progesterone content, glucose-6-phosphate dehydrogenase activity and *N*-acetyl-β-D-glucosaminidase activity (Rao & Edgerton, 1984).

Overall, these inequalities in ovarian function can be taken to reflect a potential for physiological activity in the right gonad (usually) of many placental mammals not normally matched by a corresponding activity in the left gonad. Such potential asymmetries can, moreover, be demonstrated experimentally in the male. When the left testis is removed in rats, testosterone concentrations increase more than when the right testis is ablated (Frankel, Chapman & Cook, 1989). All constituent components of the gonads from somatic and germ cell populations to the extent of innervation, vascular supply and lymphatic drainage may need to be considered when seeking a satisfactory explanation for asymmetries. Of course, the word asymmetry is used here in a simplistic and largely morphological sense; there may well be diverse physiological components of ovarian function which do not demonstrate actual or potential differences between the two sides. Even so, the genetic control of morphological asymmetry is certainly worthy of further consideration.

The list of species differences presented above could of course be extended very considerably, but it seems improbable that such a listing would bring one closer to a satisfactory interpretation. Possibly the most significant point to emerge is the apparent tendency for one gonad to try to dominate the contralateral one, at least as monitored by physiological activity after puberty. This is well illustrated in poultry, and the fact that unilateral ovariectomy in this species prompts the almost vestigial contralateral (right) gonad to hypertrophy and evolve frequently as an ovotestis is worth pondering, especially since the testicular tissue so elaborated may produce functional spermatozoa. Precisely what this tells us about the mechanisms underlying gonadal differentiation and development is uncertain in molecular terms, especially those referred to earlier in this chapter. However, a seemingly important question to ask is whether and to what extent asymmetry is reflecting chronological differences between the two sides in the loss of bipotential status in the gonad and, if so, whether this concerns differences in the extent of influence of the *Tdy* (*Sry*) gene product. Alternatively, could asymmetry simply reflect differences in cell populations at the time of loss of bipotentiality when the gonads embark upon their unidirectional evolution and growth? One conclusion that may reasonably be drawn is that there is some form of conversation between the two gonads by the time that puberty is achieved; precisely how direct is the means of communication awaits clarification.

Asymmetry in genital duct development

Instances of asymmetry in development of the Wolffian or Müllerian ducts may be primarily a reflection of the constitution of the gonads. Asymmetry can usually be demonstrated most clearly in (1) genetic females, (2) the proximal portion of the ducts adjacent to the gonads, and (3) cases of unilateral formation of an ovotestis.

An especially good example of asymmetry in duct development is found in intersex pigs (see Chapter 5) and illustrated in Plates 5.2 and 5.3. In these genetic females (XX chromosome complement), the left gonad has formed most frequently as an ovary and the right as an ovotestis. In such circumstances there is morphologically normal development of a Fallopian tube on the left-hand side and no detectable development of the Wolffian duct. By contrast, expression of the proximal portion of the Müllerian duct as the Fallopian tube ampulla on the right-hand side has been suppressed whilst the proximal portion of the Wolffian duct has been elaborated as a highly convoluted epididymal duct. As discussed in Chapter 5, these derangements can be clearly ascribed to the presence of testicular tissue in the right-hand gonad. Sertoli-like cells are presumed to have secreted anti-Müllerian hormone (AMH) in limited quantities and/or too late in embryonic development. AMH has acted to suppress full proliferation of the Fallopian tube whilst Leydig cell secretion of androgens, notably testosterone, has promoted formation of an epididymis from its Wolffian duct progenitor. Because only the proximal portion of the Müllerian and Wolffian duct systems has been detectably modified by the anomalous gonadal tissue, the influence of Sertoli and Leydig cell secretions, especially the former, is presumed to have been principally a local one. Uterine tissues remain morphologically unperturbed.

The above examples of asymmetry in genital duct development are perhaps the most extensive and dramatic seen in laboratory and domestic farm mammals. Minor differences between the two sides of the genital tract can be detected in many non-pregnant animals, either in the size of structures or in the degree of coiling. For example, one horn of the uterus in bicornuate or bipartite situations may be consistently larger than the contralateral one. Whether there is an intrinsic predisposition for the right side to be modified or whether this simply reflects the condition of the gonads has not been completely resolved. Possibly the gonads act to accentuate a trend towards asymmetry that already exists in paired structures. Transplantation surgery could offer a satisfactory means of investigating this point.

One other spontaneous circumstance concerning the ducts deserves mention. Unilateral renal aplasia has already been referred to in this chapter in the context of patients with Kallmann's syndrome. There is now an associated report of just one vas deferens in two male siblings with a kidney also missing on the same side: the right (Hardelin *et al.*, 1993). Because of the embryonic relationship between the mesonephros and its duct from which, respectively, develop the kidney (via the metanephros) and Wolffian duct, the association in this anomalous condition is not unduly surprising. However, the manner in which expression of the *KAL* or other genes – even when carrying a mutation – becomes susceptible to a unilateral influence remains an enigma. Other genetic or epigenetic factors must be required in addition to the *KAL* gene defect for expression of unilateral renal aplasia (Hardelin *et al.*, 1993).

In rats, the ACI strain shows a congenital anomaly involving unilateral agenesis of the complete urogenital tract. This defect has been noted more frequently on the right than on the left side (Deringer & Heston, 1956; Marshall *et al.*, 1982). There is also the fact that, during pregnancy in rats, only the right umbilical artery enlarges whereas the left artery atrophies after day 15 of gestation. Indeed, van der Schoot & Elger (1992) have suggested that this large and asymmetric vascular structure may contribute to the closer apposition between the right testicle and kidneys at birth. Unilateral differences may also be highlighted following endocrine therapy. For example, as a consequence of treating pregnant rats with androgens, stabilisation of the embryonic Wolffian ducts occurred more frequently on the left than on the right-hand side (Greene, Burrill & Ivy, 1939). In guinea-pigs, on the other hand, the left horn of the uterus was congenitally absent in two instances (K. P. Bland, personal communication).

Unilateral differences in gamete and embryo transport

Whilst minor differences in the relative rate of egg or embryo passage through the two Fallopian tubes into the uterus have been recorded from time to time in polytocous species, these may have been due primarily to slight inequality in the length of the left and right tubes. Whether this is true or not, much more attention has been drawn to differences in the rate of passage along the tubes between fertilised and unfertilised eggs in laboratory species such as rats and hamsters (Villalon *et al.*, 1982; Ortiz *et al.*, 1986) and also in ferrets (Mead, Joseph & Neirinckx, 1988). Such differences have been highlighted in various species of bat (Rasweiler, 1979) and most notably in horses (van Niekerk & Gerneke, 1966; Betteridge & Mitchell, 1974; Hunter, 1988, 1989).

Concerning transport in the opposite direction, asymmetries may be demonstrable as far as progression (or survival) of spermatozoa is concerned. For example, in the Japanese long-fingered bat, *Miniopterus schreibersii fuliginosus*, Mori & Uchida (1982) noted that most spermatozoa failed to reach or survive in the ampulla of the right-hand Fallopian tube, and phagocytosis by leucocytes was restricted to that side of the tract. No explanation was advanced for these left–right differences, but most probably a local influence of the single ovulatory follicle on the left ovary would need to be invoked. In rabbits, Overstreet and colleagues have noted a preferential and significant transport of spermatozoa to the left Fallopian tube. For example, in the context of rapid transport of spermatozoa in naturally mated does, Overstreet & Cooper (1978) reported an asymmetric distribution in the upper female tract. Although sperm numbers in the cervix and uterus were found not to be different on the two sides, greater numbers of spermatozoa were recovered from the Fallopian tube and its fimbriated extremity on the left side of the tract. In a separate set of experiments, Overstreet, Cooper & Katz (1978) reported asymmetry in the numbers of spermatozoa recovered from the tubes, with the left side again having the larger number. Finally, Overstreet & Thom (1982) emphasised the intriguing phenomenon of unilateral rapid transport into the left Fallopian tube, which was detected in 30 of 36 animals showing 'rapid transport' of spermatozoa. In none of these papers was any attempt made to interpret the physiological events underlying the asymmetric distribution of spermatozoa. However, speculative explanations could invoke asymmetry in the pattern of autonomic stimulation of smooth muscle on the two sides of the tract, and/or differences in receptor populations for smooth muscle stimulants such as oxytocin.

In domestic pigs, there exists a phenomenon of unilateral fertilisation (Pitkjanen, 1961; Hancock, 1962), in which the ejaculate passes predominantly into one horn of the bicornuate uterus and only eggs in the corresponding Fallopian tube are fertilised. It remains unclear whether there is any consistent trend as to the side of the unilateral phenomenon which, in any event, seems to be expressed most frequently under conditions of retarded mating and diminished myometrial contractile activity (Hunter, 1967, 1982). Patterns of smooth-muscle activity have long been accepted to differ between the two sides of the genital tract (e.g. see Wislocki & Guttmacher, 1924; Hunter, 1977). An asymmetry of sperm transport in cattle (Herz *et al.*, 1985; Hunter, Fléchon & Fléchon, 1991) may reflect a local endocrine influence of the ovulating ovary.

Distribution of uterine foetal mass

After a successful mating, the inequality in performance of the left and right ovaries in rodents finds expression in the uterine contents. This is well demonstrated in rats and mice, in which a larger number of foetuses is usually found in the right compared with the left cornua (mice: Hollander & Strong, 1950; Runner, 1951; McLaren & Michie, 1959; Falconer *et al.*, 1961; Wiebold & Becker, 1987; rats: Barr *et al.*, 1970; Buchanan, 1974).

In species with a bipartite uterus such as cows and sheep, a single foetus tends predominantly to be located in the right-hand side of the uterus, although the placental membranes will inevitably extend around the septum into the contralateral horn of the uterus as pregnancy advances. The question therefore arises as to whether this tendency towards establishment of the gravid horn on the right-hand side simply reflects the predominance of ovulations from the right-hand ovary. This would be the straightforward explanation, for the incidence of intra-uterine migration or, more correctly, redistribution of preattachment embryos in cattle and sheep is low, perhaps no more than about 10% (Scanlon, 1972; Hunter, 1980). In other words, there would be a strong tendency for the foetus to develop on the side of the tract in which fertilisation has occurred. Even so, McLaren (1982) records that in the impala, pregnancies almost always occur in the right uterine horn of an otherwise symmetrical reproductive tract, even though ovulations may often have occurred in the left ovary.

Two of the points mentioned in earlier paragraphs could also be of relevance here. The pattern of overall blood flow may be greater to the right horn of the uterus, perhaps as a reflection of asymmetries in the vascular network or of a local influence of gonadal endocrine activity, or a combination of both these possibilities. Irrespective of the underlying cause, an enhanced vascular perfusion of the right side of the uterus – if such exists – would favour development of the foetus and its placental membranes. And, as the period of gestation advances in cattle and sheep, the gravid horn of the uterus will confront the substantial dimensions of the rumen situated largely to the left of centre in the abdominal cavity. Moreover, the rumen may generate local fluctuations in temperature which could be harmful to the developing embryo or foetus if sited too closely. Physical constraints may therefore have imposed a selective advantage upon foetal development in the right horn of the uterus although, of course, a number of preferential factors may well be interacting. However, this line of reasoning is not supported by findings in camels, in which 99% of pregnancies occupied the left uterine horn (Shalash, 1965).

Concluding remarks

Whilst the unilateral trends and differences mentioned in the above discussion may not always have been demonstrated with scientific rigour, there can be little doubt that a variety of asymmetries does exist in mammalian reproductive systems. In seeking a straightforward explanation, such conditions may be viewed as an expression of the morphological asymmetry found in most types of living organisms, frequently with a systematic bias towards either right-handed or left-handed forms. Conceptually, this should not appear as a difficult point, for asymmetries in the other two planes of body development are readily accepted, and these are doubtless a reflection of the situation in ancestral species. Reference to the fossil record shows clear evidence that primitive animals were asymmetrical and that bilateral symmetry then evolved, followed by the laterality seen in modern chordates. Whether asymmetry can be related in some reasonably direct manner to underlying molecular cues expressed during early embryogenesis is a point of current discussion. However, the only gene so far identified to be involved in the inheritance of laterality is the mouse *iv* gene. In the words of a stimulating review, might there have been an ancestral left–right gene that gave rise to a family of homologous genes which specify morphological and functional handedness (Wolpert, 1991)? Also remaining completely unanswered is the question of whether abdominal lateralisation bears any specific relationship to cerebral lateralisation. Here, there is the intriguing observation that responses to hemi-castration or hemi-ovariectomy in rats seem to be based principally on cues carried asymmetrically from the central nervous system on parasympathetic fibres (Chavez, Cruz & Dominguez, 1987; Dominguez, Cruz & Chavez, 1989; Frankel, Chapman & Cook, 1989; Dominguez, 1990). Also of relevance in a reproductive context is the fact that gonadotrophin releasing hormone concentrations differ between the right and left sides of the rat hypothalamus (Bakalkin *et al.*, 1984) and, as already remarked, that unilateral renal aplasia or hypoplasia is found in Kallmann's syndrome. Indeed, there is now a clear appreciation of the fact that many hypothalamic nuclei and brain centres are paired and that, within such pairing, an asymmetric distribution of various neurotransmitters, hormones, receptors and enzymes can be demonstrated (Oke, Lewis & Adams, 1980). These asymmetries are assumed to be of functional significance, although the hypothalamo-hypophyseal portal system may act to obscure corresponding asymmetric function in cells of the pituitary gland.

Perhaps an appropriate note on which to conclude this chapter concerns what Mittwoch (1986) refers to as the basic asymmetry of normal gonadal development. It seems pertinent to question why such a situation has been retained in eutherian mammals, and to wonder what selective advantage it might have bestowed in evolutionary terms. Perhaps it offered a flexibility to ancestral species in the potential for gonadal differentiation, not least in response to changing environmental cues. However, this line of thought loses its persuasion if one focuses simply upon placental mammals. On the other hand, there does seem to be scope for inferring interaction between, for example, testis-determining genes such as *Sry* and those genes involved in promoting gonadal growth and expression of the unilateral tendency. The nature of such putative interactions and the precise manner whereby a gene imposes unilateral instructional information must await developments in the field of molecular genetics. The finding that a targeted deletion of the α-inhibin gene, generated by homologous recombination in embryonic stem cells, was frequently followed in inhibin-deficient mice (i.e. mice homozygous for the null allele) by development of unilateral tumours of the gonadal stroma (Matzuk *et al.*, 1992) may provide a useful model system for further analysis. So also may the observation that a recessive mutation in a family of transgenic mice resulted in situs inversus in 100% of the homozygous transgenic mice examined (Yokoyama *et al.*, 1993). Such an insertional mutation identifies a gene that controls embryonic turning and visceral left–right polarity.

References

Altmann, S. L. (1992). Perturbing patterns. *Nature (London)*, **360**, 545–6.

Bakalkin, G. Y., Tsibezov, V. V., Sjutkin, E. A., Veselova, S. P., Novikov, I. D. & Krivosheev, O. G. (1984). Lateralization of LH-RH in rat hypothalamus. *Brain Research*, **296**, 361–4.

Barr, M., Hensh, R. P. & Brent, R. L. (1970). Prenatal growth in the albino rat: effects of number, intrauterine position and resorptions. *American Journal of Anatomy*, **128**, 413–28.

Benoit, J. (1932). L'inversion sexuelle de la poule, déterminée par l'ablation de l'ovaire gauche. *Archives de Zoologie Expérimentale et Générale*, **73**, 1–112.

Berridge, M. J. (1993). Inositol trisphosphate and calcium signalling. *Nature (London)*, **361**, 315–25.

Berstein, L. M. (1993). Topoendocrinology. *Journal of Endocrinology*, **137**, 163–6.

Betteridge, K. J. & Mitchell, D. (1974). Direct evidence of retention of unfertilized ova in the oviduct of the mare. *Journal of Reproduction and Fertility*, **39**, 145–8.

Biddle, F. G., Eales, B. A. & Nishioka, Y. (1991). A DNA polymorphism from five inbred strains of the mouse identifies a functional class of *domesticus*-type Y chromosome that produces the same phenotypic distribution of gonadal hermaphrodites. *Genome*, **34**, 96–104.

Billington, W. D. (1965). The invasiveness of transplanted mouse trophoblast and the influence of immunological factors. *Journal of Reproduction and Fertility*, **10**, 343–52.

Bleier, W. J. & Ehteshami, M. (1981). Ovulation following unilateral ovariectomy in the California leaf-nosed bat (*Macrotus californicus*). *Journal of Reproduction and Fertility*, **63**, 181–3.

Bock, G. R. & Marsh, J. (ed.) (1991). *Biological Asymmetry and Handedness*. Ciba Foundation Symposium 162. Chichester & New York: Wiley.

Borradaile, L. A. (1955). *Manual of Elementary Zoology*, 12th edn. London & New York: Oxford Medical Publications.

Brown, N. A. (1991). Development of left and right: the worm turns. *Current Biology*, **1**, 159–61.

Brown, N. A., McCarthy, A. & Wolpert, L. (1990). The development of handedness in aggregation chimeras of situs inversus mutant and wild-type embryos. *Development*, **110**, 949–54.

Brown, N. A. & McCarthy, A. & Wolpert, L. (1991). Development of handed body asymmetry in mammals. In *Symposium on Biological Asymmetry and Handedness*, ed. G. R. Bock & J. Marsh. Ciba Foundation Symposium 162. Chichester & New York: Wiley.

Brown, N. A. & Wolpert, L. (1990). The development of handedness in left/right asymmetry. *Development*, **109**, 1–9.

Buchanan, G. D. (1974). Asymmetrical distribution of implantation sites in the rat uterus. *Biology of Reproduction*, **11**, 611–18.

Burden, H. W., Lawrence, I. E., Smith, C. P., Hoffman, J., Leonard, M., Fletcher, D. J. & Hodson, C. A. (1986). The effects of vagotomy on compensatory ovarian hypertrophy and follicular activation after unilateral ovariectomy. *Anatomical Record*, **214**, 61–6.

Casida, L. E., Chapman, A. B. & Rupel, I. W. (1935). Ovarian development in calves. *Journal of Agricultural Research*, **50**, 953–60.

Casida, L. E., Woody, C. O. & Pope, A. L. (1966). Inequality in function of the right and left ovaries and uterine horns of the ewe. *Journal of Animal Science*, **25**, 1169–71.

Chavez, R., Cruz, M. E. & Domínguez, R. (1987). Differences in the ovulation rate of the right or left ovary in unilaterally ovariectomized rats: effects of ipsi- and contralateral vagus nerve on the remaining ovary. *Journal of Endocrinology*, **113**, 397-401.

Cook, B. (1987). Endocrine asymmetry. *Journal of Endocrinology*, **113**, 331–2.

De Graaf, R. (1672). *De mulierum organis generationi inservientibus tractatus novus*. Leyden.

Deringer, M. K. & Heston, W. E. (1956). Abnormalities of urogenital system in strain A × C line 9935 rats. *Proceedings of the Society for Experimental Biology and Medicine*, **91**, 312–4.

Domínguez, R. (1990). Differential ovulatory responses of right and left ovaries to unilateral lesion and anaesthesia of the cervico-vaginal plexus. *Journal of Endocrinology*, **124**, 43–5.

Domínguez, R., Cruz, M. E. & Chavez, R. (1989). Differences in the ovulatory ability between the right and left ovary are related to ovarian innervation. In *Growth Factors and the Ovary*, ed. A. N. Hirshfield, pp. 321–5. New York: Plenum Press.

Eicher, E. M. & Washburn, L. L. (1983). Inherited sex reversal in mice: identification of a new primary sex-determining gene. *Journal of Experimental Zoology*, **228**, 297–304.

Falconer, D. S., Edwards, R. G., Fowler, R. E. & Roberts, R. C. (1961). Analysis of differences in the numbers of eggs shed by the two ovaries of mice during natural oestrus or after superovulation. *Journal of Reproduction and Fertility*, **2**, 418–37.

Franco, B., Guioli, S., Pragliola, A. *et al.* (1991). A gene deleted in Kallmann's syndrome shares homology with neural cell adhesion and axonal path-finding molecules. *Nature (London)*, **353**, 529–36.

Frankel, A. I., Chapman, J. C. & Cook, B. (1989). Testes are asymmetric in the testicular hemicastration response of the male rat. *Journal of Endocrinology*, **122**, 485–8.

Frankenhuis, M. T. & Kappert, H. J. (1980). Experimental transformation of right gonads of female fowl into fertile testes. *Biology of Reproduction*, **23**, 526–9.

Fujinaga, M. & Baden, J. M. (1991*a*). Evidence for an adrenergic mechanism in the control of body asymmetry. *Developmental Biology*, **143**, 203–5.

Fujinaga, M. & Baden, J. M. (1991*b*). Critical period of rat development when sidedness of asymmetric body structures is determined. *Teratology*, **44**, 453–62.
Fujinaga, M., Maze, M., Hoffman, B. B. & Baden, J. M. (1992). Activation of α-1 adrenergic receptors modulates the control of left–right sidedness in rat embryos. *Developmental Biology*, **150**, 419–21.
Galloway, J. (1990). A handle on handedness. *Nature (London)*, **346**, 223–4.
Geschwind, N. & Behan, P. (1982) Left-handedness: association with immune disease, migraine, and developmental learning disorder. *Proceedings of the National Academy of Sciences, USA*, **79**, 5097–100.
Greene, R. R., Burrill, M. W. & Ivy, A. C. (1939). Experimental intersexuality: the effect of antenatal androgens on sexual development of female rats. *American Journal of Anatomy*, **65**, 415–69.
Grove, A. J. & Newell, G. E. (1953). *Animal Biology*, 4th edn. London: University Tutorial Press.
Hancock, J. L. (1962). Fertilisation in farm animals. *Animal Breeding Abstracts*, **30**, 285–310.
Hancock, J. L. & Daker, M. G. (1981). Testicular hypoplasia in a boar with abnormal sex chromosome constitution (39XXY). *Journal of Reproduction and Fertility*, **61**, 395–7.
Hardelin, J-P., Levilliers, J., Young, J., Pholsena, M., Legouis, R., Kirk, J., Bouloux, P., Petit, C. & Schaison, G. (1993). Xp22.3 deletions in isolated familial Kallmann's syndrome. *Journal of Clinical Endocrinology and Metabolism*, **76**, 827–31.
Harrison Mathews, L. (1937). The female sexual cycle in the British horse-shoe bats, *Rhinolphus ferrum-equinum insulanus* Barrett-Hamilton, and *Rhinolphus hipposideros minutus* Montagu. *Zoological Society Transactions*, **23**, 224–66.
Henning, W. L. (1939). Prenatal and postnatal sex ratio in sheep. *Journal of Agricultural Research*, **58**, 565–80.
Herz, Z., Northey, D., Lawyer, M. & First, N. L. (1985). Acrosome reaction of bovine spermatozoa *in vivo*: sites and effects of stages of the oestrous cycle. *Biology of Reproduction*, **32**, 1163–8.
Heyns, C. F. (1987). The gubernaculum during testicular descent in the human fetus. *Journal of Anatomy*, **153**, 93–112.
Hogg, J. (1830). The Raid of the Kers. *Blackwood's Magazine*, **28**, 895–9.
Holder, N. & Maden, M. (1993). Mice with half a mind. *Nature (London)*, **360**, 708.
Hollander, W. F. & Strong, L. C. (1950). Intra-uterine mortality and placental fusions in the mouse. *Journal of Experimental Zoology*, **115**, 131–47.
Hunter, G. L. (1959). A contribution to the study of the problem of low fertility among Merino ewes in South Africa. *Journal of Agricultural Science*, **52**, 282–95.
Hunter, R. H. F. (1967). The effects of delayed insemination on fertilisation and early cleavage in the pig. *Journal of Reproduction and Fertility*, **13**, 133–47.
Hunter, R. H. F. (1977). Function and malfunction of the Fallopian tubes in relation to gametes, embryos and hormones. *European Journal of Obstetrics, Gynaecology and Reproductive Biology*, **7**, 267–83.
Hunter, R. H. F. (1980). *Physiology and Technology of Reproduction in Female Domestic Animals*. London & New York: Academic Press.
Hunter, R. H. F. (1982). Interrelationships between spermatozoa, the female reproductive tract, and the egg investments. *Proceedings 34th Easter School*

in Agricultural Science, University of Nottingham, ch. 3, pp. 49–63. London: Butterworth.

Hunter, R. H. F. (1988). *The Fallopian Tubes: Their Rôle in Fertility and Infertility*. Berlin, Heidelberg, New York: Springer-Verlag.

Hunter, R. H. F. (1989). Differential transport of fertilsed and unfertilised eggs in equine Fallopian tubes: a straightforward explanation. *Veterinary Record*, **125**, 304.

Hunter, R. H. F., Baker, T. G. & Cook, B. (1982). Morphology, histology and steroid hormones of the gonads in intersex pigs. *Journal of Reproduction and Fertility*, **64**, 217–22.

Hunter, R. H. F., Cook, B. & Baker, T. G. (1985). Intersexuality in five pigs, with particular reference to oestrous cycles, the ovotestis, steroid hormone secretion and potential fertility. *Journal of Endocrinology*, **106**, 233–42.

Hunter, R. H. F., Fléchon, B. & Fléchon, J. E. (1991). Distribution, morphology and epithelial interactions of bovine spermatozoa in the oviduct before and after ovulation: a scanning electron microscope study. *Tissue and Cell*, **23**, 641–56.

Hunter, R. H. F. & Nichol, R. (1986). Post-ovulatory progression of viable spermatozoa in the sheep oviduct and the influence of multiple mating on their pre-ovulatory distribution. *British Veterinary Journal*, **142**, 52–8.

Johnson, L., Petty, C. S. & Neaves, W. B. (1984). Influence of age on sperm production and testicular weights in men. *Journal of Reproduction and Fertility*, **70**, 211–18.

Knudsen, O. (1961). Sticky chromosomes as a cause of testicular hypoplasia in bulls. *Acta Veterinaria Scandinavica*, **2**, 1–14.

Laing, J. A. & Young, G. B. (1956). Observations on testicular hypoplasia in British cattle. *Proceedings of the Third International Congress on Animal Reproduction (Cambridge)*, 68–70.

McCarthy, A., Wolpert, L. & Brown, N. A. (1990). The development of the left–right axis in neural plate phase rat embryos is disrupted by α-adrenergic agonists in an apparent stage-specific and receptor-mediated manner. *Teratology*, **42**, 33 (Abstract).

McDonald, L. E. (1980). *Veterinary Endocrinology and Reproduction*, 3rd edn. Philadelphia: Lea & Febiger.

McKenzie, F. F. & Terrill, C. E. (1937). Estrus, ovulation and related phenomena in the ewe. *University of Missouri Agricultural Experimental Station Bulletin No. 264*.

McLaren, A. (1963). The distribution of eggs and embryos between sides in the mouse. *Journal of Endocrinology*, **27**, 157–81.

McLaren, A. (1982). The embryo. In *Reproduction in Mammals*, ed. C. R. Austin and R. V. Short, 2nd edn, vol. 2, pp. 1–25. Cambridge: Cambridge University Press.

McLaren, A. & Michie, D. (1959). Superpregnancy in the mouse. I. Implantation and foetal mortality after induced superovulation in females of various ages. *Journal of Experimental Biology*, **36**, 281–300.

Marsh, J. (ed.) (1991). *Biological Asymmetry and Handedness*. Chichester: Wiley.

Marshall, F. F., Ewing, L. L. Zirkin, B. R. & Cochran, R. C. (1982). Testicular atrophy associated with agenesis of the epididymis in the ACI rat. *Journal of Urology*, **127**, 155–8.

Matzuk, M. M., Finegold, M. J., Su, J. G. J., Hsueh, A. J. W. & Bradley, A. (1992). α-inhibin is a tumour-suppressor gene with gonadal specificity in mice. *Nature (London)*, **360**, 313–19.

Mead, R. A., Joseph, M. M. & Neirinckx, S. (1988). Optimal dose of human chorionic gonadotropin for inducing ovulation in the ferret. *Zoo Biology*, **7**, 263–7.
Mittwoch, U. (1976*a*). Lateral asymmetry and the function of the mammalian Y chromosome. In *Current Chromosome Research*, ed. K. Jones & P. E. Brandham, pp. 195–201. Amsterdam: Elsevier/North Holland Biomedical Press.
Mittwoch, U. (1976*b*). Differential growth of human foetal gonads with respect to sex and body side. *Annals of Human Genetics*, **40**, 133–8.
Mittwoch, U. (1986). Males, females and hermaphrodites. *Annals of Human Genetics*, **50**, 103–21.
Mittwoch, U. (1992). Sex determination and sex reversal: genotype, phenotype, dogma and semantics. *Human Genetics*, **89**, 467–79.
Mittwoch, U. & Buehr, M. L. (1973). Gonadal growth in sex reversed mice. *Differentiation*, **1**, 219–24.
Mittwoch, U. & Mahadevaiah, S. (1980). Comparison of development of human fetal gonads and kidneys. *Journal of Reproduction and Fertility*, **58**, 463–7.
Morgan, M. J. (1991). The asymmetrical genetic determination of laterality: flatfish, frogs and human handedness. In *Biological Asymmetry and Handedness*, ed. G. R. Bock & J. Marsh, pp. 234–46. Ciba Foundation Symposium 162. Chichester & New York: Wiley.
Mori, T. & Uchida, T. A. (1982). Changes in the morphology and behaviour of spermatozoa between copulation and fertilisation in the Japanese long-fingered bat, *Miniopterus schreibersii fuliginosus*. *Journal of Reproduction and Fertility*, **65**, 23–8.
Munro Neville, A. & Mackay, A. M. (1972). The structure of the human adrenal cortex in health and disease. *Clinics in Endocrinology and Metabolism*, **1**, 361–95.
Neuman-Silberberg, F. S. & Schüpbach, T. (1993). The *Drosophila* dorsoventral patterning gene *gurken* produces a dorsally localised RNA and encodes a TGF_{α} like protein. *Cell*, **75**, 165–74.
Neville, A. C. (1976). *Animal Asymmetry*. London: Edward Arnold.
Oh, Y. K., Mori, T. & Uchida, T. A. (1985). Prolonged survival of the Graafian follicle and fertilisation in the Japanese greater horseshoe bat, *Rhinolphus ferrumequinum nippon*. *Journal of Reproduction and Fertility*, **73**, 121–6.
Oke, A., Lewis, R. & Adams, R. N. (1980). Hemispheric asymmetry of norepinephrine distribution in rat thalamus. *Brain Research*, **188**, 269–27.
Olsson, H. & Ingvar, C. (1991). Left-handedness is uncommon in breast cancer patients. *European Journal of Cancer*, **27**, 1694–5.
Oppenheimer, J. M. (1974). Asymmetry revisited. *American Zoologist*, **14**, 867–79.
Ortiz, M. E., Bedregal, P., Carvajal, M. I. & Croxatto, H. B. (1986). Fertilized and unfertilized ova are transported at different rates by the hamster oviduct. *Biology of Reproduction*, **34**, 777–81.
Overstreet, J. W. & Cooper, G. W. (1978). Sperm transport in the reproductive tract of the female rabbit. I. The rapid phase of transport. *Biology of Reproduction*, **19**, 101–14.
Overstreet, J. W., Cooper, G. W. & Katz, D. F. (1978). Sperm transport in the reproductive tract of the female rabbit. II. The sustained phase of transport. *Biology of Reproduction*, **19**, 115–32.
Overstreet, J. W. & Thom, R. A. (1982). Experimental studies of rapid sperm transport in rabbits. *Journal of Reproduction and Fertility*, **66**, 601–6.

Pearson, O. P. (1949). Reproduction of a South American rodent, the mountain viscacha. *American Journal of Anatomy*, **84**, 143–71.
Pitkjanen, I. G. (1961). Doctoral dissertation. Moscow: Agricultural State Publishing House. 185 pp.
Potashnik, G., Insler, V., Meizner, J. & Sternberg, M. (1987). Frequency, sequence, and side of ovulation in women menstruating normally. *British Medical Journal*, **294**, 219.
Rao, C. V. & Edgerton, L. A. (1984). Dissimilarity of corpora lutea within the same ovaries or those from right and left ovaries of pigs during the oestrous cycle. *Journal of Reproduction and Fertility*, **70**, 61–6.
Rasweiler, J. J. (1979). Differential transport of embryos and degenerating ova by the oviducts of the long-tongued bat, *Glossophaga soricina*. *Journal of Reproduction and Fertility*, **55**, 329–34.
Rasweiler, J. J. (1988). Ovarian function in the captive black mastiff bat, *Molossus ater*. *Journal of Reproduction and Fertility*, **82**, 97–111.
Reimers, T. J., Dziuk, P. J., Bahr, J., Sprecher, D. J., Webel, S. K. & Harmon, B. G. (1973). Transuterine embryonal migration in sheep, anteroposterior orientation of pig and sheep fetuses and presentation of piglets at birth. *Journal of Animal Science*, **37**, 1212–17.
Roberts, S. J. (1956). *Veterinary Obstetrics and Genital Diseases*. Ithaca, New York: Comstock.
Runner, M. N. (1951). Differentiation of intrinsic and maternal factors governing intrauterine survival of mammalian young. *Journal of Experimental Zoology*, **116**, 1–20.
Ryder, L. (1992). Extremes of symmetry. *Times Higher Educational Supplement*, 16 October.
Salisbury, G. W., Vandemark, N. L. & Lodge, J. R. (1978). *Physiology of Reproduction and Artificial Insemination of Cattle*, 2nd edn. San Francisco: W. H. Freeman.
Scanlon, P. F. (1972). Frequency of transuterine migration of embryos in ewes and cows. *Journal of Animal Science*, **34**, 791–4.
Shalash, M. R. (1965). Some reproductive aspects in the female camel. *World Review of Animal Production*, **1**, No. 4, 103–8.
Shirley, D. G. (1976). Developmental and compensatory growth in the guinea pig. *Biology of the Neonate*, **30**, 169–80.
Short, N. (1992). Molecular neurology comes of age. *Nature (London)*, **360**, 295–6.
Sørensen, T. B. (1992). Ovarian asymmetry and follicle development in immature and mature Harbour Porpoises (*Phocoena phocoena*, L.). *Journal of Reproduction and Fertility*, Abstract Series No. 9, p. 29, No. 43.
Van der Schoot, P. (1992). Disturbed testicular descent in the rat after prenatal exposure to the antiandrogen flutamide. *Journal of Reproduction and Fertility*, **96**, 483–96.
Van der Schoot, P. & Elger, W. (1992). Androgen-induced prevention of the outgrowth of cranial gonadal suspensory ligaments in fetal rats. *Journal of Andrology*, **13**, 534–42.
Van Niekerk, C. H. & Gerneke, W. H. (1966). Persistence and parthenogenetic cleavage of tubal ova in the mare. *Onderstepoort Journal of Veterinary Research*, **33**, 195–231.
Van Niekerk, W. A. & Retief, A. E. (1981). The gonads of human true hermaphrodites. *Human Genetics*, **58**, 117–22.

Villalon, M., Ortiz, M. E., Aguayo, C. & Munoz, J. (1982). Differential transport of fertilized and unfertilized ova in the rat. *Biology of Reproduction*, **26**, 337–41.

Ward, H. B., McLaren, A. & Baker, T. G. (1987). Gonadal development in T16H/X*Sxr* hermaphrodite mice. *Journal of Reproduction and Fertility*, **81**, 295–300.

Wegenke, J. D., Vehling, D. T., Wear, J. B., Gordon, E. S., Bargman, J. G., Deacon, J. S. R., Herrmann, J. P. R. & Optiz, J. M. (1975). Familial Kallmann syndrome with unilateral renal aplasia. *Clinical Genetics*, **7**, 368–81.

Wheeler, A. G. (1978). Comparisons of the ovulatory and steroidogenic activities of the left and right ovaries of the ewe. *Journal of Reproduction and Fertility*, **53**, 27–30.

Wiebold, J. L. & Becker, W. C. (1987). Inequality in function of the right and left ovaries and uterine horns of the mouse. *Journal of Reproduction and Fertility*, **79**, 125–34.

Wilhelm, J. E. & Vale, R. D. (1993). RNA on the move: the mRNA localisation pathway. *Journal of Cell Biology*, **123**, 269–74.

Wimsatt, W. A. (1975). Some comparative aspects of implantation. *Biology of Reproduction*, **12**, 1–40.

Wimsatt, W. A. (1979). Reproductive asymmetry and unilateral pregnancy in Chiroptera. *Journal of Reproduction and Fertility*, **56**, 345–7.

Wislocki, G. B. & Guttmacher, A. F. (1924). Spontaneous peristalsis of the excised whole uterus and Fallopian tubes of the sow with reference to the ovulation cycle. *Bulletin of the Johns Hopkins Hospital*, **35**, 246–52.

Witschi, E. (1951). Embryogenesis of the adrenal and the reproductive glands. *Recent Progress in Hormone Research*, **6**, 1–23.

Wolpert, L. (1991). Introduction and summing-up. Chairman's remarks. In *Biological Asymmetry and Handedness*, ed. G. R. Bock & J. Marsh, pp. 1–2 and 316. Ciba Foundation Symposium 162. Chichester & New York: Wiley.

Yokoyama, T., Copeland, N. G., Jenkins, N. A., Montgomery, C. A., Elder, F. F. B. & Overbeek, P. A. (1993). Reversal of left-right asymmetry: a situs inversus mutation. *Science*, **260**, 679–82.

Yost, H. J. (1990). Inhibition of proteoglycan synthesis eliminates left–right asymmetry in *Xenopus laevis* cardiac looping. *Development*, **10**, 865–74.

10
Concluding thoughts and a current perspective

Historical sweep

It was noted in the opening chapter that the reproductive musings of the ancient Greeks unwittingly contained some elements of truth, for example those concerning the left–right differences between gonadal size and type in hermaphrodites, but little real progress in understanding the formation and function of the sexual organs took place until growth of the Italian medical schools in the late Middle Ages; even then, progress was slow. The dissecting room paved the way to ever more accurate descriptions and illustrations of reproductive tissues but, in terms of explaining the mysteries of generation, powerful imagination was an invariable substitute for solid facts. Only with the advent of reasonable quality microscopes were the respective gametes and their gonadal origins discovered, these giving way to general descriptions of the process of fertilisation and eventually to a visualisation of mammalian chromosomes. Although the dual nature of embryonic duct systems had been more or less appreciated since the studies of Caspar Wolff and Johannes Müller and also, in due course, the dual potential of the embryonic gonads, a rational understanding of the decisive elements in mammalian sex determination had to await the technique of karyotyping 150 years later. In effect, it was the crucial clinical evidence presented during 1959 on the involvement of the Y chromosome in imposing the condition of maleness irrespective of the number of X chromosomes, coupled with developments in the tender young discipline of molecular biology, that enabled today's grasp at the level of individual genes. However, even with the progressive identification and cloning of individual genes, much remains to be clarified before some of the anomalous sexual traits that so intrigued the Ancients will be explained in a reasonably satisfactory manner.

Sex-determining genes

The clearly demonstrable association between the Y chromosome and maleness in the higher mammals, especially man, led to the notion that there must be a factor on the Y chromosome that acts to set in train a testis-determining programme. As a consequence, there is reference to the search that proceeded increasingly at a molecular level and dominated this field during the late 1970s and 1980s. After considering various candidate factors and genes such as H-Y antigen, *Bkm* sequences and *Zfy*, each thought initially to have a key involvement in programming maleness in mammals, attention was drawn somewhat dramatically in 1991 to the tiny 14 kilobase genomic DNA fragment or gene termed *Sry*. Upon microinjection into female pronucleate mouse embryos, this fragment could induce a proportion of such XX embryos to become phenotypically male, although it did not always override the intrinsic female programme. Propelled by a wave of enthusiasm on account of this demonstrable sex reversal, *Sry* rather too promptly became known as *the* male-determining gene whereas, for elaboration of a fully-functional male, it is now appreciated that other genes on the Y chromosome, such as gene(s) for spermatogenesis, have a vital rôle to play. A more cautious and balanced view of the contribution of *Sry* was urged by Mittwoch (1992).

Although Chapter 2 accepts the dogma of a dominant testis-determining pathway in embryos bearing a Y chromosome compared with a so-called permissive ovarian differentiation in the absence of the former, there can be little doubt that specific genes must act to programme the genital ridge of mammalian females to differentiate as ovarian tissue. Hence, the nature of male 'dominance' may require a new perspective, and the same comment may apply to the previous distinction between the steps of sex determination and sexual differentiation. Initial dominance is largely in terms of timing. As brought out several times in the text, embarkation upon the testicular pathway in the presence of a Y chromosome or portions thereof appears in some critical and essential manner to be linked to precocious development in male embryos, a situation frequently demonstrable even as early as during the first few cleavage divisions and predictably so by the morula–blastocyst stage. This finding may imply quite specific interactions between sex-determining genes and genes that influence the rate of early embryonic development. In any event, if male embryos consistently show an accelerated rate of cleavage, this would suggest that they already know what sex they are, perhaps as a consequence of *Sry* transcription at the 2-cell stage. At the time of writing, three genes that appear to be critically involved in the first stages of gonadal differentiation are *SRY*, *AMH* and *WT*. Clearly, the nature of the molecular interactions between their protein products could be of very special importance, although this particular focus should not override a presumed coordinated expression of many genes at this time. Of course, it is not yet known what regulates *SRY* nor the extent to which *SRY* may act to suppress female development (i.e. as a loss of function pathway) rather than primarily to activate the male pathway (i.e. gain of function). It could certainly be argued that an *SRY* influence over a period of time is just as likely to be involved in suppressing the expression of femaleness as in inducing maleness. In reality, repression and activation programmes may be imposed simultaneously. A related possibility is that the *SRY* gene product activates *AMH*. The involvement of *WT* in reproductive differentiation also requires clarification. Because of the proximity of the embryonic gonad and kidney, it is perhaps significant that this gene that influences aberrant kidney development may also influence steps in gonadal differentiation. One suggestion is of a *WT* involvement in the expression of growth-inducing genes, for example the gene for insulin-like growth factor II.

Male precocity and timing mismatch

As a more general point, it could reasonably be argued that if the male and female gonad-determining pathways are of potentially equivalent potency, then some slight asynchrony in their imposition would indeed be required for establishment of normal gonadal differentiation. It would be immensely valuable to identify the various factors that regulate differences in growth rate between male and female embryos during the pre-blastocyst stages. Progress in this area, followed by a means of suppressing accelerated growth in male embryos, might enable the experiment of sex reversal in XY embryos by introduction of X-associated gene programmes. If anything were to come of this suggestion, then the key to maleness – in one sense at least – might be seen to be as dependent upon growth-determining factors as upon sex-determining genes. Hence, the relevance of Mittwoch's (1988) phrase 'the race to be male', and the nature of the dominance contributed by the Y chromosome might be viewed in a new light. Not completely new, however, for Brambell (1927) drew attention to the precocious development of male gonads, and 10 years earlier Lillie (1917) had stressed the need for some mechanism whereby male embryos would be protected from the sex hormones of their mother. Failure of appropriate precocity may lead to situations of timing mismatch referred to several times in the text that compromise gonadal identity (Table 10.1).

Table 10.1. *Some important recurrent themes concerned with sex determination and sexual differentiation that are highlighted in succeeding sections of the text*

Theme or topic	Referred to in Chapters
Imposing rôle of *SRY/Sry*	1, 2, 4, 5, 10
HMG domains and *Hox* genes	2, 9
Requirement for ovary-determining genes	2, 3, 6, 10
Common origin of Sertoli and granulosa cells	3, 4, 5, 8, 10
Rôle of mesonephros in gonadal differentiation	3, 4, 5, 7, 8, 10
Male gonads develop precociously	2, 3, 4, 5, 9, 10
Possible influence of primordial germ cells upon evolution and stabilisation of gonad	2, 3, 4, 5, 9, 10
Developmental flexibility (plasticity) of gonads	1, 2, 3, 5, 9, 10
Situations of timing mismatch	2, 5, 6, 7, 8, 10
Unilateral formation of ovotestis or testis	2, 3, 5, 6, 7, 9, 10
Foetal ovarian transplants resulting in sex reversal	3, 4, 6, 7
Two X chromosomes in males inevitably lead to sterility	2, 5, 6, 7, 8, 10
Importance of autosomal genes in the sex-determining pathway	2, 3, 5, 6, 7, 10
Somatic sexual differentiation in marsupials may be under primary genetic control rather than being dependent on gonadal hormones	2, 3, 4, 10

Key rôle of Sertoli cells

A recurring theme during discussion of testicular formation was the pivotal rôle of the Sertoli cell, this currently being the earliest known target of *Sry* action in the gonad. Differentiation of Sertoli cells followed by expression of their range of synthetic functions in due course imposes influences on both somatic and germ cells of the gonad and also on the neighbouring duct systems. Suppression of organisation in an ovarian direction together with regression of the Müllerian duct system are early requirements of the growing Sertoli cell population, quite apart from their direct involvement in the formation of seminiferous tubules. Reflecting on the course of events during elaboration of testicular structures, one is forced to consider precisely how and why ancestral Sertoli-like cells assumed priority and achieved such prominence, both as a target of *Sry* and as a modulator of the visible stage of confluence between the processes of sex determination and sexual differentiation. The question also arises as to the extent and timing of granulosa cell secretion of glycoproteins comparable to those of Sertoli cells (e.g. AMH), a point that receives some illumination in the context of chimaeric gonads. However, the specific gene that targets and initiates granulosa cell differentiation has yet to be isolated. One other aspect relevant to Sertoli cells and worth pondering is in situations where such cells are incompletely formed or incompetent, and are better termed Sertoli-like cells as in ovotestis development. In these circumstances, the question arises as to whether *Sry* activity or its appropriate downstream programme has malfunctioned or been in abeyance.

Establishment of primordial germ cells

The bipotential nature of the early embryonic gonads, that is their ability to develop either as a testis or as an ovary – or indeed pathologically as an ovotestis – has been remarked upon repeatedly. Whilst the precise derivation of the somatic cells from the separate lineages of coelomic epithelium, mesenchymal cells and mesonephros is gradually being clarified using specific cell markers, there is no doubt as to the extra-gonadal origin of the primordial germ cells. In part, at least, the present author views the requisite phase of migration as a means of bestowing flexibility of location during colonisation of the genital ridges and also as a potential source of differences between the left and right gonads. However, a further interpretation for the extra-gonadal origin of the primordial germ cells has been noted by McLaren (1992). Here, the original suggestion of Monk, Boubelik & Lehnert (1987) is that the primordial germ cells need initially to be sequestered away from the influences that determine somatic cell fate during gastrulation, such influences including DNA methylation. Be that as it may, the phase of migration of primordial germ cells from the epiblast coupled with their multiplication before and after entering the genital ridges might enable them to impose a form of organisation on the gonadal anlagen that would not have been possible if they had emerged there *de novo*. Although primordial germ cells are said to play no essential part in testis determination, conversations between somatic and germ cells in the genital ridge are considered vital in achieving full gonadal differentiation and function.

Despite the long period of years during which the question has been posed, and the recent highlighting of a rôle for TGF-β_1 (Godin & Wylie, 1991), there is still not a full appreciation as to the cues involved in guiding, or occasionally failing to guide, the primordial germ cells to the genital ridges. If the putative chemotactic phenomenon could be described precisely, then some exciting experimental modifications of the pattern of primordial germ cell migration should be possible. These would enable subsequent examination of the developmental fate of gonadal tissues. Even so, recent observations indicate that primordial germ cells are seemingly connected to each other by processes during the phase of migration (anonymous report, 1993). A further reference to this study in mouse embryos by Gomperts *et al.* (1993) states that migrating primordial germ cells, detectable by the SSEA-1 antigen, are in contact with at least one other primordial germ cell (see Wilkins, 1993). This finding may serve to reduce the emphasis on chemotaxis, not least because the leading cells are apparently situated very close to the target and may thus be able to guide the others.

AMH involvement in males and females

Whilst diverse anomalies of the embryonic duct system have been noted, perhaps the greatest excitement in Chapter 4 concerned the several rôles of AMH, currently thought to be one of the earliest Sertoli cell secretions in response to the influence of *Sry*. Precisely how AMH functions to achieve early regression of the Müllerian ducts in male embryos remains uncertain, for there is still no clear demonstration of an appropriate receptor molecule for this glycoprotein. But, in instances in which there has been failure of formation of the AMH molecule, which in turn engenders the persistent Müllerian duct syndrome, the defect can be levelled squarely at one or more gene mutations. Although controversial, the notion is still abroad that AMH may in some manner contribute to descent of the testes, at least in those eutherian species – the majority? – that require a scrotal gonad for successful spermatogenesis. However, this point of view would seem difficult to sustain in instances of unilateral cryptorchidism when there has been regression of both Müllerian ducts. Moreover, evidence from at least one marsupial has suggested that differentiation of the gubernaculum is hormone independent, and that there are sexual dimorphisms which precede morphological differentiation of the gonads and which are said to be under primary genetic control. As a quite distinct topic, the extent to which AMH contributes to the regulation of waves of ovarian follicular development and follicular dominance in eutherians is only just beginning to receive close attention, not least through its influence on the aromatase enzyme pathway.

On a more philosophical level, consideration has been given as to why the reproductive system has taken over the embryonic excretory ducts, that is the mesonephric and then paramesonephric ducts. The most reasonable response would seem to concern the evolution of systems of internal fertilisation in mammals and a requirement for conduits from the gonads (which develop in close proximity to the embryonic kidneys) to contain and guide the liberated gametes and in due course to support the developing embryo and foetus. Viviparity involves a high level of organisation in the genital duct system in conjunction with a corresponding development of the organ of exchange and sustenance, the placenta. In the play and interplay of evolutionary forces, who could have imagined that the ancestral paramesonephric duct system would one day embrace such a highly developed and complex organ as the eutherian placenta as one consequence of the switch from external to internal shedding of the gametes?

Aberrant sexual development

The overriding flavour that remains from Chapters 5, 6 and 7 is one of anomalous conditions and diverse syndromes that may arise in both farm and laboratory animals and in man. Thus, there is consideration of freemartins and intersexes in farm animals, with seemingly important rôles for AMH or closely related proteins in perturbing gonadal formation and function. In man, the syndromes of Turner, Kallmann and Klinefelter take thoughts immediately to the involvement of inappropriate gene programmes producing serious or even devastating reproductive disturbances. Many of the sexual anomalies portrayed in Chapters 5 to 7 were traditionally ascribed to the conditions of mosaicism or chimaerism, but it is becoming increasingly clear that extremely subtle defects in single genes – point mutations – rather than gross disturbances of the overall chromosome complement may be the source of specific anomalies. In domestic farm animals, although not necessarily in man, abnormal development of the gonads and/or adjoining duct systems would appear to be more frequent in individuals with an XX sex chromosome constitution, and this prompts the question as to whether the slightly less rapid female programme of differentiation renders XX animals more vulnerable to developmental perturbations when compared with males; this might in part be because the bipotential state of the gonad would exist for slightly longer in females. And, despite the focus on *Sry* as one probable initiator of testicular differentiation, autosomal genes downstream of – or interacting with – *Sry* must certainly make a crucial contribution to both normal and abnormal gonadal development. This prompts the thought that if the usual sequence of gene action in the developmental cascade were in some manner to be perturbed, although not involving a detectable mutation, then such a chronological derangement might offer an explanation for wayward formation of gonadal tissues.

Whereas the rôle of individual genes in provoking reproductive anomalies is possibly best understood in the case of laboratory species with the well-known examples of sex reversal, testicular feminisation and hypogonadal mice, much valuable information is accumulating for man. In both groups, one of the observations that commands attention is the association of anomalies with errors arising during male meiosis. Once again, therefore, this raises the question as to whether the process of spermatogenesis is rendered less stable in the scrotal testis with its special requirement for a temperature below that of the abdomen. Even if a certain potential instability cannot be ascribed to this need for a sensitive system of temperature control, which in man is frequently overridden by sartorial considerations, the particular disposition of chromatin during the events of meiosis in both males and females might facilitate the production of errors. Quite apart from their intrinsic value as models of such errors, studies on XX males and XY females have been of especial value in mapping the position of *TDF* on the Y chromosome. Another question brought out in the text, albeit very tentatively, is whether abnormal genetic exchange between the X and Y chromosomes in male meiosis is itself under some (rare) form of genetic predisposition rather than simply representing a completely spontaneous derangement. This is a somewhat cryptic way of wondering whether the arrangement of genes themselves could in some manner exert an influence on the frequency of mutations.

Inferences from chimaeras

The relevance of experimental work on chimaeras was seen in terms of the light they might shed upon novel interactions between germ cells and somatic cells in the embryonic gonads. There were, of course, the key questions that have appealed to previous reviewers, such as the extent to which germ cells differentiate as a function of their own sex chromosome complement compared with the extent to which neighbouring somatic cells might impose at least a provisional identity upon their incoming colleagues. And what might be the nature, for example, of interactions between germ cell gene expression and gene products elaborated by surrounding somatic cells? Can Sertoli cells be of XX sex chromosome constitution in a chimaeric gonad and if – as now appreciated – the answer is 'yes', to what extent are they functionally representative? And what of XY follicular cells? In chimaeras having both XX and XY cell populations present in the gonads, there was a specific interest in the incidence of ovotestis formation and in other forms of gonadal anomaly.

As already discussed, there appear to be 'molecular tensions' between expression of the intrinsic genetic constitution of a germ cell and influences in the local milieu in which such germ cells are arranged. The constitution of a chimaeric gonad represents a dynamic situation, the hypothetical molecular tensions being much influenced by the relative proportions of XX and XY cells. But even though transdifferentiation of cell types appears to be possible, there is a tendency towards stabilisation before the adult stage is reached, frequently involving the loss of individual cell lines. This appears significant in a true biological sense, and could seemingly offer one explanation for the small proportion of animals developing as hermaphrodites or intersexes (rather than the anticipated 50%) after generating XX↔XY chimaeras by combining cleaving embryos of opposite sex. There is an initial phase of development as hermaphrodites, but the gonads show a subsequent transition from foetal ovotestes to postnatal testes, the origin of the change residing in Sertoli cell secretion of AMH which kills oocytes and thereby provokes loss of granulosa cells. Despite this loss of female germ cells, abundant XX somatic cells would remain in the gonads of mismatch chimaeras. Moving from the situation in mice to that in farm animals, and on the basis of the preceding views, it seems improbable that a majority of XX intersex pigs could be classified as chimaeras. Moreover, despite the extensive development of testicular tissue, there was little evidence that AMH from the Sertoli-like cells had been available at an appropriate time or in sufficient quantities to kill oocytes and prevent follicular development. Indeed, ovarian tissue within an ovotestis frequently contained corpora lutea (Chapter 5).

Key factors in determining sexual differentiation in XX↔XY chimaeras, at least in mice, would seem to be the number of Sertoli cells, their spatial distribution in the genital ridges, and the extent to which the full spectrum of molecules characteristic of normal Sertoli cells is available. Plasticity in the early development of the germ cell line in chimaeric embryos can be demonstrated, although it is worth stressing that functional reversal of differentiation of XX germ cells does not occur in the mouse or in several other mammals; the XX germ cells degenerate and disappear soon after birth, and none appears capable of undergoing spermatogenesis. Studies on chimaeras indicate that there must be a potent local influence of the somatic milieu in the early embryonic gonad. Features of such molecular influences that will determine their impact doubtless include the timescale of synthesis, the quantity of molecule(s) elaborated, and the extent to which these molecules diffuse. Even so, it remains uncertain precisely how quantitative thresholds are assessed by responding germ cells. That the somatic cells should initially dominate the germ cells in the genital ridges strongly suggests that gene expression in the somatic cells precedes that of the germ cells. However, early subordination gives way to an assertion of germ cell 'rights' so that, in almost all situations examined so far, potentially functional gametes can be derived only from an association with somatic tissue of corresponding sex chromosome complement.

Musings on asymmetry

As to the penultimate chapter concerned with asymmetries in the reproductive system, it is hoped that readers may have been persuaded that the overall problem of asymmetry has some bearing on sexual differentiation, and might in particular shed light on various instances of intersexuality. The fact that the embryonic gonads are initially bipotential in terms of their possible direction of differentiation together with the initial duality of the embryonic duct system may – in a manner not yet fully appreciated – render the reproductive organs especially vulnerable to asymmetric developmental cues. And differences in rates of growth between the left and right embryonic gonads cannot readily be explained without invoking a rôle for one or more genes influencing laterality (handedness), that is to say with different and seemingly independent instructional programmes functioning at least briefly during early organogenesis. Comparable gonadal 'responses' to circulating growth hormones would have been anticipated unless genetic influences predispose left–right differences. As already suggested, there is ample evidence to indicate that embryos have some means of distinguishing left and right during formation of their organ systems.

Touched on briefly in Chapter 9 was the question of whether different patterns of primordial germ cell migration might be related to or even underlie certain forms of gonadal developmental asymmetry. If diverse distinctions can be demonstrated between left and right sides of the body, then colonisation of the genital ridges by different numbers of primordial germ cells, possibly on a slightly different timescale, remains a reasonable speculation. This, in turn, could exert an influence on the multiplication of somatic cells and thus the rate of gonadal growth and the ultimate size attained. But perhaps correlations or potential connections between cerebral asymmetries and those found in paired abdominal organs remain the most intriguing, and their various components need to be teased out in full. For the moment, one can but air the notion that left–right differences in patterns of innervation may act in a subtle way to influence organ form and function, particularly to modulate responses to circulating hormones.

A final comment

As already pointed out, an ultimate clarification of the phenomenon of asymmetry must be sought at the molecular level, and yet the many potential interactions involved during differentiation will render this a task of formidable dimensions. For example, interactions of relevance to the present treatment are those between sex-determining genes such as *Sry*, genes that influence the rate of gonadal growth, and putative genes for handedness. Nevertheless, technological progress has been so rapid and impressive in the late twentieth century that computer modelling and analysis of cellular, tissue and organ interactions against a backcloth of known molecular instructions may reasonably soon provide a quantum leap in our understanding of mammalian differentiation and development and of its many fascinating errors. If these steps are followed by convincing demonstrations at the laboratory bench in terms of being able to generate specific tissue anomalies, then the intellectual rewards will need to be set against rather substantial social responsibilities. It might be wishful thinking to imagine that the most highly developed of all the placental mammals will live up to this particular requirement but, viewed in an enlightened frame of mind, a little wisdom should not be beyond our frequently voracious grasp.

References

Anonymous (1993). Germline development. *AgBiotech News and Information*, **5**, 312–13.

Brambell, F. W. R. (1927). The development and morphology of the gonads of the mouse. Part 1. The morphogenesis of the indifferent gonad and of the ovary. *Proceedings of the Royal Society of London, Series B*, **101**, 391–409.

Godin, I. & Wylie, C. C. (1991). TGFβ_1 inhibits proliferation and has a chemotropic effect on mouse primordial germ cells in culture. *Development*, **113**, 1451–7.

Gomperts, M., Garcia-Castro, M., Wylie, C. & Heasman, J. (1994). Interactions between primordial germ cells play a role in their migration in mouse embryos. *Development*, **120**, 135–41.

Lillie, F. R. (1917). The free-martin; a study of the action of sex hormones in the foetal life of cattle. *Journal of Experimental Zoology*, **23**, 371–452.

McLaren, A. (1992). The quest for immortality. *Nature (London)*, **359**, 482–3.

Mittwoch, U. (1988). The race to be male. *New Scientist*, No. 1635, pp. 38–42.

Mittwoch, U. (1992). Sex determination and sex reversal: genotype, phenotype, dogma and semantics. *Human Genetics*, **89**, 467–79.

Monk, M., Boubelik, M. & Lehnert, S. (1987). Temporal and regional changes in DNA methylation in the embryonic, extraembryonic and germ cell lineages during mouse embryo development. *Development*, **99**, 371–82.

Wilkins, A. S. (1993). Germ line development. *BioEssays*, **15**, 699–700.

Index